CONQUISTA
DEL
ÉTER

THE ODYSSEY FROM MAGNETS TO GRAVITY

CONQUISTA DEL ÉTER

ITS HISTORY AND THE ODYSSEY

Blake Taylor

Published by

YOWAH BOOKS

YOWAH BOOKS
Australia
www.yowahbooks.com

ISBN-13: 978-0-9874729-5-3

ISBN 978-0-646-86991-9

ISBN: 978-0-646-89035-7

Disclaimer

Note that ideas expressed in this work by the author are simply theories. Whether the reader agrees with them not, they should not be used by students trying to pass examinations until they are found in their curriculum books. Until that time Yowah Books and the author take no responsibility for the consequences to those students who because of this work do not toe the party line in their examinations.

This work widens elements of my first four books.

The dyslexic author dedicates this book to spell-check.

GENERAL CONTENTS

S

Illusions

There are four really great illusions which deceived and/or deceive mankind.
The first was that of the Sun circling the Earth; which Copernicus resolved.
The second is Oersted's current/magnetism illusion; which is resolved here
and consequently penetrates
The third illusion that; 'Space is a Void.'
and the forth
The Cosmos is a Uni-One-Verse.

Part I

The History of Ether

VOIDS AINT VACUUMS

INTRODUCTION

The *ether* existed in the minds of man for at least three millennia. It was the invisible 'atmosphere' and or 'spirit' that permeated all matter and life. It marched hand in hand with both metaphysics and physics which included the phenomena of astrology, astronomy, biology, light, gravity, electrification and magnetization and more.

The perception of *ether* grew steadily with science, peaking towards the end of the 19th century. Luminaries like Swiss mathematician Leonhard Euler stated in 1760, his belief that the '*luminiferous ether*' which propagates light, also propagates electrical phenomena. British physicist James Maxwell, the father of electromagnetic wave theory, forcefully argued for a form of its existence.

It shocked and distressed many persons of science when in 1887 German born American physicist Albert Abraham Michelson (1852-1931) and American chemist Edward Morley (1838-1923) announced they had executed the *ether*. Michelson believed he had killed it with light rays and an interferometer a decade earlier while hunting the snark in Berlin, however there was a chance that the beast was only wounded. To eliminate this possibility he enlisted the services of Morley, a respected chemist and reputed perfectionist, to repeat the Berlin experiment with greater precision. At the Case School of Applied Science in Cleveland Ohio they built a rigid based, *(it was literally anchored in bedrock)*, altar, aimed their ray gun on it and fired across it.

They reasoned the since the planet Earth was flying though the *ether*, and since the *ether* supported light, then light rays fired in some directions would prosper over rays in other directions.

The rays however were found to be invariable so they proclaimed the *ether* wasn't there. It didn't exist any more! It was dead! But it was not erased.

It haemorrhaged in purgatory for about twenty years until Albert Einstein delivered the coup de grace, sealed it in a relatively heavy sarcophagus and buried it in the unkempt graveyard of toppled theories. There it suffocates beneath the suppositions of the ever vigilant *'Professorial Guards'* and their precursors.

Occasional a maverick would creep in at night and put forget-me-nots on the grave, but these were quickly withered by the *Guard's* fusty breath.

As the 19th century rolled into the 20th and the 20th into the 21st the rebellious visitors dwindled, but in the heart of hearts of many scientists and all Trekies, lonely ethereal embers still smoulder waiting for a logical breeze.

THE ETHER MILLENNIA BC

CHAPTER 1

BABYLONIA

In the west around 3000 BC three great civilisations emerged around three mighty river systems; the Egyptians along the Nile, the Indus along the Indus and the Mesopotamians along the Tigris and the Euphrates. Mesopotamian is often referred to as the 'cradle of civilisation', and one of its kingdoms was Babylonia. It began around the18th century BC, and was on and off a vassal state of the military nation of Assyria. Babylonia means 'Gate of Gods', and there were thousands of them. The minor Gods were ruled by Marduk the el supremo who finished up with so many titles that it took 10 minutes to call his name. It was far easier to call him Bel (Lord) which became the practice. His consort was Belit.

On top of their ziggurats the priests of Babylon scanned the heavens with their instruments, devising zodiac arrangements, divining prophesies and calculating astral movements. To facilitate the latter they devised the abacus which would eventually reach Japan and China three thousand years later in the 16th century AD.

By about 700 BC Babylonians had the mathematical skill to plot the paths and positions of the moon and several planets. They had tables for squares, cubes, square and cube roots, good approximations for the hypotenuse of right angle triangles, and could solve quadratic equations.

$$ax^2 + bx + c = 0 \qquad \text{multiplying by 4a}$$

$$4a^2x^2 + 4abx + 4ac = 0$$

$$(2ax + b)^2 - b^2 + 4ac = 0$$

$$(2ax + b)^2 = b^2 - 4ac$$

$$2ax + b = \pm\sqrt{b^2 - 4ac}$$

$$x = \frac{-b \pm \sqrt{b^2 - 4ac}}{2a}$$

They had a number system based on 60 which cursed the future with the 60 second minutes, 60 minute hours, and 24 hour days. They evolved the 360 degree circle, in which each degree has 60 minutes and each minute 60 seconds, sentencing all future generations to complexity. A metric time system would have 10 hours in a day, where each hour has 100 minutes, and each minute 100 seconds. There would be 100,000 metric seconds in a day instead of 86,400.

A metric second would be 0·864 of a Babylonian second. A metric circle would have 100 metric degrees, with right angles @ 25 metric degree.

Perhaps the future Spacelings will implement this system.

The Bagdad Battery

In 1936 a 15 cm high bright clay vase was discovered in excavations at a village near Baghdad. It held a smaller copper cylinder which was firstly soldered with a lead-tin alloy then further sealed with asphalt. The cylinder held an iron bar *(corroded by acid),* which was centrally suspended and fixed with asphalt, which also fixed the cylinder to the top of the jar. The asphalt doubled as an insulator.

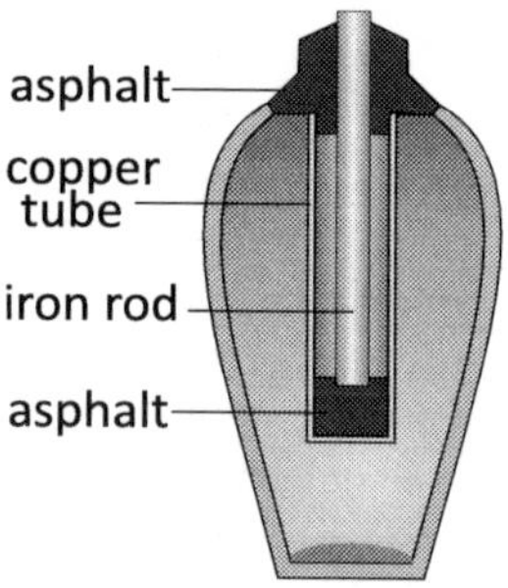

It is indisputably a battery and is at least 2000 years old, *(some say much older)*. Suggestions for its use include electroplating, acupuncture, and the electrification of metal idols so priests could zap the touch of a transgressor.

Lodestone

The magnetic properties of lodestone were known to the Babylonians, Ancient Chinese, Greeks, Egyptians, and Assyrians. Lodestone is mentioned in Chinese texts as early as 2700 BC and in Greek texts of around 800 BC.

*Lodestone is the magnetic oxide of iron Fe_3O_4. Magnets derive their name from Magnesia in Thessaly (now Manisa in Turkey) where lodestone was mined. The Roman natural encyclopaedist ‘Pliny the Elder’ however, believed magnets were named after the shepherd ‘Magnes’, who got stuck to a lodestone rock by the nails in his boots.

Pliny was also commander of the Western Roman fleet. He subsequently suffocated from volcanic fumes on his rescue mission to Vesuvius in 79 AD*

ANCIENT GREEKS

The Ancient Greek saga of creation holds that:

- In the beginning there was Chaos, the gaping void that was the foundation of all things.
- Out of Chaos came Gaea, Mother Earth, and Eros, Love. *(matter and glue)*
- Before 600 BC the Greeks believed the Earth was flat and floated on the great river Oceanus into which all the smaller rivers flowed.
- Beyond Oceanus was Cimmerii, the land of dreams and the entrance to hell.

Thales of Miletus

Thales (625?-546? BC) was instrumental in changing this mythical concept of the world into one based upon the physical substance of the Earth. He advocated that water was the essence of Earth's matter. Regarding the heavens, he thought the Milky Way was made up of millions of stars.

Legend has it that he predicted a solar eclipse on May 28/585 BC, which stopped the battle between the kingdoms of Lydia and Media.

Thales studied both the magnetic phenomena of lodestone and the effects of rubbing amber, 'ēlektron' in Greek, with various cloths, causing it to attract lightweight objects such as lint and feathers.

Thales thought ēlektron had a soul or spirit since it could affect objects via the *ether* without touching them.

Amber is a solidified translucent fossil resin from prehistoric extinct conifers. It radiates a beautiful yellow or yellow-brown colour much prized in jewellery.

Pythagoras

Pythagoras was born in 580 BC on the island of Sámos, close to present day Turkey, which at that time was ruled by Persia. He was sure the Earth was round. He theorised that the Earth, Moon, planets and Sun were harmonious globes revolving around a central fire, which couldn't be seen from Earth because it was always hidden behind another body called the Counter-Earth.

He was a semi-mystical purist who was so revolted by the tyranny of Polycrates, the king of Sámos, that he migrated in 530 BC to a Greek colony called Crotona on the heel of Italy. There he established his scholastic brotherhood or cult on very pious monastic principles of obedience, silence, abstinence, etc.

In their or his credo the soul was released from the body at death, whence it transmigrated through a series of animals before revisiting humans. To avoid cannibalism they were total vegetarians. For some really crazy reason, bed sheets had to have the impressions of the sleeper smoothed away as soon as he got up, while objects they dropped had to stay dropped.

He and/or his disciples are remembered for their number theories, which include the famous Pythagorean triangle theorem.

Pythagoras shunned fame.

When once asked what time was, he replied; "It is the soul of the world."

Anaxagoras

Towards the end of the 4th century BC some Greeks began to theorise that all things were comprised of different numbers of the same miniature atoms. Anaxagoras (502-428 BC) believed these 'atoma', meaning indivisibles, were combined by a superior being to create matter.

He also thought the Milky Way was made up of millions of stars. Each star was a world containing a sun and planets drifting in Space which had no up or down.

He was jailed by the priests for saying the Sun was a hot rock.

Leucippus and Democritus

About 400 BC Leucippus (450?-370? BC) took up the atomic baton and then passed it to his pupil Democritus (460?-370? BC) who is known as the father of atomic theory.

Atoms were now made of the same material but of different size and shape, floating around in the great void of Space. *"Nothing exists except atoms and space.*" Moreover worlds were created when these spinning 'atoma' crashed into each other and coagulated.

You smash a rock and it turns to grit; smash grit and it turns to powder; smash powder and it turns.......to 'atoma' which float around in the ether until they coagulate and re-materialize. No microscopes, no machines, no slide rulers; just a protractor, a hammer, and just right.

He also wrote "Everything existing in the universe is the fruit of chance and necessity." (In 1755 the German philosopher Immanuel Kant would reiterate this.)

Aristotle

Democritus's atoma however were not just right for Aristotle. He knew better.

Aristotle the Stagirite (384-322 BC) was Plato's most famous student but the master recognised that the pupil "needed a bridle". He was a prolific writer and by all accounts a thoroughly decent person, revered by his students and loved by his family.

His sensible philosophy still shapes society today. At one stage of his illustrious career he was the tutor of young Alexander of Macedonia.

While he got much right with philosophy, ethics, logic and even biology, he got almost everything very wrong when postulating his own conception of the physical universe:

...........the wind was not air in motion

..........sunlight was pure pristine white which when polluted by mist can become a rainbow

...........the Prime Mover *(God)* would not permit a vacuum to exist

..........evolution was impossible

the list of fallacies is long!

He held that the Earth was the centre of the universe surrounded by concentric spheres. The first contained water, the second air and the third fire. All matter sought equilibrium according to the proportions of their elements; bubbles rose through water and smoke rose through air toward fire; iron had more earth than wood so it was pulled down; etc.

Beyond the fire sphere were the crystal spheres in which the planets moons and stars rotated through a fifth element called *aither*. Here resided the 55 pure intelligences controlling the cosmos.

He effectively split the universe in two, proclaiming that earthly objects had no elements in common with heavenly objects. The Sun circled the Earth in a crystal sphere but the four base elements could only move in straight lines up or down unless they were pushed or pulled sideways. He also held that one element could change into another *(transmute)* to explain chemical action.

Fundamental to his theory of matter was that heavier weights of the same shape fell faster than lighter ones. Amazingly no one challenged this doctrine for almost 2 millenia until Galileo did at Pisa.

Stand on a cliff over a pool with a round heavy rock in one hand and a similar shaped ball of light-weight material in the other and drop them both at the same time. They will both hit the water at the same time. Start the lighter ball off an arm span lower than the rock and release them simultaneously. The sound on contact will always be plip-PLOP, not PLOIP, nor PLOP-plip. The heavy rock will never catch up with the light one.

Time the journey from different heights. 16 feet takes one second, 32 feet-1.4 seconds, 64 feet-2 seconds. Play with the results and discover that height divided by the square of time is constant, producing the ubiquitous inverse square equation:

$$\frac{height}{time^2} = constant$$

Aristotle established the peripatetic *(walking)* school in the Athens lyceum. It practiced lectures on the hoof as students strolled around the campus with their professors.

With the death of Alexander the Great in 323 BC all things Macedonian were very much on the nose in Athens, so Aristotle retired to the country, but his erroneous science unfortunately didn't.

These unscientific suppositions now piggybacked his pearls of philosophy which lent them credence and authority, becoming indoctrinating dogma which perverted the course of physics for millennia.

Theophrastus

Aristotle's devoted ex-pupil and successor was the philosopher Theophrastus *(not a Macedonian)*. For 35 years he very successfully ran the peripatetic school which sometimes had 2000 students. The curriculum was faithful to all things Aristotelian further entrenching bad physics.

Theophrastus' own writings in science were based on practical observations, and his botanic studies were still definitive textbooks as late as the middle ages

He also noticed that substances other than amber had an ēlektron soul or spirit.

Early Greek Mathematics

While physics was suffering as a branch of philosophy, the pure science of mathematics prospered. *(Mathematics is grease in the philosopher's grasp.)*

Around 300 BC Euclid published 'Elements', a thirteen book anthology of geometry which included his own number theory. This famous comprehensive work survives to this day as a geometry text book.

Eudoxus of Cnidus (around 300 BC), Diophantus (250 BC) and Archimedes are just a few of the other names that bejewelled early mathematics and geometry.

Archimedes

Archimedes (287-212 BC) of the Greek city state of Syracuse in Sicily was intentionally a theoretical mathematician, and unintentionally the applied mathematician for which he is most remembered. Investigation of the lever led to the quote "Give me a lever long enough and a place to stand and I'll move the Earth." His work in hydrostatics led to the discovery of specific gravity and the exclamation "Heurēka!"

He devised huge catapults for the defence of the city against invading ships and did extensive analyses of reflections from both flat and curved mirrors. He also charted the movements of stars about the equinoxes

The two equinoxes (equal days and equal nights) for Earth, happen when the Sun crosses the equator on March 21 and September 23.

He used Eudoxus's work on infinitesimals to prove that the volume of a sphere is $4/3\,\pi r^3$, which is two-thirds the volume of a cylinder that circumscribes it and he showed that its surface area is $4\,\pi r^2$. He was extremely proud of these two results.

The mathematician and astronomer Conon of Samos, a close friend and constant correspondent of Archimedes, is said to have been the true discoverer of formula for the 'spiral of Archimedes'.

Archimedes died tragically when the Roman army commanded by Marcus Claudius Marcellus conquered Syracuse in 212 BC. He refused to leave his work when ordered to by a soldier saying "Do not disturb my diagrams." The soldier promptly shafted him.

Aristarchus

Archimedes's notes record that Aristarchus of Sámos (310-250 BC) had believed the Earth revolved every 24 hours, and it as well as the planets revolved around the Sun. He also recorded that the Aristotelian scholars gave Aristarchus a very hard time, proclaiming that it was logically impossible for something as massive as the Earth to revolve around something as tiny as the Sun.

The only work of Aristarchus that survives concerns a method for measuring distances to the moon and the Sun. The method was right but his mathematics was poor. The rest of his works were destroyed.

Eratosthenes

In about 250 BC Eratosthenes measured the circumference of the Earth and was about 15% too large. Its true radius at the equator is 6,378 kilometres.

He knew that on the summer solstice the Suns rays fell vertically down a well at Aswān, but in Alexandria 800 kilometres away, he measured an angle of deviation of about seven degrees off vertical. Reasoning that the Sun was so far away that its rays were parallel, he made his calculations.

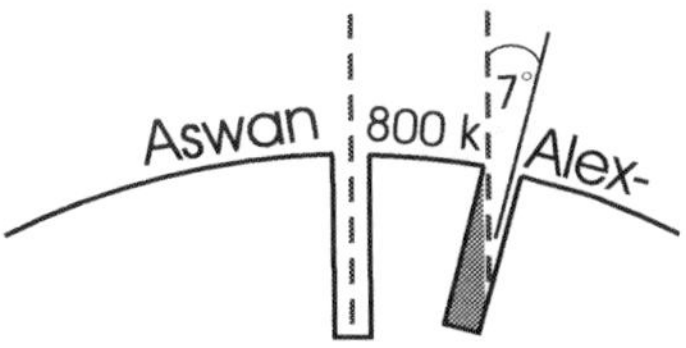

As head of the library of Alexandria in 240 BC he introduced the leap year, which demonstrates the accuracy of astral observations back then.

Apollonius

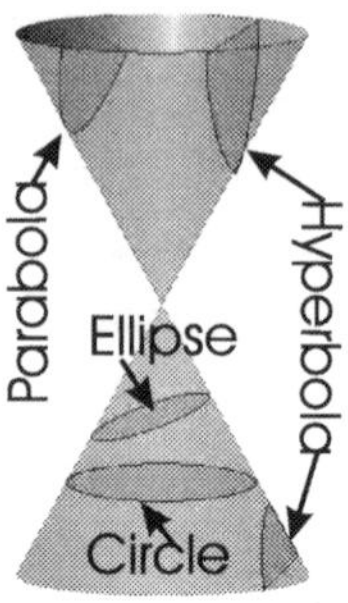

Apollonius of Perga (260-190 BC) who earned the title 'The Great Geometer', showed how the basic shapes of circle, ellipse and parabola are straight slices of one cone while a hyperbola is a straight slice of a double top-to-top cone.

Since the moving Earth model of Aristarchus had been erased by the Aristotelians, Apollonius concocted the concept of planets moving in small 'epicycles' which in turn moved on much larger 'deferents' *(orbits)* around the Earth.

Hipparchus

In around 150 BC the very brilliant astronomer Hipparchus of Nicaea, *(now Iznik in Turkey)* (? - ≈127 BC), measured the distance to the Moon in multiples of the Earth radius. He used the parallax method measuring the angle of moon either at dawn then 6 hours later at midday, or at sunset then 6 hours later at midnight. The best results were determined from observations made at two places on the same meridian. Stars near the line of the Moon are observed at the same time to help in eliminating errors of refraction and instrumental inaccuracies.

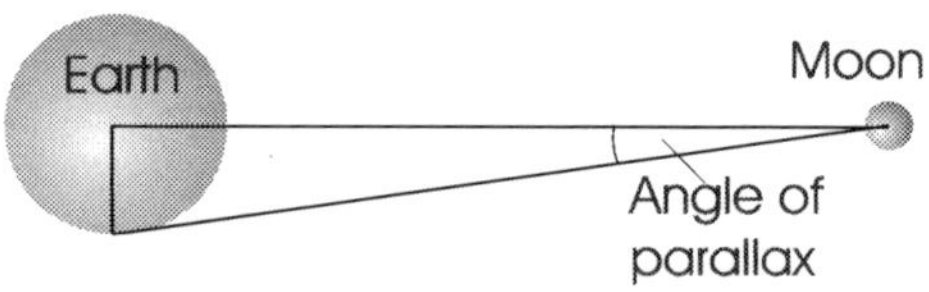

Hipparchus's results put the Moon's Parallax at 58′ *(almost 1°)* at a distance of 59 Earth radii, quite close to today's figures of a parallax of 57′02.608″ and a distance of 60.2 mean Earth radii. The mean moon distance from Earth is close to 384,400 kilometres.

If Hipparchus had used Eratosthenes' radius of the Earth in his calculations then he too would be 15 % out.

i.e.

$$\approx 6370 \times \frac{115}{100} \times 59 = 432{,}200 \ \text{kilometers}$$

The angle of parallax of the Sun across 12 hours of Earth's rotation is less than 0.01 arcsecs troublesome because of the effects of Earth's atmosphere.

Hipparchus combined the attempted measurements of the distance to the Sun by parallax with estimates of its size derived from timing the moving footprint of the Earth's umbra on the moon during an eclipse.

He concluded that the Sun was both huge and an enormous distance away.

Hipparchus is renowned for his work on equinoxes of the planets. He compared star data from his time with data from 150 years earlier recorded by Timocharis of Alexandria. Both were recorded at the equinoxes.

The principle objection to Aristarchus's idea of a heliocentric *(Sun at the centre)* solar system was that the data didn't quite fit a model with circular orbits. Hipparchus liked things to be as simple as possible, and incredible as it sounds, to him the geocentric model with its deferents and epicycles was more compliant.

His other works include the calculation of the length of the year to within 6½ minutes, and the establishment of the for-runner of all trigonometry tables, which applied to the length of chords of a circle for every ½° from 0° to 180°.

In 129 BC he produced a star catalogue of 850 stars.

Lucretius

The last word for the millennium comes from the Roman poet and Philosopher Titus Lucretius Carus (? - ≈54 BC) who is remembered for his long poem 'De Rerum Natura'; *('On the nature of things')*. Lucretius was influenced by the teachings of Epicurus (341-270 BC) who headed the other private school in old Athens. (*The garden school*)

Epicurus advocated having a good time by maximising pleasure while minimising pain without worrying too much over what the Gods might think.

De Rerum Natura contains that message as well as descriptions of atoma which differ in size shape and weight. The atoma drift and merge in the infinite void of Space. It also contains the first recorded pronouncement of the conservation of energy;

"Things cannot be born from nothing nor reduced to nothing."

The millennium closed with an essentially Aristotelian cosmos, some great mathematical tools, thousands of astral fixes, ēlektron-spirit, leap years, and the birth of Christ.

THE ETHER MILLENNIA AD

CHAPTER 2

Hero aka Heron

The 1st millennium kicked off with great promise. Hero of Alexandria (approx. 20-60 AD) was both a mathematician and mechanical engineer. He developed a surveying device called a dioptre *(a prototype of the theodolite),* a fountain that used air pressure to push up water spouts, and the incredible 'aeolipile' aka 'wind ball', the world's first steam engine. Steam entered a sphere through its side axles and exited through angled jets about its meridian causing it to rotate.

Science then was still under the influence of the Aristotelians and would be so for eons into the future. All things were held to be comprised of earth, air, water and fire. Consequently Hero interpreted the steam as being the transformation of water into air.

Whether by bad science or as a consequence of the abundant supply slave labour, the steam machine was shelved then forgotten. It would be resurrected or reinvented fifteen centuries later when the demand for energy exceeded the output that cheap human labour could produce.

He famously derived the Herons triangle formula

Area of a triangle sides a.b.c = square root s(s-a)(s-b)(s-c) ...s=(a+b+c)/2

Ptolemy

Claudius Ptolemaeus aka Ptolemy (100-170 AD) established the Ptolemaic system for astronomy with foundations set in complex trigonometric granite. This essentially mathematically based model would dominate western astronomy for the next fourteen centuries.

He continued where Hipparchus, whom he greatly respected, left off. There is even a plane geometry theorem called 'Ptolemy's theorem' which he pinched from Hipparchus.

Ptolemy established *'beyond all doubt'??* that the Earth was the centre of the universe around which the Sun, the planets and the stars circled.

- He *'clearly'?* showed that the backward motions of Mercury, Venus and Mars observed at certain periods of the year were because, as Hipparchus wrote, they moved in small circles called 'epicycles', which themselves moved around much larger circles called 'deferents' whose centre was the Earth.

- To account for discrepancies between observed data and theory, Ptolemy invented the 'movable eccentric'.

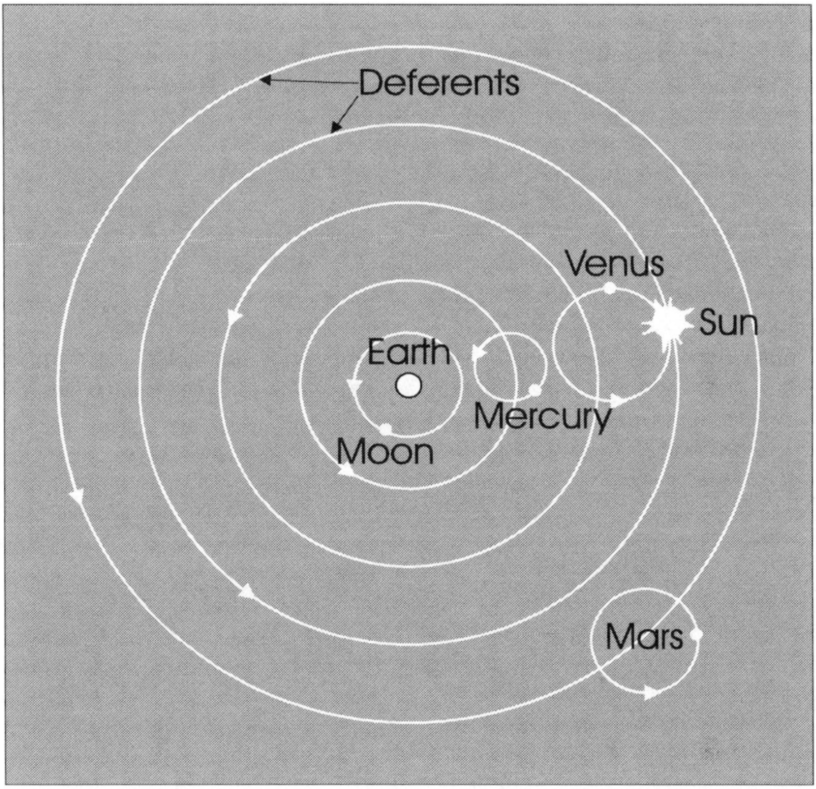

- This meant that the centres of the deferents also circled the Earth following paths that approximate to our modern equatorial satellite orbits.
- Now there was a small circle around the Earth which was the path of the centre of each very large circular deferent, which itself was the path of the centre of the little epicycles, which were the paths of the planets. Only the Sun and the Moon did not have epicycles.
- All the planets move around the Earth's ecliptic, *(a plain roughly extending out from the equator),* sometimes below it and sometimes above it.

(It borders on classic pseudologia fantastica)

The evolution of such a model from observations is a remarkable achievement of a mad genius. Imagine the abacus whizzing and protractor spinning involved in getting the facts to fit the erroneous theory, and they did almost fit it.

Subsequent observations over the next 14 centuries succumbed to this headache of a doctrine. The later astronomers must also have been mathematical savants to be able to navigate the mad maze of epicycles and deferents with moving eccentricities.

Ptolemy fared better when evaluating events restricted to the Earth which he believed was round. His studies on the magnification and refraction of lenses and glass spheres filled with water were exemplary.

Cosmas

The world was flat again by the 6th century. This was championed by heavyweights of Christianity like Cosmas of Alexandria and Constantine of Antioch, the father of 'Christian topography'. Any poor soul who contradicted it was toast. The flat Earth doctrine dominated the next three centuries,

It's hard to delineate exactly when Earth became round again, but maps of the late medieval period had it round. I don't think it was ever square but had any Pope or potentate declared it so, their scholars would have proved it.

EGYPT - INDIA - ARABIA

Indian Mathematics

From the 6th to the 12th century, while Christian doctrine was abetting Aristotelian science in the west, the works of the early Greek mathematicians and astrologers migrated into India where they flourished in the domain of Vedic mathematics.

Vedic mathematics is an ancient Hindu science dating back to the time of Buddha (563-483 BC). There are 16 basic formulas covering all calculations, which allow the practitioner to reach solutions via very fast mental arithmetic. All calculations require the use of 'zero' and often the use of the decimal point. Decimal numbers appear on Hindu temple walls in the 3rd century. In the West we are familiar with the promotional tours of Vedic adepts, some as young as seven, who amaze audiences as they race calculators to complex conclusions.

In the 6th century the Hindu mathematician and astronomer Aryabhata worked out that the moon orbits the Earth which in turn orbits the Sun. He went on to precisely predict eclipses of both. He expanded pi to 4 decimal places. ($\pi = 3.1416$).

The 9th century mathematician Mahavera wrote treatises on zero though he got division by zero wrong. He thought $N/0 = N$

The great 12th century mathematician and astronomer Bhaskara corrected this showing that $N/0 = \infty$ *(infinity).* He wrote six books on mathematics which included his own work on operating with irrational numbers. One book 'Lilavati' *(The Beautiful)* was a summary of Indian mathematics. Another was 'Karanakutuhala', on astronomical wonders.

Arab Mathematics

Indian mathematics came to Bagdad in the 8th century. Al-Khwārizmī Muhammad ibn Mūsā (780-850), made a compilation of Hindu mathematics in his treatise on 'al jabr', *(algebra),* which included his own work on algorithms.

Al jabr was translated into Latin by Gerard of Cremona in the 12th century who took it into Europe. About the same time, Euclid's Elements were translated from Greek and Arabic into Latin. Both translations woke the European dormice of the dark ages.

Over the next four centuries volumes of mathematical and philosophical works from ancient Greece, Alexandria, India, Bagdad and Persia were absorbed by Europe. The symbols + - × ÷ < > were added for simplicity, as was the equals sign '='. As well there came to Europe the astral locations of literally thousands of stars at different times of the year.

In the late twelfth century the compass arrived from China.

When a length of lodestone (magnetite) is suspended from a thread tied in its middle, it always pointed north-south. When a needle of lodestone is glued to a round float and placed on the surface of a bowl of water or alcohol it becomes a ships compass. Western ships captains could now navigate as the Chinese had done since 270 AD.

EUROPE WAKES

Petrus Peregrinus

In 1269 the French scholar and engineer Pierre Pèlerin de Maricourt aka Petrus Peregrinus *(Peter the Pilgrim),* published his 'Letter on the Magnet'. It was a summary of known magnetic phenomenon along with descriptions of his experiments.

He made a magnetic sphere out of lodestone and using a floating magnetic compass found the sphere's poles by tracking the longitudinal magnetic meridians running from one pole to the next.

He coined the term 'pole' for the magnet since he believed, as did a lot of people, that it was the Pole Star that attracted the compass needle. He also noticed, as had the Ancient Greeks, that when a magnet was broken, each segment was itself a magnet with two poles, and that opposite poles mutually attracted while like poles mutually repelled. He also observed that a strong magnet can reverse the polarity of a weaker magnet.

Nicholas of Cusa

Nicholas of Cusa (1401-1464) was a German cardinal, scholar, mathematician, scientist, and philosopher. Perhaps because he was a powerful Roman Catholic prelate, this intellectual could get away with questioning the doctrine the Earth's immobility and its geocentricity.

In 1440 he supported the dangerous idea that the universe was infinite and riddled with stars ad infinitum.

He also suggested the amendments to the calendar which Pope Gregory XIII, *(Pope from 1572-1585),* would later implement.

Cardinal Nicholas described 'the learned man' as one who is aware of his own ignorance.

Regiomontanus

Regiomontanus was the Latin name of the illustrious German astronomer and mathematician Johannes Müller von Königsberg (1436-1476). He was a wunderkind graduating from University of Vienna when 16 years old.

In 1461 he worked with his teacher, the astronomer Georg von Peuerbach on a commission from Cardinal Bessarion, the papal legate to the Holy Roman Empire. Its directive was to compose a short summary of 'Ptolemy's Almagest' and clear up some confusion that occurred in a previous translation.

Peuerbach died at the beginning of the project and it fell to Regiomontanus to finish it, which he did the following year. It was not published until 1496 when, incidentally, Copernicus was 26 years old. It contained suggestions for alternatives to the Ptolemaic model to explain the backwards motion in the epicycles of Mercury, Mars and Venus.

For a time line, the Polos were in China around 1275; Columbus reached the Americas in 1492; Vasco de Gama rounded the Horn and entered the Indian Ocean in 1497; Magellan arrived and died in the Philippines in 1521.

Copernicus

Nicolaus Copernicus (1473-1543) was a Polish astronomer and a Canon of the Catholic Church which provided his sinecure. He finished his thesis, 'De Revolutionibus Orbium Coelestium' *(The Revolutions of Heavenly Spheres)* in 1530, which demonstrated that the Earth circled the Sun. This was very much at odds with the Ptolemaic dogma in the west, and with the Church which held the Bible sacrosanct.

Joshua 10:12. Then spake Joshua to the LORD in the day when the LORD delivered up the Amorites before the children of Israel, and he said in the sight of Israel, Sun, stand thou still upon Gibeon; and thou, Moon, in the valley of Ajalon.

To the Church this was proof that the Earth was the stationary centre of the universe around which astral bodies moved.

It's a long bow yet Copernicus, who was by nature a timid person, feared its arrow. He was terrified of the consequences of uttering heresy.

Curiously it was his local peers of the frock who encouraged him to publish his thesis. Initially he thought that dedicating the work to Pope Paul III might provide sufficient life insurance, however he seems to have had second thoughts or fears regarding this. He delayed publication until he was on his deathbed in 1543, and taking more precautions still, he used a Lutheran printer in Nürnberg.

The Church at first seems a little amused by the work. It was branded 'a comedy of errors' and even a 'travesty of logic'. It was also supremely boring and the first edition of a thousand did not sell out. Ridicule was compounded by Copernicus's sums which were a little scratchy. Decade by decade however support grew, as did controversy.

In 1616 the Empire struck back. The Church of Rome reasserted its authority over reason and censured all books by Copernicus and all references to him. Most were burnt.

The suppositions of Aristotle and Ptolemy prevailed. The Sun still circled the inert Earth.

Filippo Bruno

In the town of Campo dei Fiori *(now called Brinzio)* which is about 50 kilometres North West of Milan, a statue celebrating free thought was erected in the late 19th century on the site of the martyrdom of the Italian philosopher, astronomer and mathematician Filippo Bruno (1548-1600), who was a victim of the Catholic inquisition.

Copernicus's Universe in which the Sun and all the stars and planets were contained in one heavenly sphere was finite. Like Cardinal Nicholas, Bruno believed the universe was infinite, unknowable and perfect. Unlike the Cardinal, Bruno lacked friends in high places.

After 8 years of jail and torture he still refused to recant so they burnt him at the stake.

William Gilbert

The first of the English supporters of Copernicus was William Gilbert, (1544-1603) *(physician to Queen Elizabeth I)* who postulated that the 'fixed' stars are all different distances from Earth.

In 1600 he published 'Of Magnets, Magnetic Bodies and the Great Magnet of the Earth', coining the terms: *'electric force'*, *'electric attraction'*, and *'magnetic pole'*. He repeats, either by design or by chance, Peregrinus's experiments, moving a very tiny magnetic needle over and slightly above a large sphere of lodestone.

He showed that a needle not only pointed to the sphere's north pole or from its south pole but also dipped down into the poles as well. The Earth he claimed, could be viewed as having a giant bar magnet through its axis.

He also believed that the planets were held in their orbits by some kind of magnetic force and that magnetism was itself a subtle flow of effluvium or *ether.*

Gilbert showed that ēlektron-spirit or electric attraction could be generated in glass, diamond, sapphire, amethyst, opal, jet, sulphur crystals and many other substances which he grouped in the term 'electrics'. Those that weren't electrified when rubbed he called non-electrics. He attributed electric attraction through the *ether* to the flow of electric effluvium, as distinct from magnetic effluvium.

The 'gilbert' is the unit of magnetomotive force in the c.g.s. system, i.e., The centimetres-gram-second system.

Lenses

I like to think that idea of an optical lens was inspired by the humble jellyfish. In semi-tropical waters there is a jellyfish of about 5 cm in diameter with a domed 1 cm thick top which is as transparent as good optical glass. When washed up on the beach the base flattens, and the grains of sand beneath are magnified. The earliest microscope recorded however, was made by the miniscus of a drop of water caught in a small hole.

Glass has been around for eons. Fulgurites are tubes of hard glassy material formed when lightning strikes sand. Some are 5 meters long and 4 meters wide. All are spectacular. Over time these weather into pebbles.

Before Christ the Chinese were grinding glass into lenses and making spectacles, but it was not until the 14th century that spectacles appeared in Europe.

The Assyrians were also aware of the magnifying power of segments of glass spheres. Their empire, before it collapsed in 600 BC, stretched from the top of the Persian Gulf to the western shores of the Mediterranean Sea and included lower and middle Egypt.

The first records of compound microscopes appear in the 16th century. Three Dutch spectacle makers, Hans Jansen, his son Zacharias Jansen, and Hans Lippershey simultaneously began using convex lenses in tandem which could magnify objects about 15 times. These would be improved a hundred fold in the next century by the Dutchman Antoni van Leeuwenhoek (1632-1723). Using a clever juxtaposition of focal lengths he created and marketed powerful compound microscopes.

In 1606 two children *(other sources claim that it was the apprentice),* playing with two lenses in the shop of Hans Lippershey, magnified the weather vane of the church down the street. "Look Hans no distance!" The following year Hans demonstrated his 'looker' to the Netherlands Government. Its military use was obvious. The following year little telescopes, peculiarly called 'Dutch Trunks', were sold in cities throughout Europe.

There are also reports that at that time the Roma gypsies were selling little spy glasses at county fares. These may have been the Dutch ones, or may have even predated them.

Galileo Galilei

The most infamous of the Copernican transgressors was Galileo (1564-1642). Born near Pisa he entered the University of Pisa aged nineteen to study medicine, *(by then he was a highly accomplished musician excelling on the lute).* Fortunately for science his enquiring mind soon drew him in another direction.

The life changing moment is said to have happened while he sat in the Cathedral of Pisa contemplating his future. He was absentmindedly watching the great lamp which hung from the roof swinging in the breeze coming through the open door. It seemed to him that irrespective of the amplitude of the lamp's arc, the time for each cycle was the same.

Alert now he timed the swings using his pulse and confirmed this notion. Back home he constructed his own pendulum and quickly discovered that the period of oscillation was the same for any weight irrespective of the amplitude of the swing provided the length of the pendulum arm or string remained constant. He was soon able to show that the period of oscillation was proportional to the square root of the pendulum's length. Scientific research thereafter became his calling.

Legend has it that he showed that objects with the same shape but of different weight, when dropped simultaneously off the top of the leaning tower of Pisa hit the ground at the same time. His publications however only describe experiments which time balls rolling down inclined planes.

From these results he determined that: for falling bodies.

$$\frac{height}{time^2} = constant$$

As well he determined that projectiles follow parabolic trajectories, with those leaving at an inclination of 45° having the greatest range. He described the distance 'd' an object falls in a time 't' by the equation:

$d = \frac{1}{2}at^2$ (where 'a' is the measure of how velocity changes)

Both the above formulae were contrary to the teachings of Aristotle and therefore conflicted with the teachings of his peers.

Another contentious issue concerned the natural state of bodies. It was the accepted criterion that the natural state of a body was when it was at rest. A moving body then, had a theoretical, yet to be discovered, continuous force acting on it to keep it in motion. Galileo countered this by arguing thus:

"A block when pushed will slide a certain distance along the floor. If the block is finely sanded and a lubricant applied to the floor, given the same push the block will slide further. If the block is more polished and a better lubricant applied to the floor, the block given the same push as before will slide further again.

Therefore, given a perfect polish and perfect lubricant and no friction, the block will slide for ever. No force is needed to maintain the motion, though an external force is required to change the motion." *(This would become Newton's first Law).*

Galileo heard of the Dutch Trunks in 1609, the same year they went on the market. At that time he was Professor of Mathematics at the University of Padua *(near Venice).* Due to financial difficulties he supplemented his income with proceeds from a small shop in which he manufactured, supplied and repaired scientific instruments.

Acquiring some odd lenses from a spectacle maker, he quickly assembled a prototype telescope with three times magnification. Then he analysed the principle, learned to grind glass, and built an improved model. It had a plain concave lens and a plane convex lens set at the ends of a lead tube.

A few months later he presented to The Doge of Venice a telescope with eight times magnification. The Doge was delighted. Galileo's wages were doubled and his professorship secured, which peeved many of his peers. He built for himself a telescope with 20 times magnification and saw wonders. Two moons of Mars, four of Jupiter, and the mountains and craters on our Moon. The milk of the Milky Way was made of tens of thousands of stars, etc.

He became a celebrity. A banquet in his honour was given in Rome and he was presented to Pope Paul V.

It's hard to imagine anyone better placed or equipped than Galileo was to create the first quality telescope. He had the mathematics to calculate the curvature of the lenses, the hands on workshop to grind and assemble them, and just up the road was Venice, the centre of the European glass industry where the finest quality blanks of optical glass were available.

With each of his discoveries the old guard became more vindictive and it seems that Galileo both planned and enjoyed retaliating. In 1611 he brought to Rome a little box containing a few rocks of barium sulphide called 'solar sponge' by the alchemists. Before entering the scholars' den he had saturated the rocks with sunlight. Aristotelian scholars had always believed and taught that light was a spiritual or ethereal quality of illumination. Choosing his moment Galileo removed the rocks from their cache. Astonishingly, they glowed in the dark which proved light was a transportable thing. Score Galileo 1- peers zero.

Galileo theorised that light was streams of little particles that could penetrate of bounce off matter.

In 1612 Christopher Scheiner entered Galileo's world. He was a German Jesuit who was forbidden by his superior to publish the results of his astronomical pursuits. He had observed sunspots and postulated there might be tiny planets circling close to the Sun. Via pseudonyms and other agencies this news was communicated to Galileo with the accompanying comment that the agent supposed "it was nothing new to the great master."

In fact Galileo was oblivious to the phenomena. His modus operandi was looking through a telescope avoiding the Sun which blinds anyone observing it, especially anyone with a telescope.

He received the communication in 1612 while working with his ex-pupil abbot Benedetto Castelli on a publication. He then set the Abbot to observing and recording the reported sunspots using a camera obscura.

*The camera obscura, (Latin for dark room), had been around since Ancient Greece. It was both a fairground novelty and a device used by artists to copy landscapes in perspective. Some of the former types resembled gun turrets in with peepholes substituted for the barrels. The panorama was projected via the hole onto the back wall in the darkened turret where it appeared upside down. In smaller models, the image could be projected onto a mirror which sloped at 45° to line of sight. This reversed the image again when viewed from above making it the right way up.

In 1599 Giambattista della Porta greatly improved the device by enlarging the hole and installing a lens.*

By substituting a telescope instead of a lens, the back screen clearly showed the sunspots grow and diminish (therefore not tiny planets). Because patterns repeated themselves roughly once a month they concluded that the Sun itself must rotate. Detailed drawings were taken and engravings made for publication. The book was published with the backing of the Lincean Society, (patrons of Galileo), who insisted on a preface concerning their involvement. Galileo resented this and many Jesuit colleagues of Scheiner resented both him and the Linceans for stealing their thunder. Galileo had scratched very dangerous foe.

After the book's publication in 1613 it was clear to all that Galileo was a supporter of Copernicus. He was damned from many a pulpit for his heresy.

In 1632 he published 'Dialogue on the Two World Systems'. It contained a series of conversations between an Aristotelian stalwart 'Simplicio', a Copernican character 'Sagredo', and Galileo's alter ego 'Salviati'.

Sagredo is a layman who sensibly listens to, then upholds Salviati's ideology. Simplicio is the arch conservative who is given the dialogue verbatim of what was topical for some very influential people, not the least of whom was the Pope himself. This overt lampoon of the Vicar of Christ was heretical grist for Galileo's enemies.

In 1633 he was summoned to Rome to face the inquisition and found guilty. Under threat of torture, which he didn't relish at the age of 68, and well aware of old Bruno's demise in Campo dei Fiori, he made a very eloquent if overstated public adjuration. *(methinks he confesses too much.)* He promised never again to utter write or teach that the Earth circled the Sun, upholding that Earth was the inert fixed centre of the universe. Some say that at the end of this speech they heard him utter sotto voce; "and yet it moves". His sentence to life imprisonment was commuted to house arrest by Pope Urban VIII, who as Cardinal Barberini had been his good friend and patron before being elected.

One might think that following his persecution Galileo would be hostile towards the Church, but on the contrary, as a devout Christian he was very concerned for its prosperity. He foresaw the inevitable, that in a conflict between science and rigid Holy doctrine, science would prevail and thus challenge and damage the infallibility of the Church. He ardently believed in the separation of Church and science, and considered the physical proof of faith superfluous. In his private copy of 'The Dialogue' he added the following lines for Sagredo;

"*Take note theologians, that in your desire to make matters of faith out of propositions relating to the fixity of Sun and Earth, you run the risk of eventually having to condemn as heretics those who would declare the Earth to stand still and the Sun to change position.*"

flees

He retired to the country for his twilight years writing up unpublished studies concerning the motions of objects at constant speeds and/or uniform accelerations, including their trajectories if airborne. He also worked on the breaking strains of materials. In 1638 these works were smuggled out of Italy and printed in the Netherlands under the title 'Dialogues Concerning Two New Sciences'. By then Galileo was blind and perhaps because of his infirmity, he was left in peace.

He worked on for another four years with Vincenzo Viviani, a young student as a scribe and died in his company in 1642.

*One hundred and eighty years later, in 1822 Pope Pius VII declared that the Sun was the centre of the solar system. Thus after a struggle of more than 2½ millennia, the Earth, with divine Papal approval, finally circled the Sun.

An inquiry instigated in 1979 by Pope John Paul II, née Karol Wojtyła, aka Andrzej Jawien and Vicar of Christ, concluded 13 years later that the Church had erred in castigating Galileo.*

Tycho Brahe

A little earlier, back in 1588 *(when Galileo was sixteen)* and up north in Denmark, King Frederick II, a patron of science died in 1588. In 1596 after 8 years of regency rule, his 19 year old hard-living, hard-drinking and pragmatic son, Christian IV was crowned. These events inadvertently revolutionised physics and astronomy.

In 1576, Frederick II had granted title to the island of Hven to his friend, the astronomer and nobleman Tycho Brahe (1546-1601). Hven, then called Ven is a small island of about 50 square kilometres in the middle of 'The Sound', which is the hectic passage through which ships pass between the North Sea and the Baltic. In addition Frederick provided the capital to build a very flash observatory with a three story palatial mansion to house Tycho and his assistants and students. The observatory held the finest pre-telescope equipment in the world.

There were astrolabes, *(the forerunners of sextants)*, and quadrants which measured the elevation of stars between 0° and 90°.

(The huge mural quadrant had a radius of 6 meters a scale accurate to 10 seconds of arc). There were parallactic rulers, which took readings from two locations, armillary spheres which placed stars and planets on metal rings relative to the Earth and lots more. Everything was built and decorated to the highest standards by Europe's master craftsmen.

They made it twice as nice as paradise and they called it Uraniborg, after Urania, the muse of astronomy.

There followed twenty years of dedicated astral observations, cataloguing 777 stars and accurately plotting the paths of the five known planets, particularly Mars.

Tycho concluded that the planets circled the Sun as did Copernicus, but hedged his bet by saying that they and the Sun circled the Earth.

Back in 1577 he showed that comets were astral bodies, and not the supposed atmospheric phenomena.

When Frederick died Tycho faced mounting opposition from within the Danish council and from Frederick's heir, with whom he did not 'get on'. Tycho wasn't an easy man for anyone to get on with. He was a Viking who lived and dined in the Viking drinking, wenching, boasting, bullying tradition of the mead halls.

When he was a 20 year old student, a drunken argument erupted with another student over a mathematical point which culminated in a sword fight. Instead of coming away with the obligatory elegant slash on the cheek prevalent amongst wealthy students, *(particularly in German Universities),* Tycho lost his nose. Thereafter he travelled with a little box of spare noses cast in silver or gold which were painted skin-tone. He also carried a little pot of glue to fix one when it became wobbly.

In 1597, soon after Christian's coronation, his fief was cancelled and the great observatory abandoned. He packed up all his secret and secreted data, and then drifted about Europe with his factotum/companion dwarf.

Some historians regard Tycho as a statistical niggard, but no one knew better than he that the data from Uranienborg was not only his raison d'être, but also his ticket to ride.

His travels end two years later in Prague where he acquires a new patron, the half mad Holy Roman Emperor Rudolf II, who was in lucid moments a great patron of science. He decided to build for Tycho a bigger and better observatory than Uraniborg.

In 1600 Tycho employed the promising astrologer Johannes Kepler as his assistant and their conjunction is what revolutionised astronomy.

In 1601 before the observatory was completed Tycho died 'à la grande bouffe'. He overindulged at a dinner party where it was etiquette not to leave the table until the meal was finished. Before he could reach the toilet what he called "something of importance" *(his bladder)* burst and he died eleven days later.

Kepler

Following Tycho's death Johannes Kepler (1571-1630) became Rudolf II's court astronomer and mathematician and was placed in charge of the new observatory.

Remarkably this devout Lutheran was given a senior post when all about him every other Lutheran in civil service was getting the boot or worse. The Catholic reformation was a bad time for Protestants.

Back in the 90s while Kepler was studying Lutheran theology, *(he failed the ministerial exams),* he read Copernicus's book and accepted its heliocentricity, believing *(as did Copernicus)* that God would have designed things to be as simple as possible. *(Recall that Hipparchus thought geocentricity was simpler.)*

The metaphysics of the universe was ever-present behind Kepler's real physics, and it could be argued that metaphysics held the whip. He championed the analogy that the Sun was the Father, the fixed Stars the Son, and the *ether* in between, the Holy Ghost. He also believed that comets were evil omens.

Like many stargazing physicists Kepler received some income from doing astrological charts for the wealthy and he was quite successful. In 1995 he drew up an astrology based calendar for the town of Graz, correctly predicting a peasant uprising and a Turkish attack. *(There were several during the Austrian-Turk war of 1593-1606.)*

Kepler had done much to earn his position at the new observatory. For six years prior to his appointment he held the chair of astronomy and mathematics at the fledgling Graz University where he published 'Mysterium Cosmographicum' *(Cosmographic Mystery)* in 1596. It brought observational data and calculations of circular mathematical orbits to within 5% of each other. He also included the proposition that the Sun emits a force on a planet which was inversely proportional to the square of its distance.

"There exists only one moving soul in the centre of all orbits; that is the Sun which drives the planets the more vigorously the closer the planet is, but whose force is quasi-exhausted when acting on the outer planets ...."

Following Tycho's death, Kepler not only gained an observatory, but after a political tussle with the Danes' heirs, he also acquired trunk loads of very precise data, not the least of which was the Mars data.

Kepler was the first scientist to apply mathematics to orbits. His circular orbits were not just sketches on a pad; they had to conform to formulae. To this end he 'went to war' with Mars, trying to fit Tycho's data into a circular orbit. The war lasted from 1601 to 1605 and left in its wake 9,000 pages of calculations. The data was accurate to 2 minutes of a degree but the best mathematical orbit he could devise had discrepancies of 8 minutes at certain times of the Mars year. Kepler really believed that the circle and the sphere were hallowed shapes which God would not forsake.

But the facts were facts and with great reluctance and anguish, he ditched the circle. Only then did his mind see patterns in the discrepancies occurring roughly half a Mars year apart, and thereafter the pfennig dropped.

The orbit had to be elliptical with the Sun at one focus.

*In a circular orbit a planet would move at a constant speed around its only focus, the Sun. The shorter the radius the faster a planet must go. In an elliptical orbit the planet speeds up as it nears the Sun, whips around it then slows down. The fastest and slowest points are where the orbit crosses the imaginary line drawn through and beyond both focuses of its elliptical orbit.

Our Earth is 147.1 million kilometres from the Sun on the short fast side (the perihelion), and 152.1 million kilometres away on the long side (the aphelion).

To get some idea of the difficulties faced by early astronomers consider that their observatory is mounted on Earth whose equator is rotating at a fixed 1670 kph and whose orbit velocity around the Sun varies above and below the average of 107,000 kph, depending on the time of year. From this they observe other planets moving at their respective variable speeds also dependent on their times in their years. Add to this that all the planets have orbits almost in the same plane (the ecliptic) making all the observations of them side on. Compound all this with the subterfuge created by our most prominent astral neighbour Venus whose orbit in very close to a circle, being only 1% out.*

Put all that in your abacus and rattle it!

In 1609, the same year that Galileo presented the Doge with a telescope and 24 years before his inquisition, Kepler published 'A New Astronomy Based on Causation or Physics of the Sky' which included the first two of the three sublime laws of planetary motion.

1. Planets circle the Sun in elliptical orbits with the Sun at one focus.
2. The radius of arc of all planets describes equal areas in equal time.

The second law, with its simplicity verging on innocence describes the variations in the speed of planets in orbit.

'A New Astronomy' also contained the bones of gravitation theory in its preface. *"The moon's attraction causes the tide to ebb and that attraction between bodies is proportional to their masses*." It is not mentioned in the text of the work but it was there for all to read.

There existed then in the minds of many theorists a phenomenon called 'action at a distance'. This could be summoned like a genie to explain any inexplicable attraction or repulsion between anything. Very handy but not very scientific, yet it was the mainly held reason as to why astral bodies kept in touch with one another. This theory now got the wobbles because it depended in its way on perfect circular orbits.

In 1619 Kepler published 'Harmony of the World' which included his brilliant third law amongst a lot of Pythagorean numerological, music of the spheres, quasi-religious bunkum.

The cube of any planet's mean distance from the Sun, divided by the square of its orbital period is the same:

$$\frac{d^3}{t^2} = a\ constant$$

He added more bricks to the foundations of calculus by developing a system of infinitesimals to support his optical research.

Like Gilbert he resorted to magnetism to explain the attraction that held the planets in orbit, concluding that the Sun is a gigantic magnet (*which it is*), but this model proved unsatisfactory. There couldn't be enough iron in the Sun to cause this attraction.

There had to be another force involved.

The elliptical orbit shook all who worshipped the perfection of the sphere, including Galileo who clung doggedly to the platonic ideal of circular motion. He maintained that the elliptical paths of comets were optical illusions. He refused Kepler an audience taking the counter argument to the grave. *(How the mighty falter!)* He outlived Kepler by twelve years.

Darwin

*I am reminded of a similar argument concerning the Sun that occurred two centuries later.

In 1831 the 22 year old parson, though son of an atheist Charles Robert Darwin (1809-1882), set sail on the Beagle captained by the young 26 year old Robert Fitz Roy. The Beagle's mission was to chart the South American coast.

The soujorn would stretch five years.

Because social contact between a captain and his crew was taboo, dinner companions were often hired on. Charles though very young was approved and hired. He must have been only an occasional companion because he spent much of the voyage ashore, galloping across South America frequently spending ten hours a day in the saddle, climbing every mountain he could see, digging trenches, collecting, mailing fossils to the Royal Society via passing ships, and so on. He even cruised for 300 miles in an open boat off Tierra del Fuego.

Though he had embarked as a creationist his observations and deductions en route could not support the scriptures' doctrine which dated the dawn of creation as 4004 BC.

Darwin carried with him a first edition of Charles Lyell's Principles of geology and now he was discovering first hand for himself from the evidence of fossils and geology, what some other European scientists had quantified; The Earth had to be much older.

Following an agonising battle between physics and metaphysics he disembarked at Falmouth, an atheist.

Over the next 23 years he worked at publishing his notes and developing the theory of evolution while constantly wrestling with its anti-doctrine conclusions. The first edition of 'On Origin of the species' published in 1859 met with great controversy. It also failed to honour the earlier French pioneers of evolution namely Baron Georges Cuvier (1769-1832), Jean-Baptiste Lamark (1744-1829), and Etienne Geoffroy Saint-Hilaire (1805-1861).

One of Darwin's many opponents was the creationist William Thomson Kelvin (1824-1907), the highly esteemed professor of natural science at the University of Glasgow. Kelvin was a child prodigy who graduated from Glasgow+ University when ten years old.

The age Earth had been a matter of contention for an eternity but it became more so in the mid 19th century as the science of fossils and geology gained understanding. Darwin claimed the Earth had to be over 300 million years old for certain evolutions to occur. Kelvin quite rightly, and at that time very successfully argued, that the Sun could not possibly contain sufficient coal to burn for more than 20 million years maximum. This belied the evidence of the fossils which were diagnosed as being much older than that.*

There had to be another source of fuel.

Cavalieri

Francesco Bonaventura Cavalieri (1589-1647) became professor of mathematics at Bologna University in 1629. He had been a student of Galileo's and remained devoted to him to the extent that he nobly postponed publishing his own theories on indivisibles because Galileo was working on something similar. His theories were principally concerned with evaluating the areas and volumes of geometric figures. There must have been some resolution to his dilemma because in 1635 he published his 'Geometry of the Indivisibles' which he revised in 1647 to accommodate objections to the first edition. It expanded on Kepler's work concerning the measurement of volumes using infinitesimals.

In the 17th century, Cavalieri's 'theorem of volumes between parallel planes' was celebrated.

Torricelli

Evangelista Torricelli (1608-1647) furthered Cavalieri's work on indivisibles to find solutions to the quadrature of curves. Quadrature was the title given to pursuits such as finding the dimensions of a square that had the same area as a circle. These pursuits studied the addition of infinitesimal areas, and what resulted when they approached their limits.

Torricelli had been Galileo's secretary for the last few months of the great man's life. Galileo had invented a water thermometer in 1593 and a gas one in 1607. He also had planted the seed of creating a vacuum in the mind of Torricelli, who set out not only to achieve this, but simultaneously to create a device for measuring 'the mutations of air', *(the difference between light and heavy air.)*

He filled a tube 1.2 meters long test-tube with mercury stopped it, inverted it, submerging the stopped end in a bowl of mercury and then removed the stop. A column of mercury about 96 cm high remained in the tube above the level in the bowl. Above the mercury in the tube was a 24 cm vacancy *(vacuum). (Mercury does not wet glass.)* He noticed different heights occurred on different days, and concluded that the air pressure was changing. This was the world's first barometer.

In 1644 Torricelli published his mathematical treatise 'Opera Geometrica' in which he included a chapter on the barometer.

Pascal

Blaise Pascal (1623-1662) heard of Torricelli's experiment, and built his own mercury column and conducted observations at different heights above sea level. He first carried the column to the top of a church steeple with no result. He then wrote to his brother-in-law who lived in the mountains in Auvergne, asking him to conduct the experiment up there. When the news came down that their column measured 7½ centimetres less than his, it caused a great excitement. The barometer could also be used as an altimeter. For fun Blaise built a device with a 7½ metre column of red wine.

In 1642 he invented an adding machine.

In 1850 John Speke used a crude and delicate glass altimeter to discover the source or highest point of the Nile, which he called Lake Victoria.

Horror Vacui

'Horror vacui', *(nature abhors a vacuum)*, was the title of the argument used by Aristotle to show that neither Nature nor the Gods would permit a vacuum to exist. The mercury or wine should according to Aristotle, and all scholars since him, including Descartes, flow back up the tube.

The barometer's rout of horror vacui caused a monumental kerfuffle amongst scholars of science and philosophy. It wasn't Torricelli's actual device that caused this, but like events at Pisa, this device invalidated a priori facts.

*The impossibility of a vacuum existing was certainly disproved on any scale that Aristotle could imagine, yet when taken to its limit a total vacuum can only be hypothetical. The empty space vacated by the mercury in a column is not strictly a vacuum but contains mercury vapour in such minute quantities that it is very nearly a vacuum. How does one exhaust the last atom of the previous occupant of a volume?

The so called vacuum of outer Space is also only a partial vacuum. At its least it contains starlight which is theoretically streams of photons which are held to have mass. There might be a theoretical total vacuum in a black hole or the black hole might theoretically be one itself.

Therefore on this issue Aristotle gets the benefit of the doubt. The Gods or nature possibly forbid the existence of an absolute vacuum.*

The most sensational physical devices of the early 17th century were the telescope, barometer and altimeter. The great Galileo was a play maker in them all.

Bishop Goldwin

There was yet another legacy from Galileo and that was 'science fiction'.

In 1638 Bishop Goldwin published 'The Man in the Moone', a very popular children's book inspired by Galileo's picturesque descriptions of the mountains and valleys on the moon. It was the tale of young Domingo, who accidentally flies to the moon in a chariot drawn by swans that are following their strange migratory path. He experiences weight reduction leaving the Earth, then weightlessness, then lesser than Earth weight when on the moon, where he cavorts about making enormous jumps. He breathes air all the way.

The frame of the fantasy, in particular the change in Domingo's weight, is surely drawn from Keplers work.

Domingo was to the children of the 17th century what Peter Pan was to those in the 20th. Just about everyone who could read, read it to their children.

Did this inspire Robert Hooke and Newton?

ANALYTICAL MATHEMATICS

Up til the middle of the 18th century nearly all scholastic works were written in Latin although the concise Arabian numerals had displaced the Roman ones. Despite this, in the early 17th century, mathematical proofs were very long-winded.

For example 3 x 6 = 18 would have been written in a Latin sentence as "Multiplicatio trium in sex sunt octodecim."

All this changed in that century.

Demand for algebraic solutions to volumes and areas of the geometric curves of the early Greeks increased the need for simplicity. Both Pierre Fermat and René Descartes were prime movers in this transformation.

Formulae became concisely abbreviated to quarter of a line instead of a quarter of a page. A coordinate system was established which represented curves as graphs thereby providing improved visualisation.

In 1614 John Napier published his tables of logarithms which Kepler enthusiastically adopted.

The significance of thiese transformations can be likened to that of the change over from hand tools to power tools. Not only could tradesmen work faster, but the economy of effort led to more vigorous innovative designs.

Descartes

*Saint Augustine (354-430) wrote; *"If I make a mistake, I conclude that I exist; for he who does not exist cannot make a mistake, so the fact of having made a mistake is proof that I exist."**

René Descartes (1596-1650) of *"I think therefore I am"* fame had an enormous impact on Philosophy in the second half of the 17th century.

He was a rationalist, believing action should be governed by reason. This brought him into conflict with the Catholic Church which more or less held that action should be governed by faith. Mindful of Galileo's demise by the inquisition Descartes held up publishing his pragmatic 'Le Monde' *(The World)* until he was safely established in the Calvanist Netherlands.

He applied his keen mind to all things past and present giving no quarter on any issue. Because of his religious tolerance he was branded as an atheist by the intolerant. He once said "Good sense is destroyed when one thinks too much of God."

Concerning physics he stripped the metaphysical cladding off it, laying bare both matter and forces for clear analysis.

In 1637 he published his famous 'Discours de la méthode'. It had an addendum titled 'La Géométrie' which introduced several conventions to facilitate the writing of equations. These included:

- The x,y,z co-ordinate system.
- Capitals at beginning of the alphabet e.g. A,B,C,D represent 'knowns' or constants.
- Lower case letters from the tail of the alphabet $u,v,w,x,y,z,$ represent unknowns or variables.
- Superscripts represent the number of times a character is multiplied by itself $y \times y \times y = y^3$.
- He also confirmed the prior established operating symbols (+,x,=) etc.

These conventions took a while to penetrate the education system principally because the Aristotelian scholars still held sway, and all things Descartesian were thought to be flavoured with heresy.

Descartes rejected any notion of action at a distance through empty space, propounding that matter could only exert a force on other matter through the *ether*.

Fermat

The brilliant mathematician Pierre de Fermat (1601-1665) is credited as the co-discoverer with Descartes of analytic geometry. Among his many discoveries is a method for finding the greatest or least value of formula, a method very close to differential calculus.

Fermat's principle (*circa* 1650) encapsulates reflection and refraction.

"*A light ray travelling from one point to another will follow a path such that, the time required is either a minimum or a maximum.*"

Huygens

One of Europe's most esteemed scientists of his day was the Dutchman Christiaan Huygens (1629-1695) who formulated the laws of conservation of momentum and of kinetic energy.

He invented a new way of grinding and polishing lenses which produced telescopes with sharper definition. The following year he exposed the true shape of the rings of Saturn and in 1665 he found a new moon of Saturn he called Titan. In 1666 he became a foundation member of the French Academy of Sciences and was given an apartment in the building and a very substantial income.

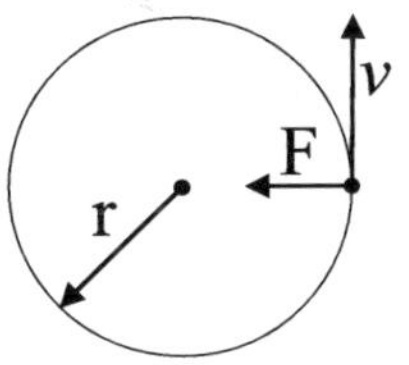

In 1673 he published 'Horologium Oscillatorium', a mathematical masterpiece with formulae for simple pendulums, centripetal force and curvature.

The centripetal force holding a mass in orbit is given by: $F = \dfrac{mv^2}{r}$

His most profound observation which bears his name went largely unnoticed for two centuries.

'Huygens principle' states that: "All points on a wave front can be considered as point sources for the production of spherical secondary wavelets that radiate with the same frequency and wavelength of their parent."

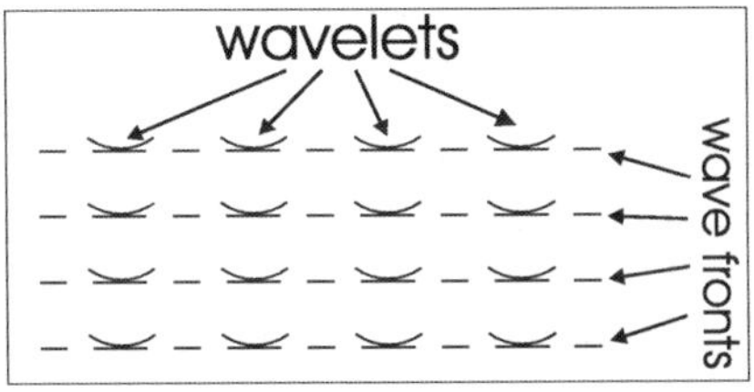

Wallis

The English mathematician John Wallis (1616-1703) was a founding member of the Royal Society. He extended Cavalieri's and Torricelli's work advocating an algebraic approach to geometry and assigning numerical values to spatial indivisibles. Thus he deduced for example:

$$\frac{4}{\pi} = \frac{3\times3\times5\times5\times7\times7\times9\times9\times11\times11\times}{2\times4\times4\times6\times6\times8\times8\times10\times10\times12\times}$$

His 'Mechanics or Tracts of Motion' was published in three parts between 1669 and 1671. It corrected many of the errors concerning motion that had persisted for millennia and included the introduction of the symbol '∞' for infinity. It also forwards the assumption that the Earth's gravity is located at its centre.

He was a cantankerous bugger warring ceaselessly with Huygens and venomously with Descartes.

His colleague at Cambridge was the first Lucasian professor of mathematics Isaac Barrow (1630-1677), who formulated the inverse relationship between tangents and areas, which is the fundamental theorem of calculus.

Hooke

Robert *(Shorty)* Hooke (1635-1703) was an Oxford scholar renowned for his law of elasticity: *('Stretch is proportional to force')*. He was appointed Curator of experiments at the Royal Society in 1662 and elected fellow the following year. He put nature under a microscope of his own design and in 1665 observed the cell structure in a thin slice of cork. He christened the little box shapes in the cork 'cells' because they reminded him of monks' cells in monasteries. That year he published 'Micro Graphia' which contained detailed drawings of magnified snowflakes, lice, fleas, weevils etc.

The same year he helped Robert Boyle develop his Boylean air pump. Boyle had been a member of the Invisible Society in Oxford before it was granted charter by Charles II in 1663 and was renamed the Royal Society.

In 1663 Scottish Mathematician James Gregory drew plans for a telescope which used a parabolic mirror at its base. The light beam was then reflected onto a focal flat mirror and bounced to the eyepiece. Hooke built one the next year and saw Jupiter and Mars rotating and defined their axes. He published much new data about the heavens which included news of the red spot on Jupiter.

He also invented the hairspring for watches and the anemometer for measuring wind speed. He supported Huygens wave theory of light.

From his thirties on, Hooke was repeatedly embroiled with Isaac Newton principally on issues relating to priority of invention.

Newton's Laws and Wars

The person who discovered the new force responsible for attraction through the *ether* was Isaac Newton (1642-1727).

He was born with an *EP&S* spoon in his mouth into what was for him a childhood hell. His father died three months before his birth and he was dumped on his grandparents at age two when his mother remarried a well off minister to whom she bore three children. Newton hated his stepfather and threatened to burn both him and his mother to death as they slept. He was no doubt thrilled when the minister died in 1653. By then Newton was eleven years old and obviously psychotic.

Psychiatric disorder is manifest throughout his life, be it as manic depression, obsessive vindictiveness or balancing on the precipice of self mutilation. He once pushed a bodkin *(a fat blunt needle)* through the corner of his eye socket and moved it about "betwixt my eye and the bone as near to the back of my eye as I could". Incredibly it did no damage.

The student or reader is strongly advised not to repeat this insane act.

With the death of the minister mummy became a land-holder, and though not rich was able to send Newton to grammar school. Though a late starter, the teachers soon recognized his potential and in his final year they encouraged his mother to allow him to sit the entrance exams for Trinity College Cambridge. He passed and entered the college in 1661 as a subsizer which meant that like Tom Brown, he had to fag for senior scholars. *(He would have just loved that!)* He got his BSc. in 1665 and his Masters in 1667.

At that time, Kepler's three laws of planetary motion were 50 years old and Descartes, who viewed reality as mechanical non-spiritual compositions of matter, was twenty years in the grave.

Even so their concepts had not yet penetrated the curriculum of the European universities including Cambridge, so Aristotelian scientific dogma still persisted.

This virulent virus still infected almost every science book and every professor that had taught their content. It was a battle hardened Trojan virus, immune to evolution.

Between his BSc. and his Masters, Newton retired to the country to dodge the plague, but he wasn't fiddling while London burned in 1666. He worked there on 'Opticks' as well as inventing calculus, both of which became controversial.

His mathematics professor, Isaac Barrow had realised the connection between differentiation and integration but had failed to find the interconnecting method of formulating it. Newton was of the new school of mathematics and using the concise Descartesian abbreviations, he found the method sparkling away in the haystack of infinitesimals. He called it the 'fluxional method' and calculus was born.

He revealed this in 1969 when he produced a manuscript titled 'De Analysi per Aequationes Numeri Terminorum Infinitas' *('On Analysis by Infinite Series')* which had limited circulation amongst the local scientificerati, but it was not published.

Barrow was so impressed by the work he recommended that Newton should replace him when he retired. This soon came to pass and Newton was made Lucasian Professor of Mathematics in 1670. *(The Lucasian presently became highest mathematical post in Europe).*

De Analysi stamped Newton as one of the leading mathematicians of the age. He credited his study of Wallis's 'Arithmetica Infinitorum' as his inspiration for discovering differentiation. *(It was prudent to throw that tiger a big bone.)*

De Analysi also contained Newton's beautiful binomial theorem:

$$(a+b)^n = a^n + na^{(n-1)}b + \frac{n(n-1)}{1\times 2}a^{(n-2)}b^2 + \frac{n(n-1)(n-2)}{1\times 2\times 3}a^{(n-3)}b^3 + \dots ab^{(n-1)} + b^n$$

Newton's Controversial Telescope

Newton is known to have studied Kepler and the entire philosophical and physics works of Descartes. Descartes believed that light was corpuscular and was refracted according to its wavelengths by a prism. It was his belief that refracted colours called chromatic aberrations could never be eliminated from a lens.

Newton agreed with Descartes, and since mirrors do not suffer aberration he built a reflective telescope in 1668 which was bigger and better than Hooke's. It was 150 mm long and 25 mm wide and magnified an image 40 times with good clarity.

In 1671 the Royal Society asked to see it, and he sent them a copy of his original. They were suitably impressed and Isaac's star was polished, just as Galileo's star was similarly boosted by presenting the Doge of Venice with a telescope. The following year Isaac was elected a Fellow of the Society.

Even though Hooke as the Societies curator of experiments, had for the last seven years been using a less precise model of the James Gregory reflecting telescope, somehow Newton gets the kudos for the invention.

Achromatic Lens

The not too distant future would prove both Descartes and Newton wrong in their belief that chromatic aberration could not be eliminated from a lens.

Even so, Newton belligerently stuck to his guns ignoring the reports of other scientists who were experimenting with prisms of different glass.

*Dispersion is a measure of the difference in the amount of refraction of red light to that of violet light at the opposite end of the spectrum. Flint glass has twice the dispersion of crown glass, so combining a converging lens of crown glass with a half as powerful lens of flint glass, evens out any dispersion. This combo is called an achromatic (no colour) lens.

In 1758 John Dollond produced the first achromatic refracting telescope.

The real worth of achromatic lenses became apparent in the early 19th century when they were used in microscopes. The lack of aberration enabled much greater clean clear magnification.*

The same year he showed the telescope, Newton presented the Royal Society with his paper on optiks which was well received by most, but it did have its critics. Not the least of these was Hooke who was more than likely a little peeved over the telescope business.

Hooke wrote a condescending review criticising the work which sent Newton into paroxysms and wild tantrums, threatening malicious revenge which included public humiliation for Hooke.

Newton retreated to the country with his 'black dog' *(depression)* where he both worked and sulked. He continued his analysis of light and in 1705 brought out a second paper with explanations of 'Newton's rings', *(the pretty rainbow coloured concentric rings which are best revealed when curved glass is placed on flat glass.)*

This paper proclaimed that white light was a collection of coloured rays, which was completely at odds with the views of Aristotelian lecturers.

Also completely at odds with the establishment, was the unique idea that an object's colour was what was left after the absorption of all the other colours. Supporting evidence of this came from his lifetime friend Edmund Halley who had been down in a diving bell. He described to Newton how sunlight streaming through a small window in the roof of the bell caused the top of his hand to have a rose tint, while the bottom had a blue tint. Newton concluded that sea water doesn't restrict red rays as much as the blue, which it tends to reflect.

Starting at a depth of 7 meters red wet-suits, fish, shells etc, lose their redness and turn through purple to grey as depth increases. Orange goes at 10 meters and yellow at 13 meters. All colours including blues and greens are grey below 30 meters, yet in torchlight all colours magically reappear.

The second paper shows Newton's clear sharp mind patiently unravelling the phenomenon of light through a prism. The coloured bands of light are meticulously measured and evaluated. One gets the impression that they were slowly and surely being rounded up like a herd of bush cattle and there was no escaping the muster.

During his retreat there had been a fair amount of intercession between Newton and Hooke by members of the Society, *(particularly by Edmund Halley),* trying to smooth out their troubles.

This all fell apart when Hooke noticed a section of the 'Opticks' paper he claimed was stolen from him. Again there was confrontation followed by intercession and an exchange of excessively polite reparatory letters. One of his over-polite letters of pacification to Hooke contains the famous *(I believe ingratiating)* quote, "If I have seen further it is by standing on the shoulders of giants."

Exacerbating the 'Opticks' controversy was the old guard which still stuck to the Aristotelian concept of light. Collectively all these objections made Newton not only more neurotic but also very reticent about publishing anything ever again. This reluctance to publish would later find him embroiled in more bitter struggles to prove priority.

It was the mouse that roared that caused Newton's grip on sanity to fail. The English Jesuits in Liège Belgium kept questioning and contravening elements of the Opticks paper. Newton wrote back answering these trivial criticisms with growing impatience until 1678, when the arrival of yet another petty letter became the last straw. He had a total nervous breakdown which lasted for six years. He spent them in retreat with his 'black dog' rarely seeing anyone.

Newton vs. Hooke

Newton and Hooke seemed destined for perpetual conflict with many of their pursuits in similar veins.

Hooke worked on diffraction, believing that the wave nature of light explained why it peeped around corners. Newton worked on colour separation by prisms concluding that light was corpuscular.

Hooke, who worked on inverse square law of gravitation in 1678 would claim that Newton, who solved the problem in 1680, ripped him off.

- Indeed he had written to Newton in late 1679 explaining his theory.
- Newton refused to reply, but somehow a sketch by Newton reached Hooke of a body's path in falling to Earth from a tower and landing slightly to the east because of the Earth's rotation. The accompanying note claimed the path would, if it could continue, spiral towards the centre of the Earth.
- Hooke corrected this, showing that the path would be an ellipse all the way down and a mirror of the ellipse out the other side. Then the object would rebound to where it started from.
- Newton didn't like being corrected by anyone but he had to accept Hooke's argument however he added a refinement by applying a constant gravitational force.
- Hooke correctly countered that, for even though Newton's figures were correct for a constant gravity, he believed that gravity fell away with the square of distance from the centre of the Earth.
- Newton claimed he had proven the falling away of this inverse square force ten years before Hooke's letters. *If so, it shows a strange malice to hoodwink Hooke with the constant Gravity charade.*
- Hooke had defined a universal cosmic force in 1679, while Newton was still working with rotational dynamics.
- Since Newton had not published anything about this, it appeared to Hooke that he had not yet discovered the concept. Yet Newton claimed he had.
- He argued that; "*In 1666 I began to think of gravity extending to the orb of the moon…… I deduced that the forces which keep the planets in their Orbs must be reciprocally as the squares of their distances from the centres about which they revolve; and thereby the force requisite to keep the Moon in her Orb with the force of gravity at the surface of the Earth and found them to answer pretty nearly.*"

- Newton did acknowledge that he was forced to rethink planetary motion as having the Sun at one focus, but Kepler had already established this *(and the inverse square idea)*.
- In 1687 Newton was encouraged, nay badgered into publishing his opus magnum, the 'Philosophiae Naturalis Principia Mathematica'. by Halley. Halley financed the publication, having recently inherited a few bob after his father was murdered. Newton deliberately wrote the Principia in Latin using scientific argot to make it as difficult as possible for any skilled layman to understand.

Prior to publication he submitted a copy of the manuscript to the Royal Society for inspection. Hooke read it and cried foul again, which re-ignited their plagiarism war.

* Scientists then and historians later, fall into two camps regarding this.
The Newtonian army passed Hooke off as a malcontent. They adopted a sympathetic attitude and used preambles like "Poor old Hooke….."
The Hooke brigade which included Christopher Wren, (it was Hooke who engineered the double dome of St. Pauls), believed Newton was the pirate.*

Midst the hullabaloo, although he had a good cheer squad, Shorty didn't really stand a chance. Before publication Newton now excised all references to Hooke from the edition. Further, he was so incensed that he refused to accept the presidency of the Royal Society until Hooke was dead. *(His vindictiveness was maniacal)*. Isaac Caligula Newton was then re-elected every year thereafter.

The Principia was a monumental work encompassing his three laws of motion and his theories of light and gravitation, all reinforced with solid mathematics including calculus. It also stamped precise concise mathematical proofs onto many incomplete theories. It was a sensational publication which crippled the old Aristotelian guard and made Newton world famous.

Newton's Laws of Motion

1. Every body continues in its state of rest of uniform motion unless it is compelled to change that state by an external force.
2. Change of motion is proportional to the force on it and it changes in the direction of that force.
3. To every action there is an equally opposite action.

- The first law is a reiteration of Galileo's sliding block analogy.

- The third law is blindingly obvious to a cave man.
- The second law does belong to Newton. Its proof requires distinguishing between an object's mass and its weight. This is embodied in his formula:

$$F = ma \quad \text{Force} = (\text{mass}) \times (\text{acceleration})$$

Newton is the man who first declares that the same force that makes an apple fall also causes attraction between astral bodies. He diligently did many tests confirming that 'inertial mass', which is what has to be moved, is the same as gravitational mass. Today this seems passé, but then it required confirmation. For proof he placed equal masses of material in the same receptacle, and weighed each of them.

Leibniz & the Calculus War

The delay of 17 years in publishing his calculus method, led to the very bitter calculus war that death did not end.

The prodigious Baron Gottfried Wilhelm von Leibniz (1646-1716) was a doctor of law at 21, an innovative philosopher, a gifted mathematician and a prince-maker, *(which provided some of his income).*

He travelled Europe on the look-out for vacant little principalities, and then went in search of documents and distant relations who could inherit them. Some of the dealings were dubious.

In Paris he'd studied mathematics under Huygens, and in 1675 for his own amusement invented infinitesimal calculus while travelling by carriage about the continent.

Over the next 8 years he developed it and polished his symbols, then in 1684 he published his differential calculus. In 1686, a year before the publication of the Principia he published his integral calculus. It is the Baron's symbols of calculus we use today. They certainly have the flavour of Germanic script. viz.

$$\int, \oint, \oiint, \Sigma$$

To speed up his work he invented a calculating machine in 1671.

The Baron lists to his credit this beautiful infinite series for pi, which is more graceful than Wallis's.

$$\frac{\pi}{4} = \frac{1}{1} - \frac{1}{3} + \frac{1}{5} - \frac{1}{7} + \frac{1}{9} - \ldots\ldots$$

When he saw the Principia, the Baron, a man of iron, German, a hustler, and totally unafraid of the push and shove of the English Royal Society, pointed the foul finger of plagiarism. Newton was incensed and the affray went very ugly. Herr Baron spoke for himself, but Newton wrote articles of attack and defence, publishing them in the names of some of his young acolytes at Cambridge, *(Super Sneaky!).*

After 1704 when Newton became president of the Royal Society, he composed the official Society's responses in favour of his priority.

He wrote, ranted and raged for yet another 25 years after the Baron's death, harassing the corpse like some bloodthirsty demon.

The Dot Rebellion

The calculus battle raged on even beyond Newton's death.

Newton's calculus symbols were not as self explanatory therefore not as concise as Leibniz's.

For example the term for differentiating y with respect to x of an equation such as:

$y = f(x)$ in Leibniz's term is $\frac{dy}{dx} =$

In Newton's language the term was called "y dot" $\dot{Y}$

This dotting eventually led to the 19th century 'dottage rebellion' by the students at Cambridge against the universities pro-Newtonian calculus format. At that time the European mathematicians had forged well ahead of the English ones principally by using a better calculus language. The rebellion succeeded and Leibniz's format was adopted.

If the dead could laugh then Leibniz would laugh last.

Series however simple are always gems. I found these while playing with Fermat's cube, but I doubt if it slipped through the net of the early mathematicians.

$$(2x+1)(x+1)x = 6\sum_{1}^{x} x^2$$

i.e. $$2x^3 + 3x^2 + x = 6(1^2 + 2^2 + 3^2 + ...+ x^2)$$

Also: $$\sum_{1}^{x} x^3 = \left(\sum_{1}^{x} x\right)^2$$

i.e. $$1^3 + 2^3 + 3^3 + ... + x^3 = (1 + 2 + 3 + ... + x)^2$$

The King James Battle

Isaac's next major battle was with King James 2nd who tried to obtain what amounted to an honorary degree at Cambridge for a Catholic priest. Newton, protestant to his toe nails, fought the good fight and won. *(He spins in his grave as honorary degrees are shovelled out to dignatries today.)*

When James abdicated after the 'Glorious Revolution' in 1688, Newton took on more public duties at Cambridge, and moved to London on the urging of a curious young Swiss mathematician house guest called Fatio de Duillier.

Nicolas Fatio de Duillier

In 1688, Newton took on more public duties. He became a patron to the next generation of Cambridge students one of whom was his young house guest Fatio de Duillier with whom he shared many interests ? ? particularly alchemy.

Fatio was a Swiss mathematician, natural philosopher (old word for physicist), astronomer, inventor who at the age of 24 had secured membership to the royal Society with his work (improvements to Italian astronomer Giovanni Domenico Cassini's explanation of zodiacal light) as being sunlight scattered by an interplanetary dust cloud that straddles the ecliptic plane .

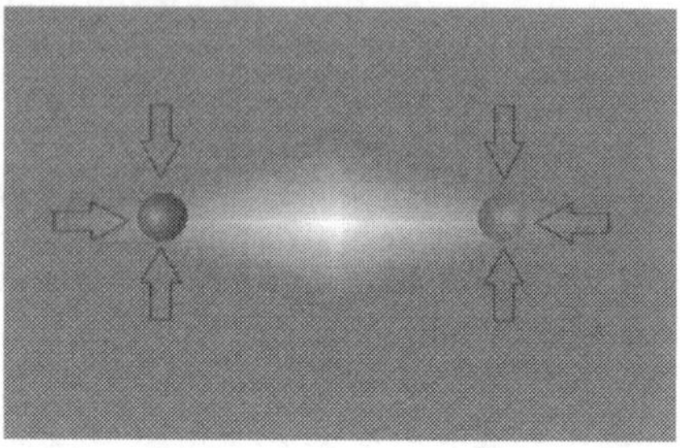

This led on to his 1690 shadow or push theory of gravitation; later elaborated by Georges-Louis Le Sage in 1750. Here Space pressure is restricted between masses.

One of his inventions was a method of drilling into jewels to create low friction bearings for mechanical watches.

He was a backbone of support for Newton in the priority Leibniz calculus war.

Unlike Newton he was diplomatic, corresponding or visiting European scientists like Huygens, Jacob Bernoulli. et al.

In 1693 Fatio became ill and his parents repatriated him to Switzerland. Newton went literally crazy with distress, suffering from force 10 mind storms and attacking his closest friends who feared for his sanity.

The storms eventually abated and he moved permanently to London securing a position via his friend Charles Montague, later Lord Halifax, as warden of the Mint on a cosy £2,000 per annum. He established serrated edges on coins to inhibit the practice of filing them down.

He took a very dim view of anyone who interfered with the money he was guarding and sent many forgers to the gallows. Perhaps these deaths appeased his hyperactive spleen.

The Flamsteed War

Another of Newton's wars that death did not end was with Astronomer Royal John Flamsteed (1646-1719), a testy sort who had catalogued over three thousand stars. Isaac, now Sir Isaac since 1705 and President of the Royal Society, wanted to use Flamsteed's observations of the moon to refine calculations.

- Flamsteed didn't like many people least of all the despot Prez who he described as "invidious, ambitious, exceedingly avid of praise, and very irritable when contradicted."
- He had had previous frosty dealings with him, so he procrastinated about releasing the data.
- Little Boots countered by getting himself appointed Chairman of 'The Visitors', a body responsible for the Royal Observatory.
- Using this position he tried to force publication of Flamsteed's catalogue.
- Flamsteed dug in.

- Newton sawed at his legs.
- For over ten years he sawed but the legs were made of the sternest oak.
- Finally Newton seized the catalogue and had Halley prepare it for printing.
- Flamsteed won it back in a court action then burnt all the proofs.

All up the resistance cost Flamsteed a fortune, but he prided himself on being one of the very few people to defeat the tyrant.

By arrangement Flamsteed's assistants published the catalogue of his life's work posthumously. Thereafter Newton took up the vindictive red pencil he had used on Hooke and expunged any references to Flamsteed from future editions of the Principia.

The Light War

There was one war that continued for over a century after Newton's death. This was the first light war.

Newton and his army, following Descartes lead and the commonly held view, forcefully propounded that light was a stream of particles vibrating though the *ether* like sound waves in air.

The other camp was headed by Huygens and the wavelets.

Both particle and wave theory found explanations for refraction. Newton believed the denser medium of glass would exert a greater attractive force on the particles of light as they approached its surface, therefore speeding them up. Huygens thought the denser medium would slow the little wavelets down.

Huygens presented a dilemma for Newton. On one hand he had incorporated Huygens's centrifugal force formula into his gravity theory and he must have admired his mathematical ability. On the other hand, not only did Huygens have the hide to contradict his particle theory, but adding the gravest insult to injury, he was the mentor of the reviled Leibniz.

Newton's particle theory carried the day because of his great prestige as the Prez, and because he outlived Huygens by 32 years, but it did not win the war. The reader will see in the section on light that, had cooler and clearer heads prevailed, this war should never have happened.

Magnetism and the Ether

It is worth noting that Newton hardly noticed magnetism.

There is one mention of it in Principia where he asserts that the force of repulsion or attraction varies with the 'cube' of distance. *(most irregular.)*

He did try to explain that gravitation was caused by differences in *ether* pressure but didn't confirm this as he couldn't find any way of demonstrating it by experiment.

When asked what caused gravity he would reply;

"I make no hypotheses."

He had an on again off again affair with the *ether*. It was present with static electricity where he believed the effluvium streamed from rubbed glass and brought back little pieces of paper with it. This was a controversial denial of 'action at a distance' theory'. *Ether* was still there in the 1675 "Hypothesis of Light", but not there in 1679, then back again refashioned in 'Opticks' published in 1704.

Eulogy

Newton was a man who few loved and many hated, *(there were no fence sitters)*, yet no one doubted his genius.

There have always been mathematical puzzles to test the intrepid. Following the advent of calculus these puzzles flourished and Newton was invariably their master.

For example Johann Bernoulli, (*who really hated Newton),* had created a problem which circulated around Europe. "What is the best shape for a track that allows a bead to roll down it and speed fastest away parallel to the Earth's surface?" Home from a hard day at the mint, Newton solved it before dinner. The next day he let the solution slip anomalously out via the Royal Society. It soon reached Paris and Bernoulli, who famously said:

"Tanquam ex ungue leonem." *"I recognise the lion by its claw."*

Newton spent more than half his life working on metaphysics and alchemy, mainly in the chemical laboratory at his country home. Much of this was in pursuit of the philosopher's stone which more often than not involved cooking concoctions mixed with quick silver. He used his sense of smell as a detector and was observed with his nose in the neck of a jar of bubbling amalgam.

There are a few biographies that attribute Newton's madness to mercury vapour poisoning. They argue that his insanity ceased after he moved to London away from his laboratory. This may explain the intensity of some events but not their spite. It does not explain the furies of his adolescence.

He died in the village of Kensington on March 20, 1727 and was buried with ceremony in St Paul's Cathedral (*beneath Hooke's double dome*). His brilliant intellect had indelibly shaped mathematics and physics.

Edmund Halley & Comets & Transits of Venus

Edmund Halley (1656-1742) was the personification of the jack of all things technical. In a long career he was a sea captain, cartographer, inventor of the weather map, the diving bell, professor of geometry at Oxford, Astronomer Royal, etc. He had many knives and some were quite sharp.

In 1682 Halley studied the path of a large comet crossing the sky and compared it with earlier recorded comet data when he could find it. This revealed that comets in 1456, 1532, and 1607 followed similar paths.

Since he was of the persuasion that the planets circled the Sun, it occurred to him that this comet might be the same one returning approximately every 75 years. He published this proposition.In 1758, 76 years later and 16 years after his death the comet returned. Thereafter it was called Halley's Comet.

This event silenced the last of the Gravity Theory sceptics

The earliest recorded sighting of the comet was in 467 BC Some say the comet of 11 BC was the Star of Bethlehem.

*Comets, *(Greek for long-haired stars)* encounter the solar wind as they approach the Sun. This is a stream of hot gases ejected from the corona at speeds of around 1½ million kph. It causes ices on the comet to vaporise creating a tail of ionised gases sometimes 300 million kilometres long. Sometimes there is a second dust tail and sometimes a third sulphur tail. The solar wind is much faster than the comet, so when the comet is speeding away from the Sun, its tail precedes it.

The period of the orbit can be stretched or shrunk if the comet passes near the gas giants Jupiter or Saturn. Computers calculate that the period of Halley's Comet has fluctuated between 74.4 and 79.6 years over time.*

It was Halley who theorised that the distance from the Earth to the Sun could be calculated by triangulating from different parts of the world, observations of the transit of Venus across the Sun.

A pair of transits 8 years apart occurs every 100 years.

In 1761 almost 20 years after Halley's death scientists from Europe sailed forth to distant parts to observe the transit. Almost all involved in this assignment failed because of either bad weather, broken equipment, or the Anglo-French war.

There was more success at the second attempt in 1769 especially from the data provided by the young commander of the Endeavour, Lieutenant James Cook. He was sent to Tahiti specifically to record the transit which he duly accomplished from a mountain top there. He then sailed off to discover the east coast of Australia.

From the combined data from both transits the mean distance to the Sun was calculated to be 150 million kilometres, stunningly close to the latest figure of 149.9 million kilometres.

Halley's concept of magnetism is bizarre sci-fantasy. Earth had an inhabited inner core. The space between inner and outer was filled with a luminous effluvium which enabled the Subterraneans to see and their crops to grow. Each surface was magnetised and the effluvium leaked out through pores in the outer skin manifesting itself as the aurora borealis.

Like Newton, Halley also picked up his annual sinecure at the Royal Mint, but only at the rank of controller.

EARTH - ATMOSPHERE - GRAVITY

Von Guericke

In the 17th century there was a score of electric-attraction generating machines using a variety of materials and designs. In 1672 German physicist Otto von Guericke (1602-1686) who was born in Magdeburg, described the one he built in 1663:

"A crank handle turned a basket ball sized sulphur sphere, which picked up electric-attraction when a hand was held against it."

Later he found that if it was turned fast enough the ball would glow, which was the first observation of electroluminescence.

He also showed that the electric-attraction could flow into substances held near the sulphur ball, temporarily inspiring them. He is also the first to show that the electric-attraction could flow through material like linen thread into distant objects and make them electrically-attractive.

Otto was a great showman. He invented the air pump in 1650, and four years later entertained Emperor Ferdinand III and his court outside the Imperial Diet at Regensburg with a demonstration of what became known as the 'Magdeburg Hemispheres'.

He connected two hollow bronze hemispheres to make a sphere of about 35 centimetres in diameter then sucked the air from it so that the halves were held together by the external atmospheric pressure. He harnessed a team of eight horses to each half and put them at pulling the hemispheres apart.

The hoofs thundered and the flints flew but the sphere didn't part. Otto then opened a valve and like Moses, magically parted the hemispheres.

Von Guericke was also an astronomer, predicting that comets would return regularly from outer Space. He showed that sound cannot pass through a vacuum but that light beams can.

The Atmosphere

One atmosphere is the pressure at the bottom of a column of air from space to sea level. It pushes mercury from a bowl, 76 cm up into a vertical vacuum tube standing in it, which is the equivalent of 10.33 metres of water. Therefore, if say the all the Earth's atmosphere was compressed to the density of water, it would be only 10.33 meters thick. Atmospheric pressure is caused by the molecules of air, which at sea level are moving at over 3,500 kilometres per hour, bouncing off each other. At 0° C and 1 atm pressure, in one cubic centimetre (1 cm^3) there are $3x10^{19}$ molecules, each travelling about 1000 molecular diameters between collisions. Even so they manage to make about five billion collisions each per second.

$$1 \text{ atm} = 1 \text{ million dynes per cm}^2 = 14.7 \text{ lbs/in}^2$$

*If you were travelling to Paris in a sports car with the cool wind in your hair at 100 kph, air molecules would strike your face at around 3,600 kph. Fortunately molecules inside your body maintain the required counter-pressure.

Only in the fantasy of Hollywood do bodies explode when ejected from Spaceships. The reality is that there are several exits through which body gas can easily escape which make pressure explosions impossible. In fact, though animals lose consciousness after 15 seconds in a vacuum, if they are removed from it within 2 minutes, they can usually be revived. If a Spaceship's airlock failed to connect properly with another Spaceship's airlock, the brave could fly the 15 second vacuum gauntlet. Many mice and monkeys died for science establishing this limit. Incidentally don't take your pet canary into Space with you. It will starve to death because it can't swallow without gravity.*

Boyle

A less dramatic but just as sensational demonstration as that of the Magdeburg hemispheres was put on by Robert Boyle (1627-1691) who showed the British that sound can't pass through a vacuum.

In 1660 he suspended a bell in a vacuum chamber with a glass window in it. Though one could see the bell ringing, no sound was heard. Observers found the Zen experience *(like one hand clapping),* maddening.

Speed of Sound	
Material	kph
Ether	0
Rubber	180
Air	1200
Hydrogen	4630
Water	5220
Copper	13,000
Iron	18,500

Aristotle got some physics right when he supported the credo that sound was transmitted from a vibrating source via little pockets of air vibrating in chain reactions, and he correctly hypothesised that if there was a vacuum *(which there wasn't because the Gods forbade it),* then sound would not go through it.

The speed of sound through compressible material follows different rules from those of incompressible materials. As a general rule, the greater the density of the incompressible material is, the faster the speed of sound through it.

Boyle is renowned for his gas law. Boyle's law states: "If the temperature is held constant, the volume of a gas varies inversely with its pressure."

$$P \propto 1/V$$

This he proved with a very simple yet elegant experiment.

A glass tube of known diameter was shaped into a 'Jay'. The short end was closed and mercury was poured in through the high top end. Measurements were taken after the bottom of the 'J' was full.

Additional mercury compressed the air caught in the short end. Both heights were measured then more mercury added and the measurements repeated. The relationship between pressure and volume soon became apparent.

Slavery, Newcomen and Watt

The second half of the 18th century saw the birth of the industrial revolution which began in England then spread to Europe and America. This was powered by steam.

There had been no further development of the steam engine since Hero of Alexandria's invention in 50 AD.

In 1698, English engineer Thomas Slavery built a water pump. By alternating steam pressure into two copper cylinders it could push water up 6 meters.

In 1712 Thomas Newcomen built a water pump using steam to push pistons. It was used to pump water out of mines.

In 1769 James Watt improved this by incorporating and patenting a condenser which brought used steam back to water for re-use.

Lavoisier

"It took them only an instant to cut off that head, and a hundred years may not produce another like it."

This was said about the father of modern chemistry as his head fell into the bucket under the guillotine of the French revolution, joining the heads of 26 other Tax Farmers. They had lent money to the crown *(Louis XVI)* in return for the privilege of collecting salt and tobacco tax among other excise. *(privatisation of taxation)*

Five years earlier Lavoisier (1743-1794) wrote; "*We must lay it down as an incontestable axiom, that in all operations of art and nature, nothing is created; an equal quantity of matter exists both before and after the experiment…… and nothing takes place beyond changes and modifications in the combinations of these elements*." This was the first 'modern' pronouncement of the conservation of energy, a law of nature which eluded both Galileo and Newton.

He also discovered the elements of water, and named them hydrogen and oxygen. He showed that the weight materials gained when they burned was due to the absorption of oxygen. This was contrary to the accepted theory that materials when they burned gained weight as they released 'phlogiston' which had a negative weight like antimatter.

When he was 28 Lavoisier married a 14 year old Tax Farmers daughter called Marie Anne Paulze who became an adept assistant and later a chemist of renown. The marriage was childless. She reappears in the history books later as Count Rumford's second wife.

Jacques Alexander Cesar Charles

In 1772 the Montgolfier brothers invented the big hot air balloon. On September 19th 1773 the first test pilots, a sheep, a rooster and a duck flew three kilometres across Paris. On 21st November 1783, the first manned untethered flight took of from Versailles piloted by Jean-François Pilâtre de Rozier and François Laurent, the Marquis d'Arlandes. They sailed over Paris for 25 minutes and traversed about 9 kilometres.

Balloon fever erupted in Europe. Cities vied to make the prettiest balloons which outperformed the flights of their rivals. The military advantage of having a spy in the sky was quickly realised.

Lavoisier was involved in designing a device which would produce huge quantities of the gas. This was achieved by ripping the oxygen out of steam as it percolated through a gun barrel filled with hot iron rings *(which rusted).*

$$Fe + H_2O \rightarrow FeO + H_2$$

In 1783 the physics teacher Jacques Charles (1746-1823) piloted the first hydrogen balloon. He made several flights sometimes going above 3 kilometres.

Charles is renowned for Charles's law which in fact really belongs to the more rigorous scientist Joseph Gay-Lussac: At a constant pressure gasses expand equally over the temperature range of 0°to 100° C.

*In the mid 1850s this was refined to:

"At a constant pressure, the volume of a fixed number of particles of gas is directly proportional to the absolute (Kelvin) temperature."

Therefore doubling the absolute temperature doubles the volume when pressure is constant.

e.g. A volume @ 30° C = 303° K, doubles @ 606° K, i.e. @ 333° C.
*Absolute zero = 273.16° Centigrade**

Measuring the Earth

Newton and others held that the spin of the Earth would cause the equator to bulge. This was eventually confirmed by the 1760s. The polar radius is 6357 kilometres, which is about 1/300th less than the equatorial radius of 6378 kilometres; clearly not very ellipsoid. The surface velocity at the equator is about 1672 kilometres per hour.

The force of gravity stops us and our atmosphere from being flung off into Space.

Newton had proposed a method for calculating the mass of the Earth. This firstly required determining the value of Newton's '*G*', the universal gravitational constant. The method required a pendulum of known mass to be placed near a mountain of estimated mass and the pendulum's deflection noted. From these results one could calculate '*G*' and subsequently the mass of the Earth.

In 1774 this experiment was conducted on Shiehallion Mountain in the Scottish Grampian Mountains. A team surveyed the mountain over the summer months. The copious volume of trigonometric data was processed producing a best approximation for the mountain's mass. The angle of deflection this mass caused to the pendulum was measured and the mass of the Earth calculated at 5,000 million million tons or about 4.5 x 10^{15} kilograms.

John Michell

The English clergyman, physicist and geologist John Michell (1724-1793) was the world's first seismologist. He also researched astronomy, magnetism and gravitation.

He was the first person to suggest a good estimate for the distance to the stars.

He theorised the existence of binary stars which orbited their centre of mass. *(Half the stars are binary.)*

Almost two centuries before Einstein he propounded that light would be bent as it passed a massive body!

His research into the cause of the massive 1755 Portuguese earthquake and its consequent tsunami suggested the epicentre had been out in the Atlantic Ocean and that high pressure vents in the ocean floor were responsible.

This work led to him being elected a member of the Royal Society in 1760, and appointed Cambridge professor of Geology in 1772.

Michell wanted a delicate tool to measure gravitational forces associated with earthquakes. He soon discovered that apparatuses with springs were far too insensitive so he devised the torsion balance. There is some evidence that this occurred as early as 1768.

Later he became interested in measuring the mass of the Earth. For some reason the experiment stalled and he died in 1793 before he could conduct it.

The torsion balance designed for the experiment however had is own momentum. It went first into the care of Francis Wollaston, the Jacksonian professor at Cambridge who passed it on the Lord Cavendish who then built a sturdier copy.

Cavendish

Henry Cavendish (1731-1810) was utterly eccentric. He lived modestly with his father *(both of them doing scientific experiments)* until the age of 40 when he inherited millions of pounds. *(Both his grandfathers were Dukes.)* When his father died two years later Henry became a Lord.

Lord Henry appears to be the shyest man in the universe. The only time he ever went out was to attend a speech or lecture on science. Many of these were scientific soirees conducted at the home of the botanist Sir Joseph Banks, president of the Royal Society from 1778 until his death in 1820. Guests here were requested neither to speak to, nor even glance at the shy Lord. At home he only communicated with his female servants, who were banished from his sight, through notes. Some people thought him a loopy taciturn misogynist.

In 1776 he discovered the properties of hydrogen and determined its specific gravity. He went on to discover the composition of water and conducted many electrical experiments.

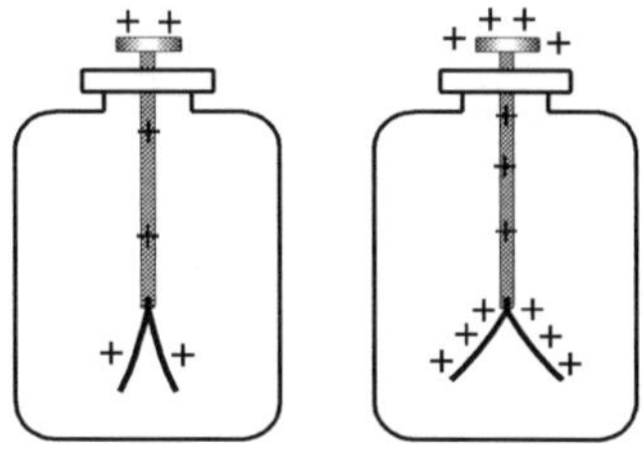

Before Coulomb, he applied the inverse square law to charge and anticipated Faraday by evaluating the amount of charge capacitors could contain with a variety of substances sandwiched between them. He used an electroscope to determine the amount of charge an object held, measuring the angle of spread tof he gold leaves.

He anticipated Ohm's law showing that the potential difference across a conductor was proportional to the amount of charge going through it. To achieve this he became a human voltmeter, holding the charged wire in one hand and the earth wire in the other then observing how far up his arms the shock travelled.

The student or reader is strongly advised not to repeat this experimental method.

The Cavendish Mitchell Experiment

Between 1797 and 1798, using his copy of Michell's apparatus, Lord Cavendish measured the density of the Earth from which its mass was calculated. It was Cavendish's last known experiment.

The apparatus was set up in a garden shed of his house. It was almost hermetically sealed in a 3 metre square heavy wooded box which was built around it. Ropes through the box wall could manoeuvre the internal elements. Observations were made through telescopes mounted in the wall and motion was timed with a John Harrison 1761 Chronometer. The box had a door which sealed draught-tight.

A 40 inch long fine steel wire stretched from a cross spar down to the centre of the 6 ft long dumbbell whose lead balls were 2 inches in diameter, each weighing 1.61 lbs or 730 grams. The resulting super sensitive torsion balance was then calibrated by timing its oscillations.**Details of an inner timber draught proof skin enveloping the dumbbell are not included in the diagram; nor is the vernier scale which measured displacements.*

The device was so sensitive that it never completely stopped twisting. In early experiments Cavendish used copper wire with periods around 15 minutes long but these proved to be too variable. The steel wire though stiffer was more consistent. It produced a period of oscillation of 7 minutes. With the period of oscillation known, the torsional constant was calculated.

The rotational inertia *'I'* of the light dumbbell is given by:

$$I = \sum ml^2 = 2m\left(\frac{l}{2}\right)^2 \text{ kg}\cdot\text{m}^2$$

where: l is the distance between the ball centres of the dumbbell
m is the mass of one dumbbell ball

With *'I'* and the period *'T'* known, the torsional constant k is given by:

$$T = 2\pi\sqrt{\frac{I}{k}} \qquad \text{therefore} \qquad k = \frac{4\pi^2 I}{T^2}$$

The second phase of the experiment involved loading a central structure with two drop down wires, each holding a heavy 12 inch diameter lead ball weighing 348 pound or 157.8 kg to a height level with the dumbbell. This structure could be turned from outside the box by a rope thereby allowing the large weights to be manoeuvred to specific distances *(around the 9 inch mark)* from the stationary *(mean oscillation)* position of the dumbbell.

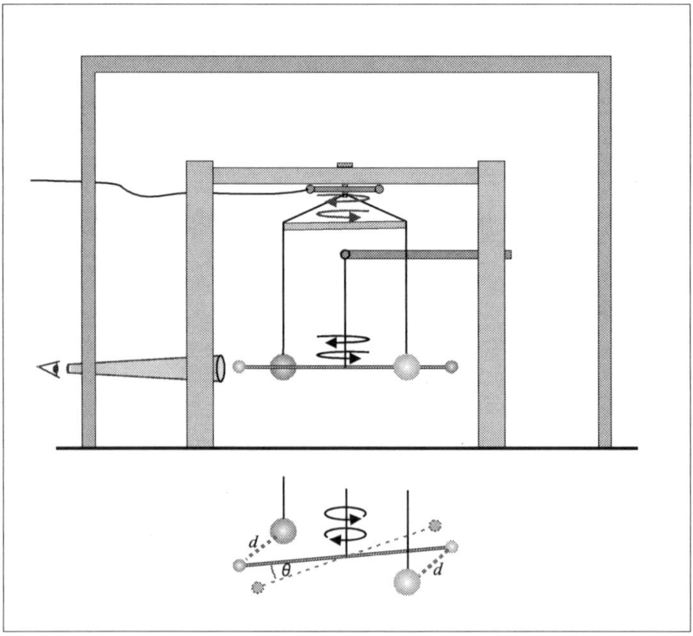

Attraction between the small and large weights caused the smaller weighted dumbbell to move through a measurable arc observable through a telescope.

The delicate experiment was repeated many times over the year to accommodate changing temperatures and levels of humidity and the mean result established.

The result Cavendish published was that of the density of the Earth. His calculations compared the forces of attraction between objects with their known weights. He did not pause in the middle of them to generate a result for Newton's universal gravitational constant *'G'*.

It would be more than 80 years later before anyone made another reference to 'G'.

He found the average density of Earth to be about 5.4 gm/cm^3 or 5.4 times the density of water. The latest estimate is 5.5268 times the density of water.

This is considerably denser than the average surface rocks which lead him to the obvious conclusion that the Earth has a heavy core.

All this was achieved in a garden shed using four lead balls of known weights, a few sticks, a few metres of wire and a clock.

From wave data collected from earthquakes and underground nuclear explosions scientists in the 21st century view the core as being a crystal structure within the molten outer-core. They perceive that its 'axis of anisotropy', is offset by 11° from Earth's geographic axis and that it rotates with, but 0.25% faster than the Earth.

The Universal Gravitational Constant *G*

The force between the opposing dumbbell balls is $$F = \frac{2GMm}{d^2}$$

And the torque from this force is given by:
Torque = (force) x (momentum arm length) $$\tau = 2\frac{GMm}{d^2}\frac{l}{2}$$

With k known the torque is given by $$\tau = k\theta$$

Therefore $$G = \frac{k\theta d^2}{Mml}$$

Cavendish would have found Universal Gravitational Constant G to be 6.754 x 10^{-11} nt-m^2/kg^2 if he had wanted to, which is very close to recently established value of 6.670 x 10^{-11} nt-m^2/kg^2.

The Density and Mass of the Earth (from Cavendish's results)

Known:

- Universal Gravitational Constant G = 6.754 x 10^{-11} nt-m^2/kg^2.
- Acceleration due to gravity 'g' at the surface = 9.80 meters / sec ^{2}.
- Radius of the Earth R_e = 6370 kilometres

If the Mass of the Earth = M_e and a mass on the surface = m_s

Then $$F = \frac{Gm_s M_e}{R_e^2} = m_s g$$

therefore $$M_e = \frac{gR_e^2}{G} = \frac{(9.80\ \text{meters/sec}^2)(6.37\times10^6\ \text{meters})^2}{6.754\times10^{-11}\ \text{nt}\cdot\text{m}^2/\text{kg}^2}$$

$= 5.89 \times 10^{24}$ kg

The latest estimate $= 5.97 \times 10^{24}$ kg

If bypassing *'G'* then: $$F = \frac{\left(\frac{k\theta d^2}{Mml}\right)\times m_s M_e}{R_e^2} = m_s g$$

The universal gravitational constant G is a scalar.

$$G = \frac{v^2 r}{M} \text{ has the units } M^{-1}L^3T^{-2},$$

Whereas **g**, the acceleration on a body caused by mutual attraction with the Earth is a vector with the units LT^{-2}.

Measuring the Solar System

There were now in place the four elements required to measure the solar system.

The mass of the Sun could be calculated from:

- The distance from Earth to the Sun.
- The mass of the Earth.
- Newton's universal gravitational constant *'G'*
- Thereafter Kepler's laws of planetary motion reveal the masses of the planets and their distances from the Sun.

(Here the French Scientist and mathematician Simon Laplace was at the forefront of establishing them.)

Moon and the Satellites

The force holding a satellite orbiting the Earth is given by: $F = \dfrac{GMm}{d^2}$

$$\text{Then} \quad F = ma = \frac{mv^2}{d} = \frac{GMm}{d^2} \quad \text{or} \quad v^2 = \frac{GM}{d}$$

where: M is the mass of the Earth, m the satellite mass
d is the distance from the centre of Earth
v is the velocity of the satellite

*Note that 'm', the mass of the satellite is eliminated from the final equation. A ping-pong ball could occupy the same orbit as the moon if its speed was the same.

A satellite of any mass orbiting 35,919 kilometres above the Earth's surface at the right speed will remain directly above one Earth location in a so called 'stationary orbit.'*

ELECTRICITY & MAGNETISM

The First Conductors

Von Guericke's conduction of electric-attraction along a thread seems to have gone largely unnoticed for half a century. The phenomenon was rediscovered in England in the early 1700s by Stephen Grey. He found that it flowed readily through the so called electrics like copper, silver and gold whereas the non-electrics tended to resist flow.

The metal group became known as conductors and the latter group comprising amber, glass, sulphur, etc., became known as insulators.

About then in France, Charles Dufay found that there were two types of charge, vitreous and resinous. He observed that when two electrified corks dangled on strings near each other, like charged ones repel and unlike charged ones attract each other.

In the early 1700s an Englishman called Hawkebee substituted a glass ball for Otto's sulphur one in an electric generator and achieved better results. In the 1740s the sphere was replaced by a cylinder and pads were used instead of hands while insulated wires were attached to the machine. This dog could hunt.

Leyden jar

Interest in the electric attraction changed dramatically in about 1745. Simultaneously Pieter van Musschenbroek, at the University of Leyden *(Holland)*, and Ewald Georg von Kleist of Pomerania *(North Prussia),* invented what became known as the Leyden jar.

The first jars were glass with a wire which went through a stopper into the water inside. One held the jar in one hand and touched the wire to an electric-attraction generator then moved the device away breaking contact. If the other hand touched the wire one got a severe electric shock. This was the world's first capacitor and it contained a large amount of charge capable of delivering a big jolt to anything or anyone that completed the circuit. The first recipients of the big jolt were of course the inventors and von Kleist was belted so hard he abandoned the research. The Dutchman proceeded with extreme caution.

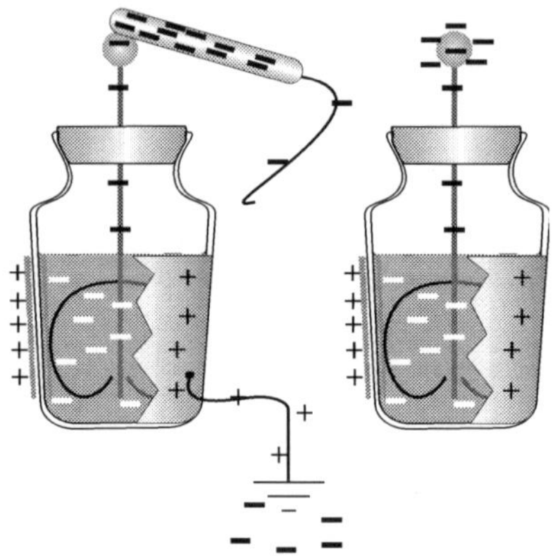

The device was quickly improved by the English physicist William Watson who added metal foil coating on the inside and outside surfaces substantially increasing its capacity to store charge. In 1747 he transmitted an electric spark through a wire strung across the River Thames at Westminster Bridge.

The jar was the launching pad for electricity and electricians, though they weren't known as electricians then. Many of these were showmen and they travelled Europe with a jar and its companion generator giving big jolts to all and sundry. Big money was earned on the carnival circuit where people paid to enter tents and see birds and mice zapped dead by the invisible electrics. In France the abbé Jean-Antoine Nollet zapped a line of Carthusian monks connected to a kilometre length of iron wire. There was great wonder and merriment when they were all simultaneously jolted, causing them to jump as one into the air.

Benjamin Franklin

One jar and electric-attraction generator crossed the Atlantic to Philadelphia, where in 1747 Benjamin Franklin (1706-1790) and three friends began playing with 'electrical fire'. With the aid of an instrument maker they conducted several experiments the results of which were sent into Europe via the Royal Society in London.

They discovered that electricity was not two sorts of effluvium but one. A substance was neutral, or electrically abundant (*vitreous*), or electrically deficient (*resinous*). When two substances were rubbed together electric fluid flowed from one to the other.

They found that the electric fluid flowed readily onto metals and water.
(not distilled water)

Franklin believed clouds were giant Leyden jars, and designed and announced an experiment to show the electric nature of lightning. It was conducted in England and France before a building could be built tall enough in Philadelphia to get near storm clouds. Exasperated with waiting he suddenly thought 'kite'! He was influenced in this modus operandi by his childhood experience of floating on his back in a lake while the kite he was flying dragged him along.

In 1752 he sheltered from the rain under a shed while flying a kite made of a cedar wood cross and a silk handkerchief. A pointed iron rod was tied to the kite frame and a long twine attached to it. The lower end of the twine held an ordinary old fashioned door key. The link from the key to Franklin was a length of silk twine which remained dry under the shelter.

When the kite flew into the storm clouds he noticed the strands of the wet twine trying to part and when he held his hand near the key, sparks jumped across the gap as the current went through him to Earth. Metal, i.e. the key, was necessary since the damp thread could not accumulate very much localised charge.

Next, a Leyden jar was held to touch the key and it became charged. It was well known that lightning could kill both man and beast, so the risky kite experiment created sensational headlines around the world.

The following year he was elected a fellow of the Royal Society.
(No credit for any discoveries ever went to his colleagues.)

The student reader is strongly advised not to reproduce this experiment.

In the heat of the moment Monsieur Reihmann tried it in St Petersburg and was toast. Many thought him a martyr. Herr Boze, a German inventor tried to burn for science but he seems to have failed in his efforts. His heroic gesture however was applauded by many in the fraternity.

Storm clouds contain huge updrafts which cause ionisation. Positive 'ions' (Greek for wanderers) are lifted to the top of the cloud leaving the lower cloud heavily negatively charged. This in turn pushes electrons in the Earth below away, leaving a zone of positive charge. The system gradually builds up channels of charged particles between Earth and cloud through which, when the paths are well defined, electrons surge and lightening flashes. (In certain conditions it's the negative ions that rise to the top of the cloud.)

Franklin found that a charged sphere could discharge with a spark to Earth through an air gap of 25 mm but at the same voltage, a sharp pointed rod could discharge to Earth through an air gap of 200mm, without a spark.

He realised that St Elmo's fire was a discharge that diluted the channels about a steeple causing the lightening to feed elsewhere. This led to the construction of pointed lightning rods, taller than the buildings they protected, which soon festooned America and Europe.

*With little encouragement, the 20 million lightning strikes to ground per year in the U.S., kills about 100 people and wounds another 1000. Each bolt contains about 30,000 amps @ 28,000° C.

St Elmo's fire is a glow often seen during thunder storms emanating from the steeples of churches. (Today include the masts of iron ships and even the propeller tips and nose tips of aeroplanes. It is now called brush discharge.)*

In 1755 Franklin became the first person to observe that there wasn't any electric charge on the inside of an insulated can.

"I electrified a silver pint cann, on an electrical stand, and lowered into it a cork ball, of about an inch in diameter, hanging by a silk thread, til the cork touched the bottom of the cann.

The cork was not attracted to the inside of the cann as it would have been to the outside, and though it touched the bottom, yet drawn out, it was not found to be electrified by that touch, as it would have been by touching the outside. The fact is singular. You require a reason; I do not know it."

Back in his teens Franklin became influenced by the writings of a sect of British Pythagoreans who like the original Pythagoreans believed most fervently in the transmigration of souls. By eating meat one might eat a fellow human from a past life, and that would be cannibalism. Franklin religiously adhered to this doctrine.

He is accredited with ascribing the words positive charge to vitreous charge and negative charge to resinous charge.

He is also accredited with saying; "Beer is proof God wants us to be happy."

One of his inventions was the bifocal lens.

Priestly

Franklin had notified the clergyman and physicist Joseph Priestly (1733-1804) about the absence of charge inside a can. In 1767 Priestly remarked that the inverse square law relating to charge at a distance could be concluded from the experiment.

It was known mathematically via calculus, that as a result of Newton's inverse square law of gravitation, if a planet were a shell, i.e. a crust without a core, then any mass inside the shell is pulled equally in all directions.

The mass might be close to the shell wall at one point and strongly attracted to the near side, but the accumulative pull from the larger surface area behind it, *(the far side),* is exactly the opposite of the near side attraction.

Priestly, saw that a similar result was happening with charge on a closed container, and therefore postulated that charge too must obey the inverse square rule. He also showed therefore that the charge distributed itself evenly over the exterior of an object.

Galvani

Another Leyden jar and electric charger found its way into the laboratory of Professore Luigi Galvani (1737-1798) of the anatomy school at the University of Bologna. This is today, the oldest university in the world still in existence.

The early electricians had often observed spasms and twitching in the birds and rats they privately or publicly electrocuted. Post-mortem probing with electrified scalpels also produced muscular spasms and this phenomenon motivated Galvani's research into 'animal electricity' which he began in 1765.

In the early 1780s when poking about with scissors in dead frogs legs, *(as you do)*, he noticed that they twitched during an electrical storm. Sometime later in 1786 he was again dissecting frogs legs when, while touching a nerve with scissors a visitor cranked the handle of the electric charger in the room. There was no connection to the scissors yet the frog's leg again twitched. He concluded that the probe was picking up an induced charge in both occurrences.

Again in thunderstorm in 1786 he noticed that frogs hung by their spinal column on a copper hook which itself hung on an iron rail, twitched whenever their legs touched the rail. Later he noticed that even though there was no storm or no electric charger being excited, the legs still twitched when they made contact with the iron rail. He knew this twitching was the result of electric stimulus, but wrongly concluded that the energy had been stored in the frog's muscles. This was however the modern world's first battery.

Volta

In the 1790s, the Italian physicist Alessandro Giuseppe Antonio Anastasio Volta (1745-1827), an old friend of Galvani, was further researching animal electricity. When he placed coins of different metal over and under his tongue with a wire connecting them, he noticed a bitter salty taste.

In 1796 he discovered that he could produce an electric current by substituting a piece of cardboard soaked in brine for his tongue.

Next he stacked alternating discs of copper (or silver) and zinc, separating each with brine soaked discs of cardboard or leather.

The animal electricity lobby opposed this new metallic-electricity but Volta's demo of his voltaic pile in 1800 using cardboard in the sandwich gagged them.

He showed it in Paris in 1801 to Napoleon Bonaparte, who thereafter made him a Count of the Kingdom of Lombardy.

Napoleon, a patron of scholars, lived by the maxim that 'knowledge is power', particularly scientific and geographic knowledge.

A steady current was now a reality launching a thousand experiments across Europe. It was soon realised that the current from a voltaic accumulator was plentiful but at a much lower 'tension' than the high tension produced from the charge stored in Leyden jars.

To test the potential difference between the top and bottom of voltaic piles constructed from a variety of metals. he invented the electrometer *(volt-meter)*

Volta had shown that a chemical reaction could generate electricity. It was just a little hop to conclude the reverse reaction, namely that an electric current could split a compound into smaller compounds or molecules.

Now scientists had a new toy and a new field of electrochemistry opened up. Before the year ended Nicholson and Carlisle had passed a current through water and broken it down into hydrogen and oxygen.

Coulomb

Yet another Leyden jar and electric charge generator was in the laboratory of French physicist Charles Augustin de Coulomb (1736-1806) who was more of an engineer than a physicist.

In 1777 he invented, *(the English say copied)* the torsion balance and used it to confirm with experiment Priestley's inverse square assumption.

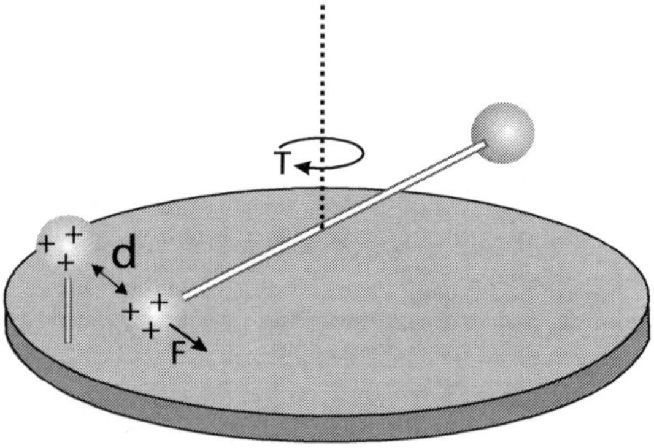

The result is known as Coulombs law:

'The force between two charges is inversely proportional to the square of their distance of separation.' $F \propto \frac{q_1 q_2}{d^2}$

1 coulomb = the combined charge of 6.24 x 10^{28} protons or electrons.

Romagnosi

In 1802 Gian Domenico Romagnosi observed that an electric current affected magnets, but he was only an Italian Juror of little standing so everyone ignored him. Progress then lost another eighteen years until a professor of the establishment accidentally rediscovered the same phenomenon. This time the world reacted with fervour.

Ørsted

Hans Christian Ørsted (1777-1851) was the professor of physics at wonderful wonderful Copenhagen University. During a lecture in 1820, while demonstrating some of the properties of electricity, he caught a flicker of movement out of the corner of his eye. It came from a compass that had somehow accidentally strayed onto the table. Further investigation showed that a compass needle placed anywhere about the conductor pointed along the tangent of a circular path whose radius was perpendicular it to the wire. When he turned off the current the needle returned to magnetic north.

The ēlektron-spirit and the magnetic spirit were connected.

There is an old Croatian saying that: "Before you put your socks on in the morning, the news is overseas". As the news raced across Europe predatory scientists raced to their batteries trying to win immortality in the Guinness book of priorities.

Pole position went to French physicist Dominique Arago who showed that magnetised needles and unmagnetised iron filings were attracted to a conductor. *(derr!)*

Ampère

Second pole position on the greedway went to the professor of mathematics at the École Polytechnique, André Marie Ampère (1775-1836). He had been an 'enfant prodige' who by the age of 12 had mastered all the known mathematics of that time. He realised with a flash of *derr-sight* that electric currents produced magnetic fields. Within a week of the news reaching him, he had published several papers on electromagnetism containing mathematical analysis.

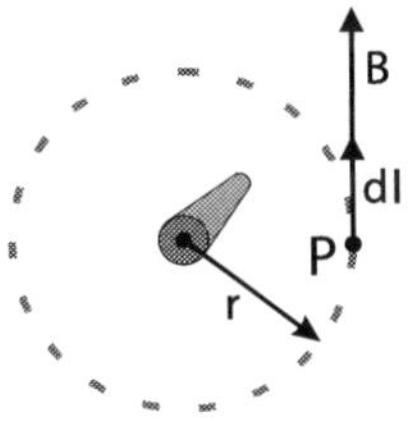

One of these contained Ampère's law:

"The magnetic force at a distance r from a straight conductor is directly proportional to its current divided by that distance."

$$B \propto \frac{i}{r}$$

Ampère belonged to the 'action at a distance' school so his formulae don't have references to, or elements of, magnetic or electric fields. Maxwell would soon expound this law to incorporate fields and produce the contemporary form of Ampère's law:

$$\oint \mathbf{B} \cdot d\mathbf{l} = \mu_0 i$$

To Ampère, magnetism was a by-product of electricity so he tried to eliminate it, thus reducing electromagnetism to purely electrical ingredients. This suggestion was not welcomed by the establishment, principally because of the apparent perpendicular nature of electricity to magnetism.

He showed that when currents flow in the same direction through parallel conductors, the conductors are attracted to one another like poles of a magnet. When the currents are opposing, the effect is repulsion.

Two years after Ørsted's discovery Ampère placed a needle within a current carrying coil and the electromagnet was born. By placing a spring held needle between two coils he could measure both the direction of a current and its intensity on a scale. He called the device a galvanometer in honour of Galvani.

Schweigger

In 1825 Johann Schweigger (1779-1857) modified the galvanometer to its preferred embodiment by fixing the needle to the coil and allowing it to pivot between two fixed permanent magnets.

Gauss

Carl Friedrich Gauss (1777-1855) was arguably one of the world's top ten mathematicians of all time. He was a wunderkind with an appetite for complex mental arithmetic coupled with a gift for languages. His parents were extremely poor.

His school teachers were astounded by his ability and when he was 14 years old they recommended him to Der Herzog von Braunschweig *(The Duke of Brunswick)* who became his patron. Gauss was always grateful and very loyal to his patron.

The following year he put his first stamp on world mathematics when he showed that the long standing puzzle of how to draw a heptagon *(7 equal sides and angles)* with only a ruler and compass was impossible to solve.

By factorising polynomial equations, he presented a proof that the only odd sided polygons that could be drawn in this manner were those of the series 3,5,17,257,and 65,537 or multiples of two or more of these.

His mathematical gift provided him with a steady occupation, firstly in astronomy and later as a surveyor.

In 1780 Giuseppe Piazzi (1746-1826) was appointed professor of higher mathematics at the Academy of Palermo. He then became the driving force in the establishment of the famous Palermo Observatory where he produced a catalogue of over 7½ thousand stars.

One of these astral bodies was the new asteroid Ceres which soon disappeared behind the sun. There were many astronomers vying for the triumph of being the first to find it again. In 1800 using a device called 'the method of least squares' Gauss relocated Ceres.

He spent many years as astronomer to Der Herzog von Hanover calculating planetary orbits.

He then spent 14 years surveying the State of Hanover. To speed things along he invented the heliograph which reflects and directs the Suns rays to distant points. The curvature of the planet presented trigonometric difficulties, so he invented the required geodetic mathematics to overcome them.

His investigation into electromagnetism began in the 1830s. He was involved in the world wide survey of the Earth's magnetic field for which he invented the magnetometer. In 1832 working together with Wilhelm Weber they put a galvanometer on each end of a cable and one of the first electric telegraphs was a reality.

Gauss saw a point in space as being a very tiny circle with a perpendicular axis *(like a minute drawing pin)*. A surface was then the sum of all these points, each with its own axial orientation vector pointing out. The orientation of all points on a flat plane surface is the same. When a surface is submerged in a flowing fluid, the force or drag of the surface varies with the angle between the orientation vector of the surface and the direction vector of the current. If that angle is zero, the drag is at a maximum and if that angle is 90°, the drag is zero.

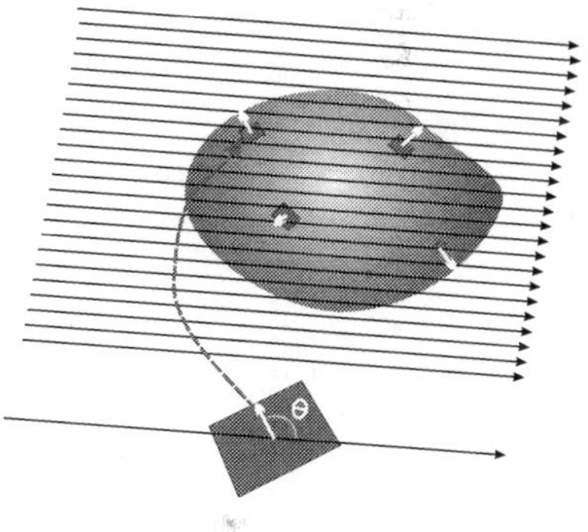

If the angle is θ, the drag is given by:

$$\text{drag} \propto \text{area} \times \cos\theta$$

($\cos 90° = 0$, $\cos 0° = 1$, $\propto$ = proportional to)

The flux of something through an area is the number of hypothetical streamlines that flow through it. The symbol for flux is 'Φ'. If the flow is an electric field the symbol is Φ_E. Acknowledging a flux exists somewhere denies the idea of 'action at a distance' there.

If **E** is the electric field flowing and *d*S is the tiny element surface area of an object *(the minute head of the drawing pin)*, then the electric flux passing over the surface is the sum of the fluxes through all the little elements of surface area. This is determined by each ones orientation vector ***d*S**. *(note the convention for vectors of the bold type, i.e. of the **S**.)* It represents both the area *d*S and the orientation of its perpendicular to surface barb.

Then the flux over the surface is given by:

$$\Phi_E = \oint \mathbf{E} \cdot d\mathbf{S}$$

i.e. The sum of all the little areas at their respective orientations to the flux flow.

In 1832 he established what are known as Gauss's Laws:

1. "The electric flux across any closed surface is proportional to the net electric charge enclosed by the surface."

$$\varepsilon_0 \Phi_E = q$$ where ε_0 is the permittivity constant

i.e. $$\varepsilon_0 \oint \mathbf{E} \cdot d\mathbf{S} = q$$

2. "The magnetic flux across any closed surface is zero." *(Therefore an isolated magnetic pole is impossible).*

The latter means that all of the magnetic field which enters a closed surface becomes zero while inside, and all of it must exit on the far side. Since nothing is missing when it exits, every north pole must have a south pole.

By contrast the former means that electric fields can begin and end on a charge.

Gauss was not a good communicator and consequently neither a good teacher nor a good professor. He always was jumping from one horse to the next. When someone would explain what a new mathematician was about, Gauss would often comment that the parvenu was about a third of the way there. It was only after his death that the great bulk of his work was accessed, and indeed it showed that he had the jump on a fair slice of the recent mathematics. There were many revelations concerning number theory.

Units

**In mathematics, when one thing is proportional to another it also means that one thing equals a constant times the other.*

The units of the constant allow the units on the right side of the equation to balance those on the left.

In dealing with gravity,
its constant is the universal gravitational constant G, with its units $M^{-1}L^3T^{-2}$

$$F = G\frac{Mm}{d^2} \qquad \text{units } MLT^{-2}$$

$$G = \frac{Fd^2}{Mm} \qquad \text{units } M^{-1}L^3T^{-2}$$

$$= 6.670 \times 10^{-11} \qquad \text{nt-m}^2/\text{kg}^2$$

In dealing with static electricity,
its constant is the permittivity constant:

$$\varepsilon_0 = 8.85 \times 10^{-12} \text{ farads/meter } (M^{-1}L^{-3}T^2Q^2)$$

In dealing with induced electricity (via magnetism) its constant is the permeability constant:

$\mu_0 = 1.26 \times 10^{-6}$ henrys/meter (MLQ^{-2})
or $= 4\pi \times 10^{-7}$ newtons/square ampere

Note: G is a stand alone constant but ε_0 is more than often accompanied by the numerical constants 4 and π

e.g. the force between two charges: $$F = \frac{1}{4\pi\varepsilon_0}\frac{q_1 q_2}{d^2}$$

and $$\frac{1}{4\pi\varepsilon_0} = 9.0 \times 10^9\, nt - m^2/coul^2$$

For an example of μ_0:

The magnetic induction at a distance 'r' from a long wire is given by:

Ampère's law $$\beta = \frac{\mu_0 i}{2\pi r}$$

Weber

In 1846 the German physicists Wilhelm Weber (1804-1891), and Friedrich Kohlrausch (1840-1910) *(the authority on electrolytes)*, deduced the ratio of electrostatic units to magnetic units and found it to be equal to the speed of light.

It is known as Weber's constant: $$\frac{1}{\sqrt{\varepsilon_0 \mu_0}}$$

$$\frac{1}{\varepsilon_0 \mu_0} = \frac{1}{(8.85 \times 10^{-12})(1.26 \times 10^{-6})} \quad \text{units} \frac{1}{(M^{-1}L^{-3}T^2Q^2)(MLQ^{-2})}$$

$= 1/11.151 \times 10^{-18}$ farads henrys/ m^2
$= 8.967 \times 10^{16}$ units $L^2\, T^{2.}$
$= (2.995 \times 10^8 \text{ m/sec})^2$
$= c^2$ (c = *speed of light)*

$$c = \frac{1}{\sqrt{\varepsilon_0 \mu_0}} \text{ m/sec} \qquad \text{units } LT^{1}$$

The weber is the unit of magnetic flux in the international system of units (SI).

1 weber (Wb) = 1 joule per ampere = 1 volt-second.

Thompson aka Rumford

Benjamin Thompson (1753-1814) was born in Woburn which is just above Boston Massachusetts. He began investigating chemistry while working as a shopkeeper's apprentice in Salem. At the age of 19 he married a wealthy widow and moved to Rumford (now Concord N.H.) where he continued his research and invention.

In the beginning of the American War of Independence (1775-1783) he was a spy for the British. When discovered in 1776, he managed to escape to England leaving his wife and daughter behind. He rose through the ranks of British civil and military service and at one stage even commanded a British regiment in New York.

He was knighted by George III in 1784, then with his permission, worked for 11 years in the Bavarian civil service, rising to War and Police Minister then Grand Chamberlain to the Elector of Bavaria. He was created Graf *(Count)* von Rumford of the Holy Roman Empire in 1791.

Thompson's leitmotif was heat. He studied gunpowder, introduced Watt's steam engine to Bavaria, improved chimneys and fireplaces, invented a kitchen range, the double boiling saucepan and the drip coffeepot.

He redefined heat, changing the perception that it was a material substance into the acceptance that it was created only by motion. He was influenced in this conclusion by observing the considerable quenching required over many days to counteract the heat generated by friction when a canon barrel was honed out. The barrel was continually topped up with water as it boiled off. In 1798 he published 'An Experimental Enquiry Concerning the Source of the Heat Which is Excited by Friction'. It contained data on the equivalence of heat and mechanical energy.

He was elected to the Royal Society in 1799 and the same year joined forces with, among others, Sir Joseph Banks. This body was instrumental in influencing George III to establish the Royal Institution.

The huge Royal Institution of London, whose main lecture hall held thousands, was opened in 1801 with a charter to bring science education to the populace. On Thompson's recommendation British chemist Sir Humphry Davy was appointed assistant lecturer.

Davy

Humphry Davy (1778-1829) was born and educated in Penzance Cornwall, a stone's throw from Lands End. He was a bright popular middle class lad with a zest for investigating everything be it scientific, poetic, or of nature, which involved long exploratory walks and much fly fishing.

When he was 19 years old and a doctor's apprentice, he befriended Davies Gilbert who had both a very good library and a friend who had a well equipped chemistry laboratory. For Davy, a visit to the laboratory was the equivalent of showing a penguin water.

Before the end of the year he was delving into chemical theory, conducting experiments, and producing nitrous oxide *(laughing gas)* in his own small laboratory.

His advancement in the scientific world was meteoric. Gilbert saw to it that he was appointed chemical superintendent of the Pneumatic Society in 1798, whose charter was to discover the therapeutic properties in any of the many new gases being discovered. Much of this research involved guinea pigs which often included Davy and his associates. He had his literary friends such as Coleridge and Southey 'researching' laughing gas. His experiment of inhaling water-gas *(the gas produced when steam is blasted through coal)* was near fatal. The carbon monoxide in it causes only a mild headache and possible some nausea before unconsciousness. However by then severe damage is done and it is not easy to recover from the deep coma.

His research results were published in 1800. In 1802 he was appointed professor of chemistry at the Royal Institution, and in 1803 elected as a member of the Royal Society. He was 25 years old.

He researched the chemistry of electric cells and their electrolytes publishing 'On Some Chemical Agencies of Electricity' in 1806. In 1807 he created the arc lamp using 2000 cells. The spark jumped across carbon sticks 10 centimetres apart. Between 1807 and 1808 he used a powerful battery with 250 cells to break down through electrolysis some liquid compounds of metals, isolating sodium, potassium, calcium, strontium, barium, and magnesium.

In 1807 he was awarded the Napoleon prize from the Institute de France even though England and France were engaged in the Napoleonic wars, which ran almost continuously from 1799 to 1815. A few years later during the same war, he toured France carrying a miniaturized chemistry laboratory and investigated 'x' gas *'iodine'* and its sister chlorine.

Davy was blessed by miners for his invention of the miner's lamp. He was knighted in 1812 and raised to a baronetcy in 1818, becoming president of the Royal Society in 1820.

In his fruitful career he discovered many new elements and gases, but some say his greatest discovery was Michael Faraday, but really it was Faraday who found Davy.

Faraday

Michael Faraday (1791-1867) the son of a poor blacksmith, learned to read and write in Sunday school. He delivered papers for a book binder for a crust and became apprenticed to the same in 1805. He read the books he worked on and was intrigued by an article in the third edition of Encyclopaedia Britannica on electricity. Using old bottles and scrap timber and wire, he constructed an electrostatic generator and began a career of experimentation.

In 1812 someone gave him a ticket to attend one of the free to public science lectures at the Royal Institution. The lecturer here was Sir Humphry Davy. Young Michael wrote up the notes of the lecture, bound them in his master's shop, and sent them to Davy with a letter asking for a job. When a lab assistant position became vacant that year, Davy hired Faraday.

The following year he accompanied his new master on an extended tour of Europe. Faraday was very much the flunky here dining in the servant's quarters etc.

In 1820 Faraday finished his second apprenticeship with a reputation of being an excellent analyst. That was the year Ørsted set science alight with his discovery that currents affect magnets. Faraday then began his part time experiments hoping to change electric energy into mechanical energy.

In 1821 he showed how moving a magnet towards a conductor created a current flowing in it. *At that time a host of other scientists were doing the same thing.*

After being put up twice for membership of the Royal Society, *(and oddly twice rejected by Davy's veto),* he was third time lucky and elected in 1824. The next year he was appointed director of the laboratory of the Royal Institution.

Faraday spent from 1821 to 1828 trying to get a current to cause a mechanical effect. One day he wound an insulated cable around an iron ring and hooked it up to a voltaic pile and a switch. Another wire wound around the other side of the ring was connected to a galvanometer. When the current was switched on the galvanometer needle jumped and with it rushed the euphoria of success. When the current was switched off the needle surprised him by jumping briefly again, but this time the other way.

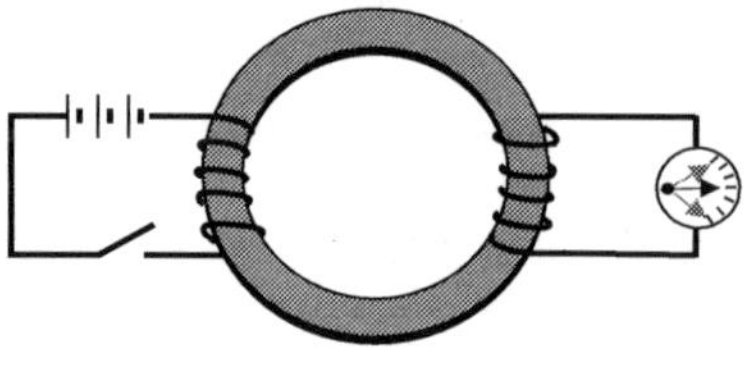

Next he started pushing a permanent magnet in and out of a coil and again a current flowed back and forth across the galvanometer and the faster magnet moved, the greater the charge. He had harnessed a mechanical effect to electricity and magnetism.

Thereafter he built a machine with a crank handle to rotate a thin copper disc. It spun between two magnets arranged so that north faced south. The rim section of the disc moved faster through the magnets than the more central sections. A current flowed along a lead from the brush at the rim to the brush near the axle. This was the dynamo or electric generator and father of the electric motor.

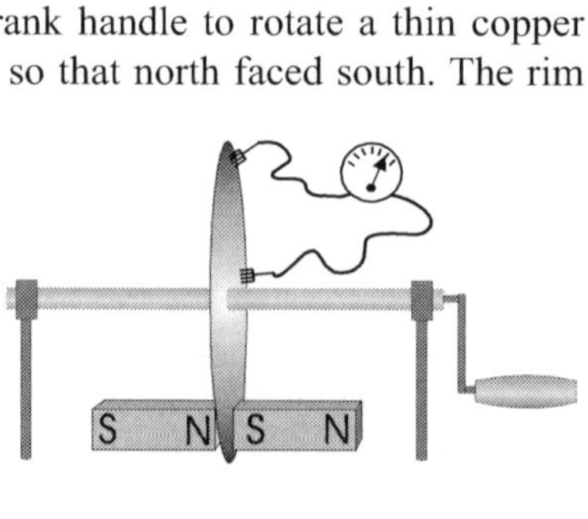

Satisfied that a current flowed he stuck the gadget in a corner and continued other experiments.

Legend has it that in 1832 when the Prime Minister visited the Institution, he inquired about the purpose of the peculiar bread slicing contraption in the corner. Faraday is famously said to have replied "I know not, but I wager that one day your government will tax it."

To Faraday, the lines of force revealed by sprinkling iron filings onto a page suspended above a magnet were very real. Magnetic strength was determined by their density through the *ether*.

His law of induction states that:

"The magnitude of the force induced in a circuit is proportional to the rate of change of the magnetic lines of force *(flux)* that cuts across the circuit."

This law was the foundation upon which the future rules of electromagnetism were built.

The flux is also called the field.

In 1833 he was appointed Professor of Chemistry at the Royal Institution.

In 1839 he had a nervous breakdown which lasted for six years. Returning to work in 1845 he found the Faraday Effect. Polarised light, when flowing through a magnetic field parallel to the lines of force, is rotated in the direction of the current that set up the magnetic field.

Next he looked for, and found that all matter responded to magnetic fields. One group of substances which he named 'paramagnetic', like iron, nickel and cobalt etc., aligned their crystalline axes to the lines of force and headed for the strong zone, whereas others, which he named diamagnetic, like bismuth or benzene for example, lined up perpendicular to the lines of force and headed for the weaker zone away from each pole. He concluded that the para-group, conducted magnetism better than the surrounding space, but the dia-group conducted it less.

Faraday made considerable contributions to chemistry using electrolysis.

When he was thirty years old he married and joined the church of his new wife, the Sandemanians. These were extreme Christian fundamentalists who frowned on aggrandisement, seeing everything from saving money to laughter as sinful.

It was a childless marriage. Work was one of the few things not frowned upon, and it was here that he got his stimulation.

To really prove that all excess charge resided on the outside of a conductor as Franklin had shown, he electrified an insulated large metal covered cube using a powerful electrostatic generator, and in his words:

"*I went into the cube and lived in it, and using lighted candles and electrometers, and all other tests of electrical states, I could not find the least influence upon them, though all the time the outside of the cube was powerfully charged, and large sparks and brushes were darting off from every part of its outer surface.*"

By obeying a command to dine at the palace with Queen Victoria he had to forgo his Sunday church service. As a consequence his church membership was, unhappily for him, suspended for many years. Twice shy and terrified of Sandemanian Coventry, he declined a knighthood, preferring to remain in his words, "plain Mr. Faraday".

He continued to give the Christmas youth lectures at the Royal Institution as long as he was able. In his dotage he was given a cottage in Hampton Court by Queen Victoria.

At all times Faraday believed that Space was not 'nothing'. It was a medium, an ether, capable of supporting electromagnetic forces.

Dynamos

There were several types of steam driven dynamos among the 26,000 exhibits at the Vienna Expo in 1873. One design invented in 1869 by the Belgian Zenobe-Theophile Gramme (1826-1901), was a continuous current dynamo which put out much higher voltages than its competitors. There was a great hullabaloo when the sales technician mistakenly hooked up the output power leads of one working dynamo to the output leads of a second idle one. Its dynamo now became a motor which started driving its pistons.

The event raised the profile of electric power, driving the electric age forward. The first taxes were levied on generated electricity in 1880.

Sturgeon

Back in 1823 when Faraday was three years out of his apprenticeship to Davy, William Sturgeon (1783-1850), a self taught electrical engineer built the first large scale electromagnet. It was a horseshoe magnet weighing only 200 grams with current carrying wire wrapped around it, yet it could lift a weight of 4 kilograms.

In 1832 he invented the commutator which would later be incorporated into generators to maintain the direction of current, and into electric motors to reverse the current.

Henry

In 1829 the American physicist Joseph Henry (1797-1878) was building electromagnets capable of lifting huge weights by insulating the winding rather than insulating the iron core. He built one for Yale which temporarily held the world record by lifting 2,086 pounds. By suddenly interrupting the current in his big transformers he could induce colossal voltages as the field collapsed, which in turn could induce magnetism in iron objects at a distance.

In 1831 Henry, assisting Samuel F.B. Morse, built the world's first practical telegraph which was 1·6 kilometres long. The following year he discovered self-induction and three years later built the first electric motor.

He was by 1832 well aware of induction but did not publish, so Faraday got the priority. In 1833 however, Henry showed how far along the induction road he was when he published the laws relating to transformers. At that time he was investigating self inductance and here he beat Faraday to the Guinness, but only just.

In trying to decipher inductance Henry was the first scientist to hint at the wave structure created by a changing electric current. In 1842 he hypothesized that when a Leyden jar discharged rapidly, the charge flowing out would over shoot in one direction. It would then have to back up but would now overshoot in the reverse direction requiring it to reverse again etc. therefore behaving like a dampened wave.

When he was 22 Henry was elected to the Royal Society. In 1846 he became the first director of the newly formed Smithsonian Institution.

In 1893 'the henry', symbol H, became the standard electrical unit of inductive resistance.

The Smithsonian Institution

In 1846, by an act of Congress, the bequest of British physicist James Smithson (1765-1829) was formalised, which established the Smithsonian Institution.

What began in a moderate sized castle, today is comprised of 16 museums and the National Zoo. It is smack in the heart of Washington D.C. stretching between the U.S. Capitol building and the Washington Monument. Smithson's bequest stipulated that it be "an establishment for the increase & diffusion of knowledge," which was more or less the directive of London's Royal Institution.

Smithson was a mineralogist, and smithsonite is a mineral of zinc carbonate named after him.

Lenz

One of the other scientists working on induction in 1834 was the Russian, Heinrich Lenz (1804–1865). He was the first to concisely define the well observed phenomenon with what is known as Lenz's law:

"*When a bar magnet approaches or retreats from a coil, a current is induced in the coil whose direction opposes the direction of the current which would have created the magnetism in the bar.*"

If there wasn't any resistance in the material of the conductor, once the surrounding magnetic field had adjusted to the change, the current could flow for ever.

Ohm

George Simon Ohm (1789-1854), was two years older than Faraday. He was a professor Mathematician at the Jesuits College in Köln Germany when in 1827 he published 'The Galvanic Circuit Investigated Mathematically'. It included Ohm's Law:

$$V=IR$$

Volts = the product of current (amps) and resistance (ohms).

At first this law coming from such a minor institution was insultingly ignored, however after he moved from the cultural backwater into Nürnberg polytechnic in 1833 it was universally recognised.

In 1942 Ohm was elected as a foreign member to the Royal Society.

THE ATOM

Dalton

John Dalton (1766-1844), a weaver's son, is the modern father of atomic theory even though how he reached his conclusions remains an enigma.

Just when the ancient Greek concept of the atoms was beginning to be taken seriously Dalton put a curve in it. He claimed that each element had a unique *(not a common)* atom and different atoms combined in the simplest way. This was difficult to prove but each objection was progressively overcome.

When he was 21 years old Dalton started a meteorological data collection which amassed over 200,000 observations during the next 57 years.

He showed that it rained when the temperature fell, which was not necessarily the result of variations if any in atmospheric pressure.

He proved that the total pressure exerted by a mixture of gases is equal to the sum of the separate pressures that each of the gases would exert if each alone occupied the whole volume.

This is 'Dalton's Law of partial pressures':

$$P_{(1+2+3+\ldots)} = P_1+P_2+P_3+\ldots$$

Avogadro

Amedeo Avogadro (1776-1856) proposed in his 1811 hypothesis that: "At the same temperature and pressure, equal volumes of gas contain equal numbers of molecules."

*Avogadro's number is 6.022137 x 10^{23}, and is the number of elementary units (atoms molecules ions electrons etc.) contained in one mole = one gram molecular weight of a substance.

The carbon isotope C^{12} which has an atomic weight of exactly 12 grams was chosen as the standard for the mole. It has 6.022137 x 10^{23} atoms.

The molecular weight of oxygen O_2 is 32 gm/mole.

Though it is named after him Avogadro did not participate in evolving this number. This was done in the late 19th and early 20th centuries. As technology improved so did the estimate.

The law was formulated for gases only but it holds true for all substances.*

Cannizzaro

The Italian chemist Stanislao Cannizzaro (1826-1910) was an aficionado of Avogadro's who worked tirelessly after his idol's death to have his hypothesis accepted as law by showing how it confirmed Dalton's atomic theory. He was himself an excellent chemist and is credited among many other things with rigorously defining the difference between atoms and molecules.

Newlands

In a paper he read to the Royal Chemical Society in 1866 John Newlands showed that every eighth element had similar properties. Unfortunately for him and progress, he compared his results to music and particularly notes and octaves on a piano. This piano analogy was too far fetched for the conservatives of the Society who slammed the lid on the fingers that had recorded what were essentially astute chemical observations.

Mendeleyev

In 1869 the Russian chemist Dmitry Ivanovich Mendeleyev (1834-1907) fared much better when he published his periodic table grouping the known elements into eight columns. He announced errors in some established atomic weights and predicted new elements would fill the gaps in his chart. When a few of these elements were located Mendeleyev was celebrated and the stampede to fill the rest of the vacant slots began.

Late in life Mendeleyev published his conception of the *ether*. He thought *ether* belonged to the zero group of elements. Each element was a millionth the size of a hydrogen atom. Here was one scientist who believed Michelson and Morley only grazed the *ether* with their light rays.

When Mendeleyev attended conference of heavyweights in London, nobody recognised who the funny little man was. He had to introduce himself about by saying "Ich bin Mendeleyev".

Meyer

In 1870, the near identical periodic table of Lothar Meyer was published in the 'Liebig's Annalen'. He just missed the fame train. *(Who ever heard of Lothar?)*

Maxwell

James Clerk Maxwell (1831-1879), was the doyen of 19th century theoretical physics. He was a meticulous, adroit, gifted mathematician who implicitly trusted mathematical footprints to lead to discoveries. He let 'the figures do the talking.'

He arrived at a time when most of the primary properties of electricity and magnetism were known but not all had a tight mathematical foundation, and was particularly interested in embedding Faraday's discoveries into mathematical concrete, thereby proving that changes in electric fields produce magnetic fields.

In 1862 Maxwell imagined that electromagnetism was a honeycomb of whirlpools of magnetic *ether* separated by cells of electric *ether*.

By 1864 however, he had abandoned the concept of a honeycomb of cells or vortices. His mathematics now defined electromagnetic waves as string like vibrations with repeating electric sections and magnetic sections. If the electric section's wave vibrated up and down, the magnetic section's wave vibrated left and right.

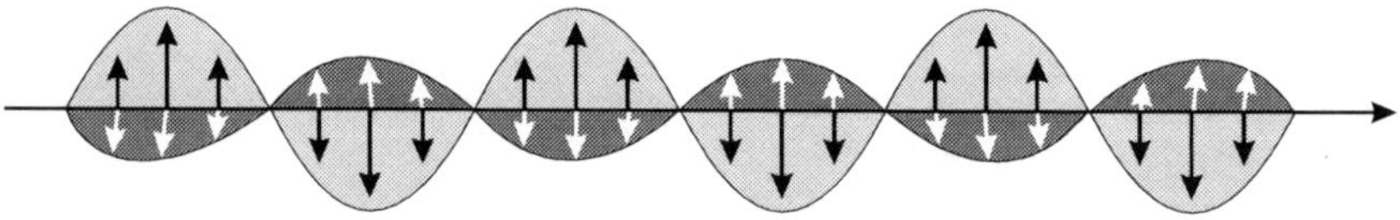

Changes in electric fields caused respective changes in magnetic fields which in turn caused changes in electric fields.

The result was Maxwell's four 'field' equations which were quite complex.

Later Hertz simplified these into their familiar form:

1. Electric field diverges in all directions from an electric charge.

   Coulombs law $$\varepsilon_0 \oint \mathbf{E} \cdot d\mathbf{S} = q$$

2. There are no isolated magnetic poles. You can't have a north without a south.

   Gauss's law $$\oint \mathbf{B} \cdot d\mathbf{S} = 0$$

3. Magnetic fields are produced by changing electric fields.

   Ampère's law $$\oint \mathrm{B} \cdot d\mathrm{l} = \mu_0 \left(\varepsilon_0 \frac{d\Phi_E}{dt} + i \right)$$

4. Electric fields are produced by changing magnetic fields.

   Faraday's law of Induction $$\oint \mathbf{E} \cdot d\mathbf{l} = -\frac{d\Phi_B}{dt}$$

With these four commandments, Maxwell stamped the lines of force into the cerebral cortex of science. No one really knew whether electric lines of force ran from positive to negative objects or vice versa nor whether magnetic lines ran from north to south poles or vice versa. Even today no one knows or cares, as long as the radio and TV work.

From the equations Maxwell deduced that the speed of the wave would be equal to the ratio of the current, (measured in electrostatic units) to the same current, (measured in electromagnetic units). This was Weber's constant which equalled the speed of light. That led Maxwell to the profound revelation that light itself was an electromagnetic phenomenon.

$$\frac{1}{\sqrt{\varepsilon_0 \mu_0}}$$

The most out of character thing Maxwell ever did was to conduct a Faraday type try-anything experiment. He tried to magnetise pieces of iron by rotating them rapidly.

Hertz

Between 1883 and 1888, Heinrich Hertz (1857-1894) confirmed all of Maxwell's electromagnetic predictions.

The device he used was a rectangle of wire with a gap in one side whose ends were capped with brass balls. The opposite side of the rectangle was adjustable. An oscillator on a distant discharging Leyden jar provided Maxwell's theoretical waves which would be a proven reality if a spark jumped across the gap, which it did.

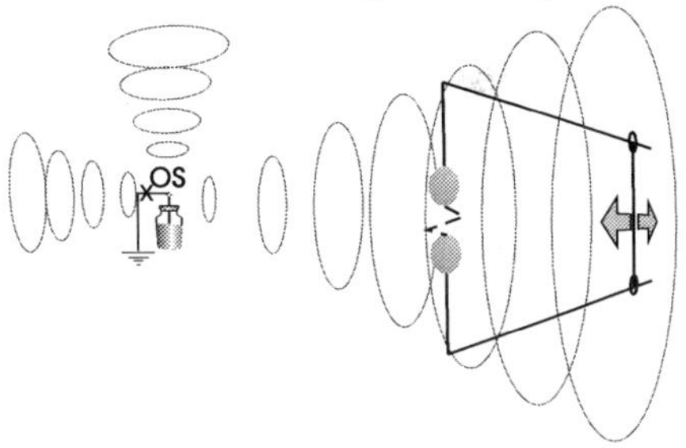

By moving the rectangle to and from the oscillator he showed that the intensity of the spark varied in a way consistent with a waves amplitude. He also demonstrated that the waves could like light be reflected and polarised.

Hertz was not impressed with the phenomenon. He even wrote to the Dresden Chamber of Commerce stating that research into Herzian waves had no practical future benefits and therefore would be a waste of effort and money.

It's one of those ridiculous statements like the one famously uttered by the head of the US patent office around 1900; "Everything that can be invented has been invented."

The Hall Effect

Prior to the execution of the *ether* and the discovery of the electron one of the most asked questions was; which way does the charge in the conductor flow? Does it flow from the positive terminal and neutralise the ions at the negative terminal, or vice versa?

In 1879 by American physicist Edwin Hall resolved this dilemma by discovering the Hall Effect.

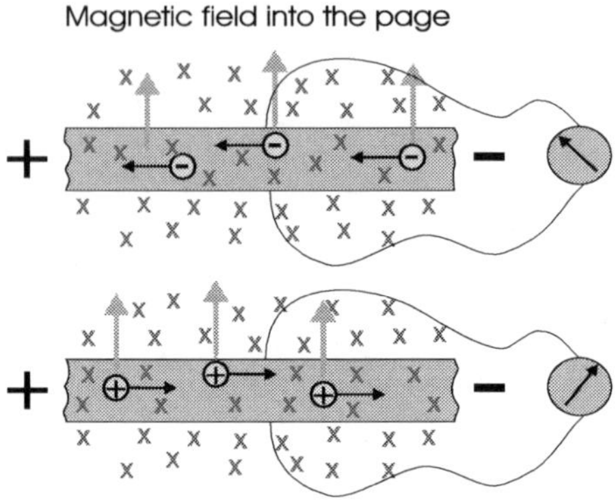

He sandwiched strips of different metal conductors between two strong magnets *(north to south)* and measured the potential difference across the strip with a voltmeter. If the charge carriers were positive they would drift to one side showing a voltage across the strip which was the reverse of that of negative charge carriers.

He found that in most metals the carriers were negative but in a few metals such as beryllium, zinc and cadmium, the carriers were positive.

The sideways drift is not unlimited i.e., all the charge carriers don't run down one edge of the tape because there is now an electrostatic attractive force between the positive and negative sides, and a repulsive force between charges already on the edge and those trying to get there. A balance is reached.

In both cases, irrespective of the carrier, the metal strip itself is forced in the same direction.

The Hall voltage is directly proportional to both the voltage and the strength of the magnets but inversely proportional to the number of charge carriers. Therefore Hall voltage is lower in a good conductor like a metal than it is in a semiconductor.

Several metering are devices based on the Hall Effect, since it requires very little energy to monitor a current.

Cathode Rays

The demand for glass scientific instruments had been increasing steadily since the advent of the thermometer. Skilled artisans were employed to blow glass onto a variety of spectacular shaped vessels, many of which had to withstand the pressures of an internal vacuum.

One such master glass blower was Heinrich Geissler, who blew glass for the German physicist Julius Plücker. In 1854 Plücker had two electrodes incorporated into each end of a highly evacuated tube, *(the Geissler tube)*. When the electrodes were connected to power a current flowed through the space between the electrodes causing a green glow on the walls of the tube which emanated from the cathode. These were called cathode rays.

In 1876 Eugen Goldstein (1850-1930) showed that cathode rays cast sharp shadows and that they left the cathode in a direction perpendicular to the surface. *(Future cathodes would have slightly conical faces to focus the rays.)* He also discovered canal rays, comprised of positive charged ions going from the anode through to the cathode.

Crookes

In 1878 the English scientist Sir William Crookes (1832-1919), *the discoverer of the element thallium*, created the Crookes tube to study the effects of the cathode rays passing through a variety of gasses. They were also directed at a variety of shaped objects of different material structure which cast shadows in the phosphorescence at the end of the tube proving that the cathode rays travel in straight lines.

Even though the anode is quite low in the tube, the cathode rays in the main pass over the top of it because they are ejected from the cathode like a stream of high speed bullets whose direction is perpendicular to the plane of the cathode.

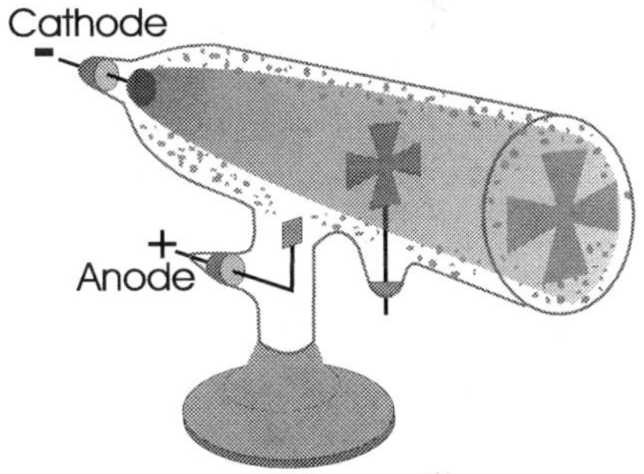

There were two schools of thought as to what a cathode ray really was. In general the English and the French believed it to be a stream of particles but the Germans thought it was a manifestation of the *ether*.

Crookes invented the radiometer for measuring the heat that mechanical or electric devices produce. It was essentially a paddle type wheel in a semi-evacuated bottle. Its few vanes were polished metal on one side and black on the obverse. The black side absorbed more heat than the shiny side causing the wheel to rotate slowly and steadily.

In later terms, the black absorbed the most heat radiation which excited free electrons near it in the rarefied atmosphere. This put pressure on the black side of the blade which was greater than the pressure on the polished metal side.

Crookes had more faith in Hertzian waves than Hertz. In 1892 he suggested they could be used to send Morse code messages through the air.

He was knighted in 1897 and received the Order of Merit in 1910.

Marconi

The twenty year old Guglielmo Marconi (1874-1937) didn't stand behind the door when they were giving out opportunity.

Like Crookes he saw the potential in Hertzian waves and quickly developed a better receiver. This 'coherer' was a jar of loose iron filings which weakly conducted a current. In the presence of Hertzian waves however its conductivity increased ten fold, which he boosted further with the addition of an antenna.

In 1896 the Italian Government of 1896 thought little of it, so Marconi hastened to England where he got a patent granted for wireless telegraphy.

LIGHT

Much was known about light before it was used to kill the *ether*. Its speed had been measured in a variety of ingenious ways and its pathways through lenses and prisms extensively explored. Each discovery brought science closer to unravelling its complexity.

The Speed of Light

Galileo's intelligent layman Sagredo, tells Simplicio how one might measure the speed of light at night by flashing a lantern beam at his assistant perched on a distant hill top, who on seeing the light, flashes acknowledgement with his lantern. Here Galileo was having a little joke.

Parallax

The difficulties in using the parallax method for the Sun were mentioned in the earlier section on Hipparchus. Astronomers found a way around this by determining the parallax of and distance to other planets when they were more or less the same distance as the Sun from Earth.

In 1672, pre-Newton's gravitation laws, a French team working both

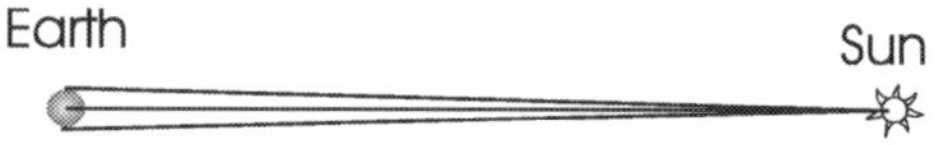

at Cayenne in French Guiana and in Paris produced the first fairly accurate measurement of the solar parallax of 9.5 seconds (9.5″) of arc. This produced a smaller than true astronomical unit of 138,439443 kilometres. One hundred years later this distance was refined during the transits of Venus.

The latest figure is for solar parallax is 8.79414″ of arc.

Once the distance to the Sun was established the distances to the stars could be estimated. Instead of having a base line of the diameter of the Earth astronomers used a base line of the diameter of the Earth's orbit.

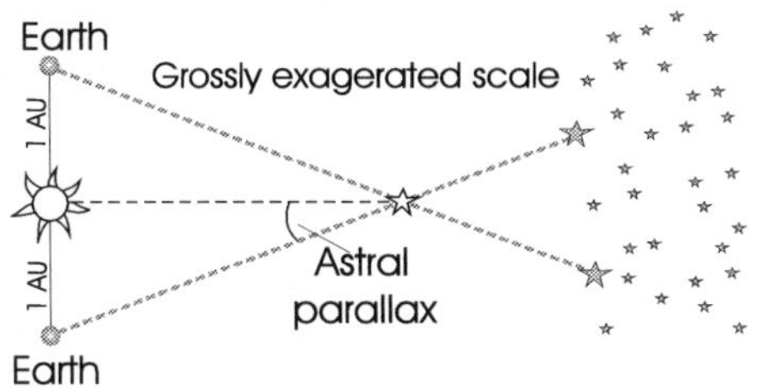

The bigger the astral parallax is, the closer the star. The closest star to Earth or the Sun is Alpha Centauri which has an astral parallax of 0.76″ (seconds).

All the rest of the stars have smaller parallax. If a hypothetical star were closer with a parallax of one second it would be one parsec from Earth which is a distance of 3.26 light years or 30,000,000,000,000 km.

The average distance from the Earth to the Sun is called one astronomical unit (1AU) and in its latest refinement this is 149,597,870 km or 92,955,808 miles.

Roemer

In 1676, the Danish astronomer Olaf Rømer *(or Roemer)* (1644-1710), working in the Royal Observatory in Paris, published his calculation that the speed of light was 140,000 miles per second or 225,000 kilometres per second. This result though 33% too low was still staggeringly fast, and showed that light speed was finite not infinite, as most people had believed.

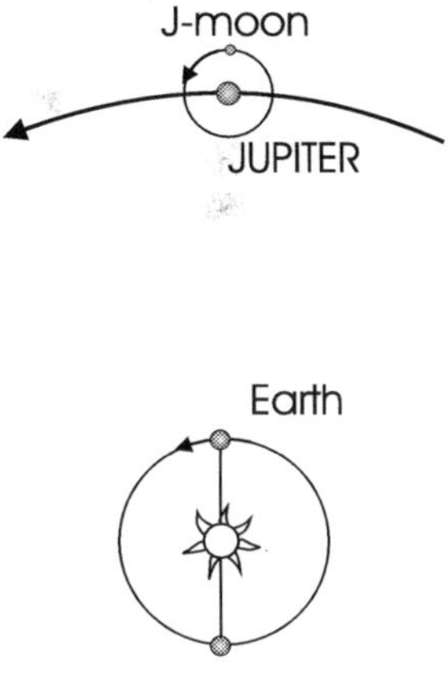

The moons of Jupiter discovered by Galileo were at that time well observed and documented. Roemer noticed that when the moon Io, which orbited above the Jupiter's equator every 42.28 hours, should have disappeared behind Jupiter, it remained visible for a short time, and when it should reappear, there was then a short time delay. Jupiter takes twelve Earth years to do a lap, making it possible to observe the time gains and lags when the Earth is firstly between the Sun and Jupiter, and secondly 6½ months later when the Earth is far-side of the Sun from Jupiter. The difference in the time delays he observed to be 22 minutes.

He calculated the speed of light to be 140,000 miles (225,000 km) per second. Today's figures are 186,282 miles (299,792 km) per second.

To achieve his results Roemer invented the 'transit instrument' to which a telescope could be mounted. This was the ultimate measuring device that could pan along arc frames from east to west at any angle. It had microscopes focused on scale rulers on the up-down and left-right arcs and a timepiece that electrically recorded transits of stars when the observer pressed a Morse key. The clock was accurate to 1/100th of a second.

The calculation of the speed of light made Roemer an instant celebrity. He visited London and met all the heavyweights like Newton, Flamsteed and Halley before being appointed in 1681 Royal Mathematician and Professor of Astronomy at the University of Copenhagen.

Bradley

In 1728 Jamesw Bradley (1693-1762) got much closer to the mark when he calculated the speed of light at 183,300 miles per second or 295,000 kilometres per second.

Bradley had worked with British astronomer and politician Samuel Molyneux (1689-1728) improving reflecting telescopes. They had limited success when trying to observe the stellar parallax of stars. Shortly after Molyneux's death, using the same method Bradley discovered the aberration of light.

Hooke had observed the giant star Gamma Draconis aka Eltanin or Etamin in 1669 but failed to resolve the observations with regard to parallax.

When Bradley repeated Hooke's work, he observed a large one second shift of arc of the giant every three days. This was not only in the wrong direction for parallax, but was also impossibly large. So what caused it?

He went sailing on the Thames, as you do when in a quandary, and while watching the boat's weather vanes, noticed them adjusting their orientation to accommodate the wind and the boats velocity. 2+2 = the phenomenon of the aberration of light.

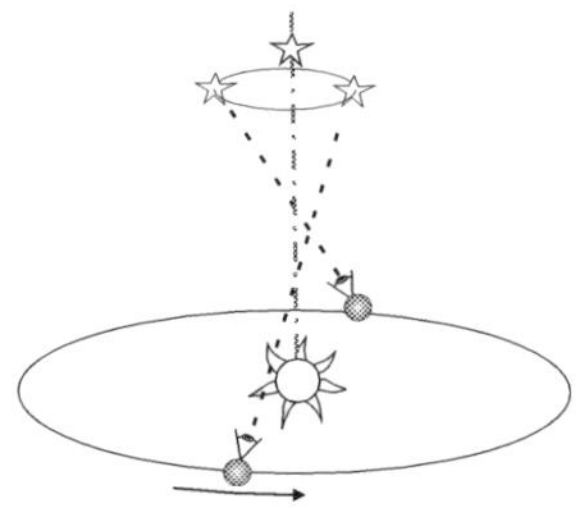

When walking into rain which is falling vertically, one angles the umbrella slightly forward to compensate for one's speed. Start running and one must tilt the umbrella further forward.

In observing Gamma Draconis near the North celestial pole, Bradley had tilted the telescope slightly in the forward direction of the Earth's orbit to see the star in its apparent position. Repeating the observation six months later the same tilt showed the apparent position displaced to the other side of the real location of the star. If one took observations every day the apparent star would be somewhere on a tiny ellipse that mirrored the Earth's elliptical orbit. The real location of the star is somewhere on a line from the Sun through the centre of the tiny eclipse.

Bradley found that the angle between the real and apparent position was 20 seconds of arc = 0.00555°

The speed of light '*c*' then is the velocity of Earth divided by the tan 20″.

Light aberration was an immensely important discovery. Apart from finally and absolutely confirming that the Earth circled the Sun, it had consequences that would lead to the theory of relativity. For the time being however the phenomenon of the aberration of light affirmed the existence of the all permeating *ether*.

The maximum aberration occurs when the light ray arrives normal to the earth's elliptical plane. This maximum is called the constant of aberration, and it is 20.47 seconds of arc. It is based on the sum of the Earth's velocity in orbit plus the speed of its daily rotation plus the speed of the solar system through the aether. The total velocity is about 1/50 the speed of light.

The Royal Society awarded Bradley the Copley Medal in 1748.

Fizeau

The desire to rigorously define the *ether* kept pressure on scientists to produce an evermore accurate measurement of the speed of light. One hundred and twenty years after Bradley, the regally named Armand Hippolyte Louis Fizeau (1819-1896) gave it his best shot.

He conducted an experiment using a variation of Galileo's idea. He placed a fixed mirror on the distant hill and shone a light at it through the gaps in the cogs of a spinning wheel on the home hill.

When the reflection was seen at its brightest coming through the second (next) cog, calculations were made. The problem here was that human error was involved in the judgement of when the reflected light was brightest when perceived through the location of the stationary position of the second cog.

Foucault

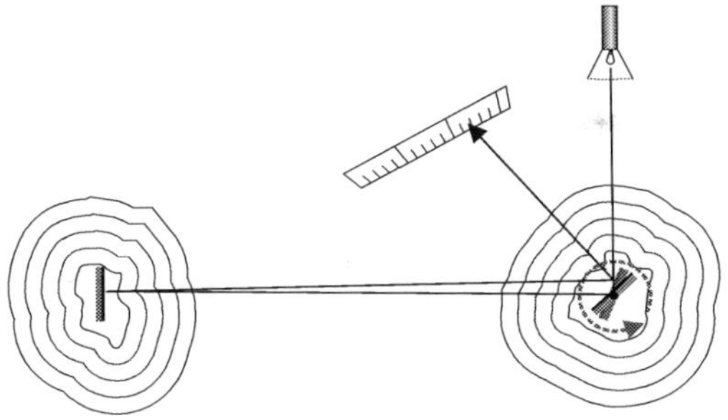

(this is not a clown's scooter)

Fizeau's co-worker was Jean Bernard Léon Foucault (1819-1868) who replaced the cog wheel with a rotating mirror. The light now shone first on the rotating mirror, then across to the distant mirror which reflected it back to the revolving mirror which had by then moved on a bit and now reflected the light onto the scale. The distance between mirrors was known, as was the speed of rotation of the mirror and the angle of deflection.

Not only was a highly accurate result achieved but in 1850, with variations, the experiment was repeated under water. The results were sensational.

Light travelled at about three quarters of its air speed.

Foucault's name is familiar to many people because of his pendulum experiment. Foucault's long heavy pendulum was a 28 kilogram plum bob on a 70 metre long steel thread hung from the central dome of the Panthéon, *(originally the Church of Sainte-Geneviève),* in Paris.

The pendulum experiment was conducted in 1851, presumably just before the church was reconsecrated. It would be awkward conducting Mass in the presence of a higher mass, roaring up and down like sword of Damocles over the heads of the kneeling faithful.

If the pendulum was suspended at a height above the North or South Pole, the ground underneath, i.e. the planet Earth, would rotate every 24 hours. A person standing at the Pole looking up would see the arc of the pendulum above him, rotate each 24 hours.

If the pendulum were located at the equator, the ground would not rotate under it so a person there looking up wouldn't see any change in the direction of swing.

In Paris the pendulum rotates in a length of time in between 24 hours and never. Foucault timed that a 360° pendulum rotation here took 32 hours.

Thus in a huge wind proof space, using a lump of iron and a length of wire, it was proven without going outside that the planet Earth rotates.

The Panthéon was secularised during the French Revolution. It twice reverted back to a church from 1828 to 1830 and from 1851 to 1870, and is now a civic building.

Foucault is the discoverer of electric eddy currents, sometimes called 'Foucault currents', which occur in copper plates rotating in magnetic fields.

In 1852 he built the world's first gyroscope.

In 1855 he received the Copley medal.

The End of the First Light War

The measurement of the speed of light through water temporarily terminated the two century old controversy of whether light was a stream of particles vibrating in the *ether* or a wave through the *ether*.

Huygens and the wavelets were the victors. Newton's theory that light slowed down in denser mediums was on the rocks but not washed up. His faithful diehards continued to support his particle theory despite this revelation and Thomas Young's earlier double slit experiment

*There followed a grudging peace for about a century until the emergence of two new revolutionary armies of physics, the relativists and the Alices in quantumland who today battle for laureates on the fields of Cosmos where they wage the Second Light War.

All the battles so far are bogged down in the mud of duality. If you look for a wave it's there but the particle isn't and if you look for a particle, it's there but the wave isn't.*

Refraction

Refraction occurs when light passes from one transparent medium to another. The apparent bend in a half submerged straight stick demonstrates refraction.

In 1621 Dutch physicist Willebrord Snell (1591-1626) discovered the relationship between the angle of incidence of a light ray and its angle of refraction, but did not publish the result. Descartes, in 1637 *(independently....?)* discovered the same with a simpler trigonometrically formula than Snell's geometric one.

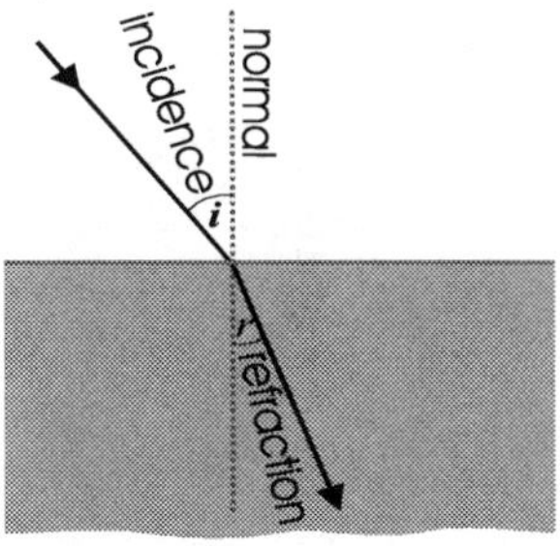

This version is known as Snell-Descartes law: The constant *'n'* is the index of refraction.

$$\frac{\sin i}{\sin r} = n$$

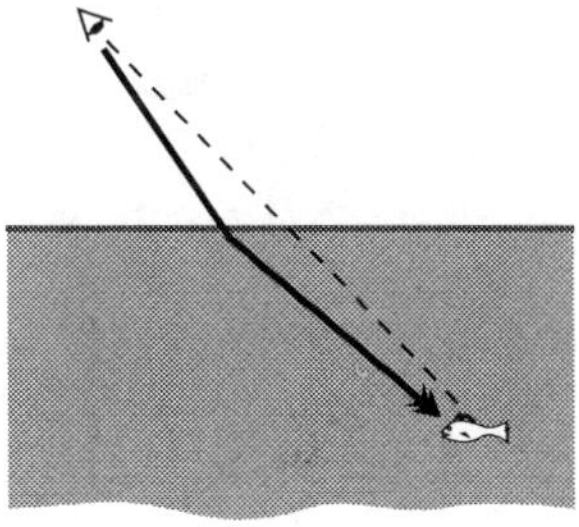

'n' for water is 1.33; glass is 1.5; and diamond is 2.42.

To spear the fish one has to aim below its image. The spear when under water appears to bend upwards.

Foucault's under water results supported Huygens' ideas on refraction. As the wavelets in the wave front penetrated the optically denser medium, those wavelets in the material slowed causing the wave front to wheel like a rank of soldiers.

Alternately one can visualise that the wavelength, *(the distance between successive wave fronts)*, gets shorter, but the frequency is the same.

Since speed = frequency × wavelength,

$$c = f\lambda$$

the speed of light must slow.

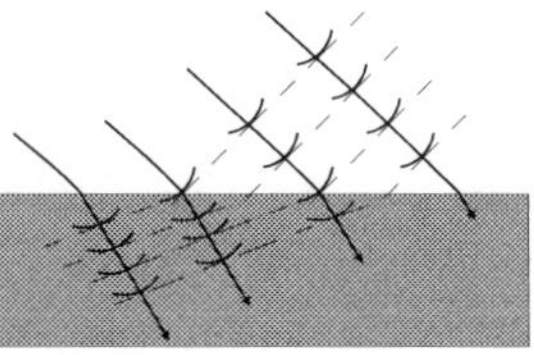

The greater the optical density of a material is, the greater its refractive index and the slower the speed of light through it.

(This is opposite of what sound does passing through incompressible materials).

It was soon discovered that the velocity of light in a medium:

$$v = \frac{c}{n}$$

where c is the speed of light in vacuo,
n is the refractive index of the medium, *($n > 1$)*

The speed of light		
Medium	Kilometres per second	Miles per second
In vacuo	299,792	186,000
In water	199,558	124,000
In diamond	123,919	77,000

This change in a wave's speed of light causes its refraction. With the exception of reflection, light travels along the shortest time path, which is not necessarily the shortest distance between two points, in a similar way to events in this illustration.

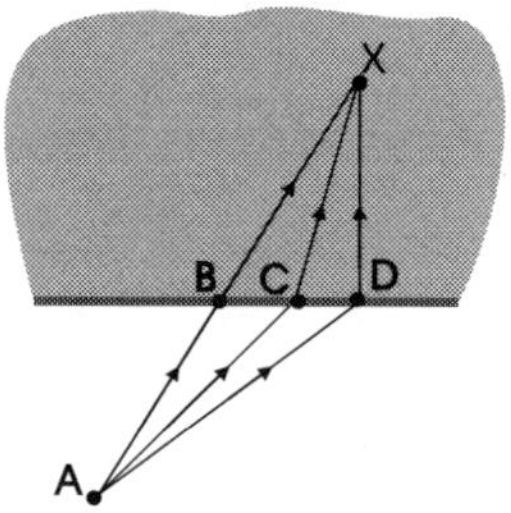

A person at point 'A' needs to run to the shore and swim out to rescue 'X'. He can run faster than he can swim. Route ABX requires too much swimming; Route ADX has the least swimming but because of the greater distance to travel, it still takes a little longer than ACX.

ACX splits the difference and is the quickest rescue path.

Interference and Thomas Young

In 1803 Thomas Young (1773-1829), English scientist physician and Egyptologist *(he helped to decipher the Rosetta stone)* conducted the double slit experiment.

Light emerging from the first slit in the first screen, spreads out according to Huygens's principle that each point on the wave front generates wavelets. The process is repeated twofold at the second screen where new secondary wavelets continue on to the back screen.

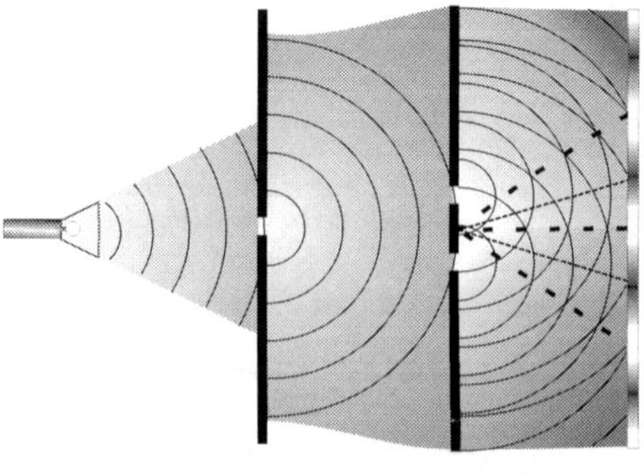

When these wavelets collide, there is constructive interference, where the crests and troughs of each are in phase *(bright zones)*, and destructive interference, where the crests and troughs of each are half a wavelength out of phase *(dark zones)*.

The brightness fades and the darkness lightens towards their respective outer sides.

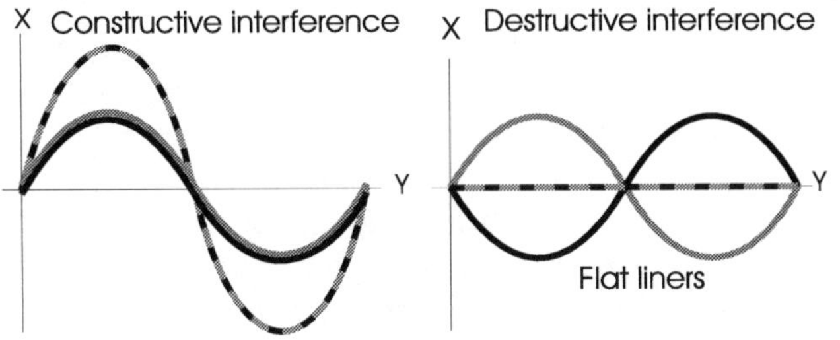

Interference can be seen when ripples collide after two stones have been dropped in the pond. Amplified waves with higher peaks and lower troughs radiate between lines of flat water.

Only coherent light will produce regular interference patterns. Coherent light rays all have the same frequency and wavelength and are usually produced from the same source. When light rays from different sources meet each other after each passes through a slit, the resultant display on the backboard is incoherent. *(hotchpotched)*

In 1807 Young showed that red, green and blue mixed in the right values could create any colour.

Diffraction and Grimaldi

The Italian physicist Francesco Maria Grimaldi (1618-1663) was the reason why the Newton-Huygens light war should never have started. In 1665 he directed sunlight through two consecutive holes and noticed that though the second hole was the right size for rectilinear propagation, the light had expanded and didn't quite fit through the hole, evidenced by a ring of light around that hole.

It was clear that light had diffracted.

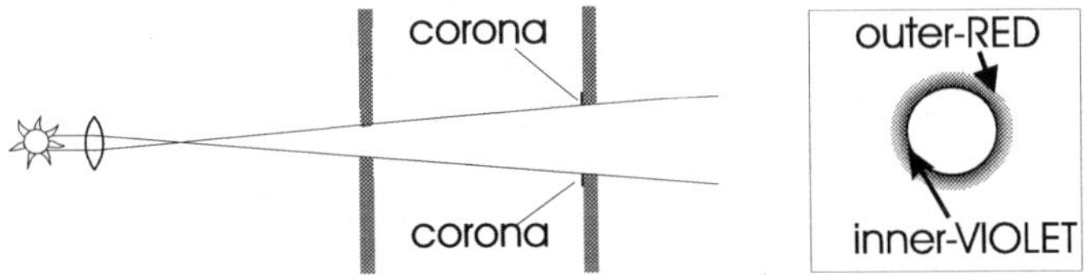

Moreover on closer examination he found that the inner edge of the resultant ring was tinged violet while the outer edge was tinged red.

The phenomenon of the waves spreading out after passing an opening is called diffraction. It is the reason why sounds made in one room after passing through the doorway can be heard in all corners of the next room.

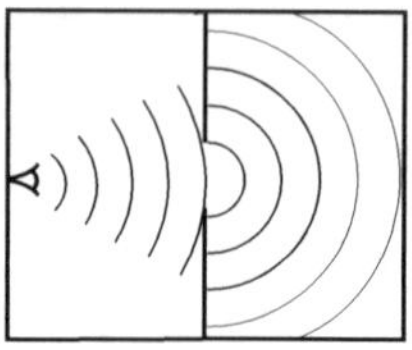

Diffraction is clearly seen when water waves pass through a gap, but light is fast and fine and appears on casual examination to leave sharply defined shadows, which encourage the belief that there is no diffraction.

For different reasons Newton and Huygens both chose to ignore Grimaldi's experiment.

Newton, who repeated the experiment, thought the result too trivial to affect his particle theory and Huygens simply did not believe in diffraction of light. His wavelets were only instrumental in propagating the wave front so only the tangential fronts of the wavelets were held to be relevant. Both scientists guarded their dogmas like crocodiles.

Fresnel

In 1814 *(the same year Napoleon escaped from Elba)*, French scientist Augustin Jean Fresnel (1788-1827), member of the French Academy of Sciences and the Royal Society of London, independently discovered interference patterns. He went on to put Huygens wavelets on a sound mathematical base that could concisely describe diffraction.

Each point on a wave is itself a tiny wavelet with the same wavelength as the master. As two side by side wavelets go forward, they expand sideways as well. This sideways motion from one wave is moving out of phase toward the approaching sideways motion of its neighbour. Consequently there is mutual self destruction leaving only the front of the wave intact, which is the wave front, going forward.

In Grimaldi's experiment, at the first hole the wave is chopped, leaving the round central wave front moving forward whose circular rim has nothing outside of it to cancel its expansion. This expanded edge forms the visible fringe around the second hole. Violet rays with the shortest wavelengths are the least diffracted, while red rays are the most.

Fresnel also designed a compound lens called the Fresnel lens which produces the penetrating parallel beams that emerge from lighthouses.

He discovered how to circularly polarise light, and researched the optics of motion which later became pertinent to the theory of relativity. Most of his work was known only to a group of scientists during his lifetime. A lot was published posthumously.

Fresnel believed the thrill of the discovery far outstripped the pleasure of acclaim from his peers. He was a quiet achiever

Diffraction Gratings

Fine straight lines scratched on a sheet of glass stop the sideways destructive interference of the incident wavelets. Thus the whole very tiny wavelet goes forward, not just the wave front. Many scratches produce a 'grate' with many whole lines of wavelets going forward. A diffraction grating might have 10,000 to 20,000 lines per centimetre which are constructed either by being pressed onto a polished surface by a diamond tipped tool or cut in with a laser. Light travelling through each gap between two scratches forms a diffraction spectra.

Each individual diffraction spectrum combines with its neighbour's spectrum and amplifies it. When 10,000 or more spectrums combine and are therefore amplified and the resulting spectrum is very sharp and clear. Also the infrared and ultraviolet outside edges of the spectrum are clear. Lenses and prisms have difficulty showing these far edges of the spectrum of light.

There are transparent or mirror diffraction gratings, both producing good spectra with sharp lines.

Doppler

In 1842 the Austrian mathematician Christian Doppler branched into physics with his publication "Über das Farbige Licht der Doppelsterne", *(Concerning the Coloured Light of Double Stars)*. In it he explains what is now known as the Doppler Effect.

The pitch of sound emanating from a moving body rises as it approaches the listener and falls as it recedes from the listener.

Doppler theorised that starlight should emulate sound by raising its frequency if the star was approaching an observer and lowering it if it was receding from the same. He argued that light from approaching stars should be blue-shifted and light from receding stars should be red-shifted. Initially however he proclaimed that the visibly blue stars in the heavens *(the blue giants)*, were approaching and the visibly red stars were receding. The analysis is flawed because it doesn't take into account that stars emit huge amounts of radiation in non-visible wave lengths like infrared.

In 1848 Fizeau came to the rescue, proving the effect by observing the Fraunhofer lines. The lines from a particular element are known from laboratory research to be located at specific points on graph paper designating their position in the light spectrum. If these same element lines from starlight move across the graph paper towards the violet end of the spectrum, the star is approaching. Conversely if they move toward the red end, the star is receding.

The effect is known as the Doppler-Fizeau shift, and the rest is Hubble.

*Thousands of stars have now been analysed and found to be either approaching or receding at between 35,000 kph and 140,000 kph. On the other hand all but a few close galaxies are receding. The furthest away and dimmest galaxies have the greatest red-shift.

Focusing the telescope on one side edge of the Sun then on the other side edge, *(not using eyes)*, readily shows by the Doppler-Fizeau shift that the Sun is rotating.*

POLARISATION OF LIGHT

Bartholinus

In 1669 Dutch physician Erasmus Bartholinus (1625-1698) made an accidental discovery which confounded both Newton and Huygens, although Newton did erroneously suggest a possible magnetic cause.

Erasmus was playing with a crystal of calcium carbonate *(Iceland spar)* on his blotter when he perchance noticed that while there was only one ink spot under the crystal, looking at it through the crystal revealed two spots. Moreover, as he twiddled the crystal around, one spot remained stationary while the other moved. He called the event 'double diffraction' and named the steady ray the 'ordinary ray = o-ray', and the other the 'extraordinary ray = e-ray'.

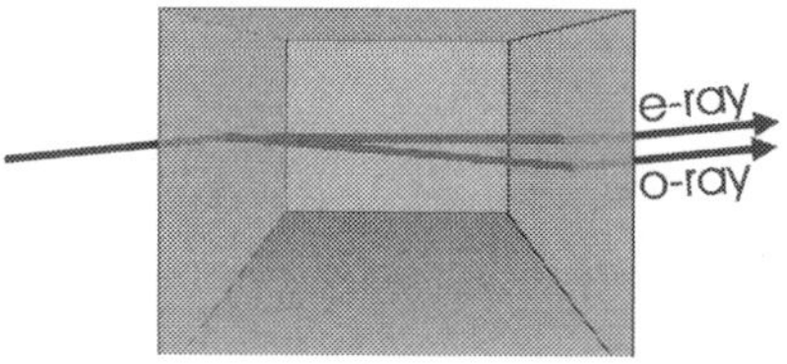

Huygens further investigated the phenomenon in 1678. Directing the diffracted beams through another crystal, he found that each beam was plane-polarized perpendicular to each other. The ordinary ray obeyed Snell's law and the extraordinary ray therefore couldn't. More perplexing was the fact that the extraordinary ray did not lie in the plane of incidence and refraction.

Malus

One and a half centuries later another accidental event revived and reinvigorated interest in polarisation. In 1808 Louis Malus (1775-1812) to pass the time squinted through his double diffracting crystal paperweight at sunlight reflected from a distant window. Wonder of wonders there was only one miniature Sun in the crystal, whereas at all other times before there had been two. Somehow the window had cancelled out one ray.

These were heady days for optics and Erasmus's records were hastily disinterred, brushed off and analysed. This time an explanation was soon forthcoming.

Newton's particle wave was a longitudinal wave like a sound wave. A longitudinal wave can not be polarised, according to conventional logic, but certain transverse waves can.

A water wave bobs up and down perpendicular or transverse to the wave direction.

Electromagnetic waves are three dimensional and bob in and out to all points of the compass perpendicular or transverse to the direction of the wave.

The electric field component of artificially generated electromagnetic waves like radio waves or microwaves oscillates in a regular manner, whereas the electric field component of light waves and other electromagnetic waves of nature, behave in a more random way

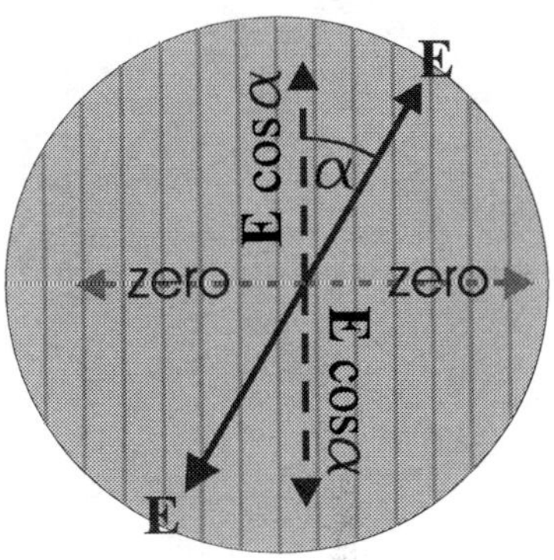

When light is polarised by either reflection or via passage through a calcite crystal or a polarising screen, only the electric field vectors of the light which are parallel to the constraints of the polariser can pass through. Those vectors which are perpendicular to the constraints of the polariser are blocked.

The amplitude 'A' of the light that gets through the screen is given by:

$$A = \mathbf{E} \cos \alpha,$$

The intensity of the light that gets through the screen is proportional to the square of the amplitude, and is given by:

The Law of Malus $I_i = I_p \cos^2\alpha$

where I_i is the intensity of the incident ray

I_p is the intensity of the polarised ray

Light cannot pass through two polarising screens whose polarisation slots are perpendicular to each other.

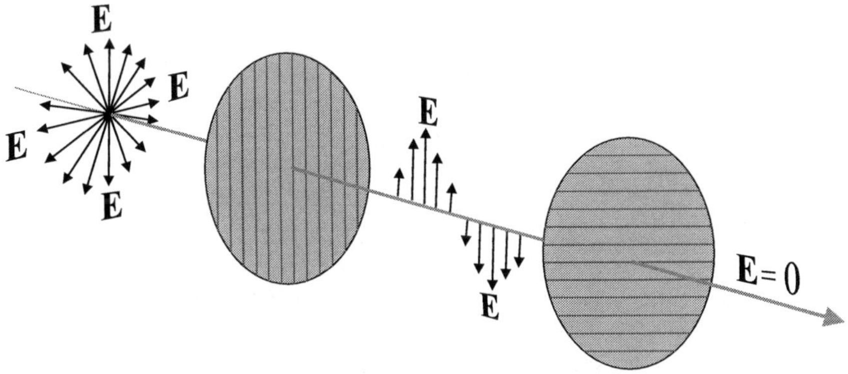

The diagram below is said to be a representation of a plane-polarised light ray. '**E**' is the electric field vector and '**B**' is the magnetic component.

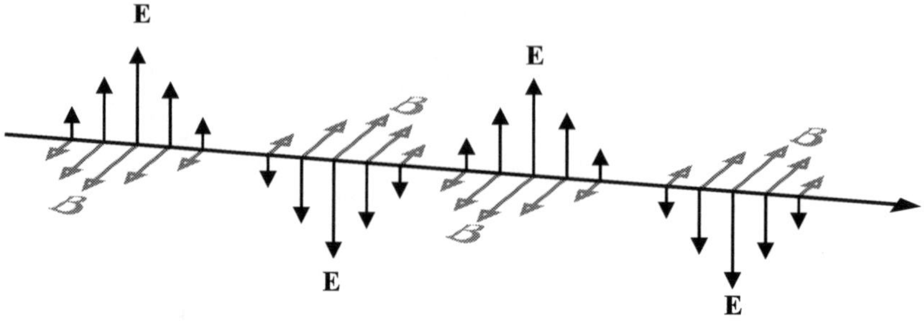

As Malus also discovered, when light is reflected from a transparent medium some of it is converted to polarised rays.

Brewster

In 1815 Scottish physicist David Brewster (1781-1868) the inventor of the kaleidoscope, found that the polarisation level of reflected light increased up to a specific angle and then decreased again.

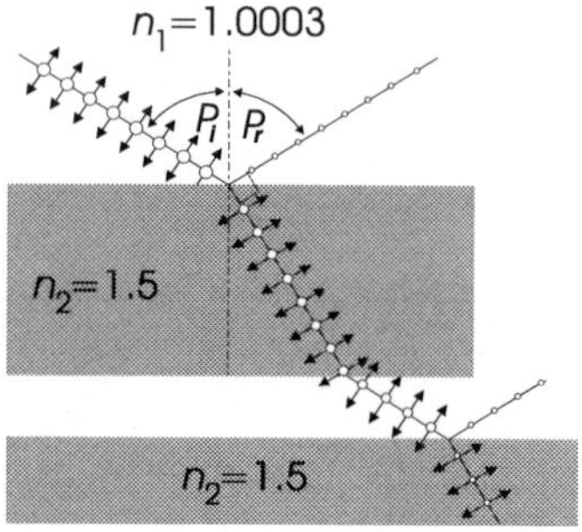

This angle is defined by Brewster's law that states that: "The angle between the refracted ray and the maximum polarised reflected ray is 90°."

Also it states that: "The tangent of angle of maximum reflection P_r equals the index of refraction of the medium that contains the refracted ray n_2, divided by the index of refraction of the medium that contains the incident ray n_1."

$$\tan P_r = \frac{n_2}{n_1}$$

The index of refraction for air $n_1 = 1.0003$ and glass $n_2 = 1.5$

Then $\tan P_r = 1.5/1.0003$,
Therefore $P_r = P_i = 56.3°$

The generation of the polarised reflected ray depletes the refracted ray in that plane. When another slab of glass is placed after a gap beneath the first, another polarised reflected ray emerges from that leading surface and the refracted ray is further depleted in that plane.

After a few more layers of glass, all that remains of the light ray, now totally depleted of light in one plane, is another polarised ray which is polarised in the plane perpendicular to the reflected polarised rays that were bled off.

Water has an index of refraction of 1.33, therefore the polarised reflected ray is at its maximum when tan P_r = 1.33/1.0003, i.e. when P_r = 53°. For glass P_r = 56.5° which was the Sun's angle of incidence at the window when one of its reflections disappeared from Malus's crystal.

The commercial polarising screen called Polaroid is made from microscopic crystals orientated in the same direction. Polaroid sun glasses are made with the screen orientation vertical. When one looks at water through the glasses the polarised reflected rays parallel to horizon are screened out leaving the refracted ray dominant. Water appears most transparent when looking down at an angle of 53° from the vertical or 37° from the surface.

A flat stone skips off water when thrown with the flat side parallel to the surface and dives if thrown face vertical. Similarly the surface parallel vectors of light tend to bounce off a surface while the vertical vectors tend to dive into it.

Nicol

In 1828 British scientist William Nicol (1768-1851) invented a device called the Nicol Prism which greatly simplified the production of plane-polarised rays.

A crystal of Iceland spar is cut into a rhombus then sliced in half through the short diagonal then glued together with a layer of Canadian balsam. The base is painted black to absorb the ordinary ray. Half the light that enters the system emerges as the plane-polarised e-ray. The o-ray before it was absorbed was the half of the incident ray polarised perpendicular to the e-ray plane. One still doesn't know in which plane the e-ray is polarised. Passing the e-ray through a second identical prism system resolves this. By rotating the second system about the axis of the first emerging e-ray, the intensity of the second prism system's e-ray varies between being not effected to nonexistent (zero) over a range of 90° of rotation. Half way between zero and zero over a range of 180° determines the polarity plane of the first e-ray.

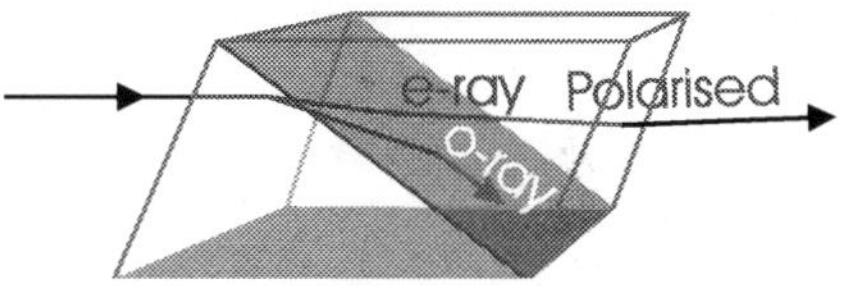

The first prism system is called the polariser and the second prism system is called the analyser. The combination forms the 'polariscope'.

(Foucault substituted air for Canadian balsam to polarise ultraviolet light)

Certain chemicals polarise light in various ways. When a container of these is placed between the polariser and the analyser their left or right handedness can be determined. Many chemical reactions were a mystery before the advent of the polariscope.

Circular Polarisation

Circular polarisation is created when a plane polarised ray passes through a thin polarising crystal disc. The ray is both normal to the disc and offset 45º from the optical axes which are in the plane of the disc. At the correct thickness of the disc, two beams emerge sharing the same axis but 90º out of phase with each other and plane-polarised perpendicular to each other.

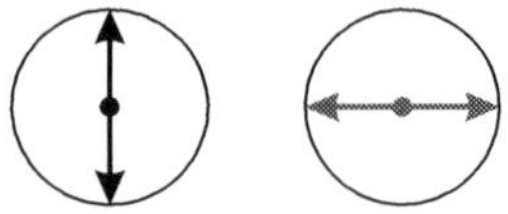

The phase difference sets up a sequence so that when the ray is viewed head on as it emerges, the electric field vector is rotating. In this instance rotation is clockwise therefore the emerging light is *right-circularly polarised.*

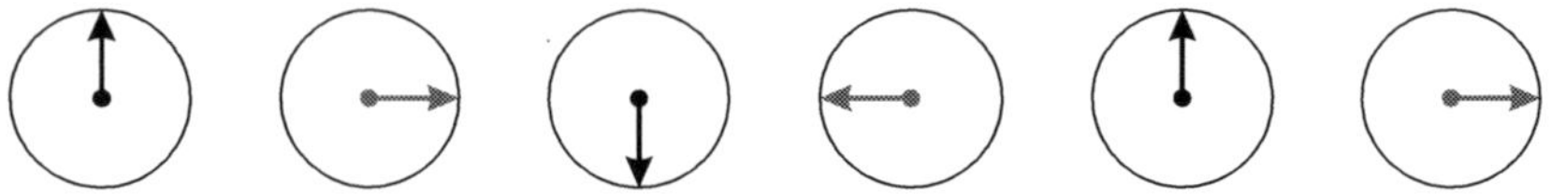

Rayleigh Scattering

When light hits gas molecules it is scatters in all directions. Blue light rays scatter more easily than red rays hence the sky is blue when the Sun is overhead. However at sunrise and sunset much of the short wavelength blue light has been scattered or bled out leaving red light dominant. Hence the Sun and the Moon when near the horizon have a red or orange tinge.

This was discovered in 1871 by John William Strutt aka Lord Rayleigh (1842-1919) and the process is called 'Rayleigh scattering'.

In 1887 he published the 'Theory of Sound'; a monumental work covering all sounds through all materials. Even today it is regarded as the Bible of sound.

Scattering from gas also causes polarisation. In the atmosphere sunlight and moonlight is plane-polarised with its electric field parallel to the horizon. Therefore in the morning and in the afternoon part of each ray is polarised north-south as they hit the Earth's surface. This phenomenon is used by 'polarisation compasses' to determine north-south. Bees' eyes have the same capability as they contain polarising filters for their navigation.

Polarisation compasses are particularly useful around the Earth's magnetic Poles where magnetic compasses are useless.

The Earth's magnetic poles, which are not necessarily directly opposite one another, are known to have drifted about 250 kilometres in the last 100 years. Every ½ million years the poles completely reverse in a process that takes from 1000 to 5000 years. There have been 171 reversals so far.

Coherent Light

The best way of creating two beams of coherent light is by passing the light through a split mirror where half the beam goes straight through while the other half is reflected. A split mirror is a sheet of polished glass with a very thin film of silver on it.

Interferometer

An interferometer is an arrangement where a light beam is split in two by a half silvered mirror and each beam after a journey is reunited and directed at a photographic plate, (*nowadays some electronic screen or receptor),* where an interference pattern forms. The fringe patterns produced reveal miniscule changes if any caused by any activity that might have influenced either or both rays on their individual journeys.

Shock Wave

For at least 2½ thousand years the *ether* had existed. It was as real as the air we breathe and the more science unlocked the secrets of the universe, the more existent *it* became. There were no heretics here; the *ether* was a friendly, benevolent thing. Suns, stars, planets moons, comets, magnetic fields, electric fields, cathode rays and electromagnetic waves including light all travelled harmoniously through it. So imagine the consternation amongst the scientific world, when it was snuffed out.

In August 1983 the Island of Krakatoa exploded pulverising 35 square kilometres of its total of 45 into fine dust which travelled around the globe in the upper atmosphere for the next three years creating countless spectacular sunsets. The sonic boom from the explosion reverberated for 4,800 kilometres.

Just as the last of the ultra-fabulous sunsets abated, another explosion rocked the planet when *ether* got the chop. The shock wave from this also reverberated around the globe leaving scientists reeling in its wake.

A true friend was dead and its dominion was dark.

Michelson and Morley Experiment

Since the Earth was travelling at 106,000 kph around the Sun, it should also be travelling at 106,000 kph through stationary *ether*. If the *ether* were moving relative to the stars then the Earth would be travelling either more or less than 106,000 kph through it. Scientists thought that a light beam should travel slower going into the *ether wind*, and faster when running with it.

In 1887 Albert Abraham Michelson the professor of physics at the Case School of Applied Science, and Edward Williams Morley, professor of chemistry at Western Reserve University share the same campus.

They believed they had a device sensitive enough to confirm or deny the effect of ether on light. It required enormous sensitivity and rigorous method since 106,000 kph is only about 1/10,000 the speed of light.

Michelson had tried the experiment years in Berlin and couldn't find any fluctuation in light speed. This time with Morley the experiment would be more precise and the results more final.

The experiment was conducted in the basement of the Western Reserve University dormitory (Cleveland), on a rigid table which was anchored into bed rock to avoid any external movement. Mounted on the table was a top plate which could swivel about its centre. The light source, mirrors and receptor were fixed to this top plate.

The split mirror M_s, mounted at 45° to the incident light beam, split the beam along the pathways shown. Mirror M_2 could be adjusted to tune in sharp interference fringe lines at the receptor. Having done this M_2 was fixed.

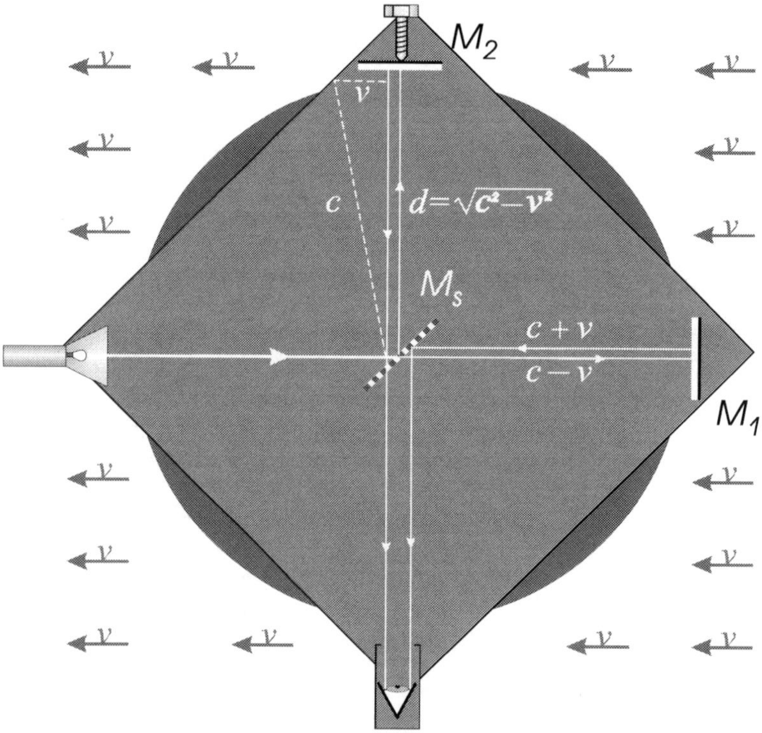

Observations were made at various rotational orientations of the top plate at different times over a period of many months. This span covered the possibility that the *ether* might at some point be moving parallel to the Earth's orbit.

The time t_1 to go from M_S to M_1 and back to M_S is

$$t_1 = \frac{d}{c+v} + \frac{d}{c-v} = \frac{2dc}{c^2 - v^2}$$

The time t_2 to go crosswind from M_S to M_2 and back to M_S is

$$t_2 = \frac{2d}{\sqrt{c^2 - v^2}}$$

and

$$\frac{t_1}{t_2} = \frac{c\sqrt{c^2 - v^2}}{c^2 - v^2} = \frac{c}{\sqrt{c^2 - v^2}}$$

Dividing top and bottom by c

$$\frac{t_1}{t_2} = \frac{1}{\frac{\sqrt{c^2 - v^2}}{\sqrt{c^2}}} = \frac{1}{\sqrt{1 - \frac{v^2}{c^2}}}$$

In every instance on every rotation of the table, there was not the minutest change observable in the interference fringe patterns at the receptors.

In July 1887 after thousands of trials, Michelson and Morley (M&M) announced that the *ether* was extinct. *The Ether Era was over.*

There were of course many who disagreed with the result. The principal objection was that all the equipment shrunk or expanded simultaneously with the *ether wind* which therefore distorted the result. Einstein would soon silence this

There were also die hard dissenters like DC Miller who repeated the experiment with variations and more sophistication from 1902 to 1924. He announced to the American Physical Society in 1925 that he had indeed found a variation the speed of light through the *ether*. Few were convinced.

A very sophisticated version of the experiment was performed in 1979 by Alain Brillet and John Hall of the Joint Institute for Laboratory Astrophysics (University of Colorado and National Bureau of Standards). They used a fixed reference laser and a second laser mounted on a turntable. Their results found no variation in the speed of light in any direction. The accuracy here was 6 parts in 10^{15}, which soon silenced the Miller crowd.

The speed of light in vacuo is 2.998×10^8 meters per second.

or c = 1,079,252,849 kph

Epitaph

The *ether* had been known as the omnipresent, transparent, weightless, frictionless medium which permeated and entwined all nature.

With the *ether* gutted, science resorted to using the pretentious word 'fields' to presume understanding of how magnets or static electric charges communicate.

In a bizarre twist,since the 'fields' had no medium, like the phoenix,
'action at a distance' was 'born again'.

ELECTROMAGNETISM CIRCA 1900

CHAPTER 3

Ferro-Magnetism

- In ferromagnetic material, usually iron, cobalt or nickel, inter-atomic forces cause the magnetic moments of neighboring atoms to align in the same direction. These become domains that resemble the assorted sized, snug fitting blocks.
- Each domain has an inherent but random magnetic orientation when the bar is unmagnetised.
- If finely ground iron oxide is dusted onto a bar magnet it will collect on the borders between the domains. These can just be seen with the naked eye, but are clearly visible through a magnifying glass.

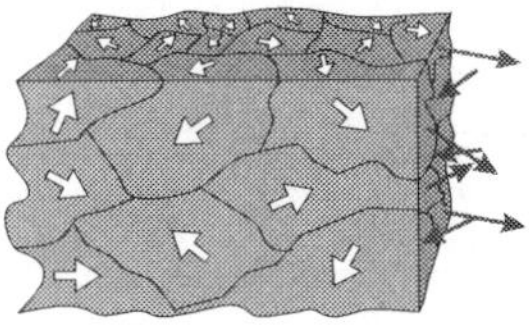

- A current flowing through a coil generates magnetism. When the coil is wrapped around a ferromagnetic bar the domains with a sympathetic magnetism to that of the coil grow at the expense of those not simpatico which shrink. The bar becomes directionall ymagnetised, increasing its magnetism as much as 10,000 fold that of the isolated coil.
- When the coil current is turned off the coil's magnetism collapses but many of the domains in the bar do not revert to their pre-magnetised condition. These directionally magnetised domains linger for a time. The harder the steel is, the longer they linger and the more permanent the magnet.
- Various environmental disturbances such as heat can cause magnets to revert to their random domain alignment.

The Curie point is the temperature at which heat breaks magnetism down. The Curie point for iron is 770° C. Combo iron-neodymium-boron magnets are more resilient and more powerful.

Magnetic Fields

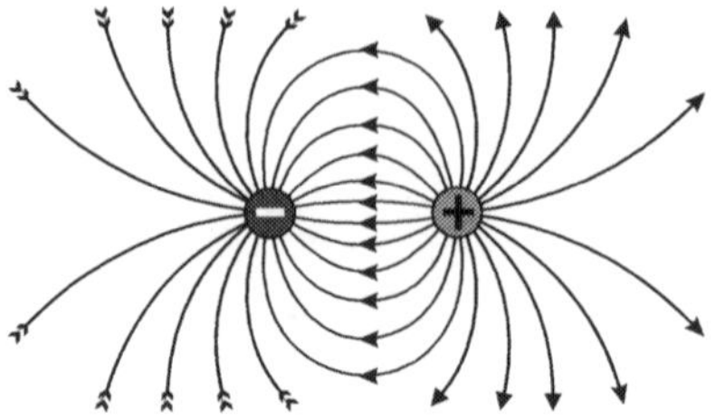

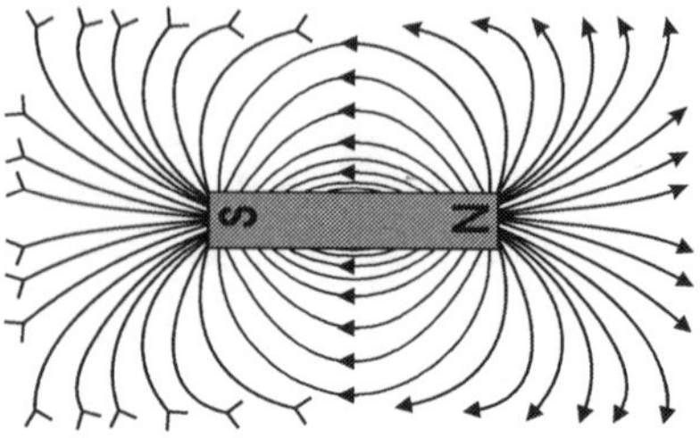

By convention a magnetic field flows from a north pole to a south pole.

Hans Ørsted in 1820, and the earlier never recognised Gian Domenico Romagnosi in 1802, showed that a current caused deflection in a compass, therefore a current creates its own accompanying magnetic field. Following convention, the direction of the field is the direction to which a north pole of a magnet points. If a right hand grips an insulated conductor with the thumb pointing in the current direction, the fingers curl in the conventional direction of the magnetic field.

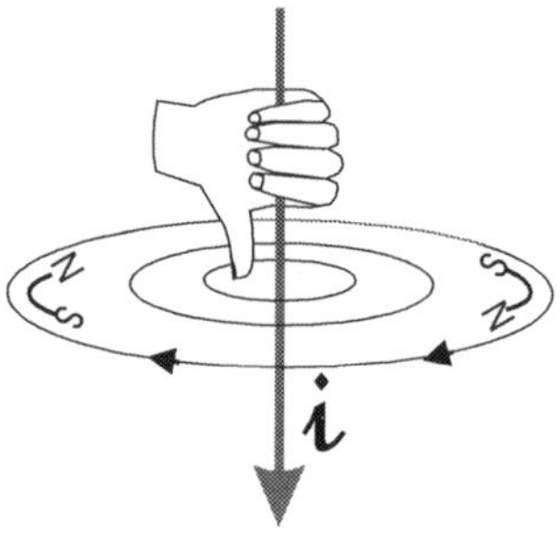

The right hand grip, also establishes the *(conventional)* direction of the magnetic field through and around a current loop or a solenoid.

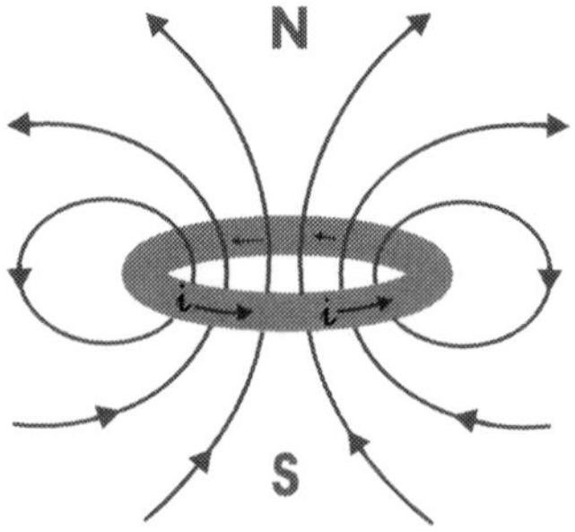

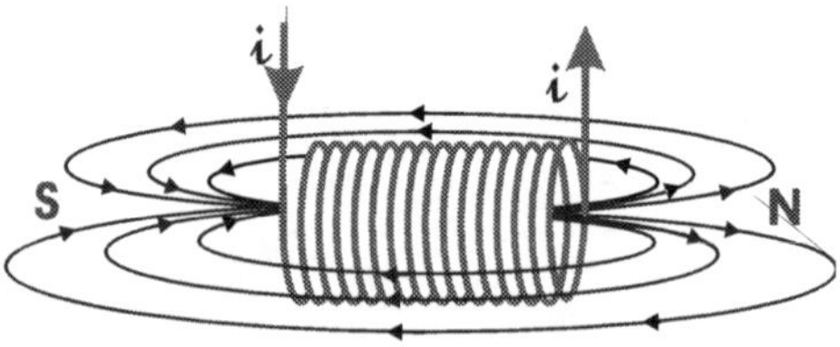

Ampère showed that parallel conductors attract each other when travelling in the same direction and repel each other when in the reverse direction.

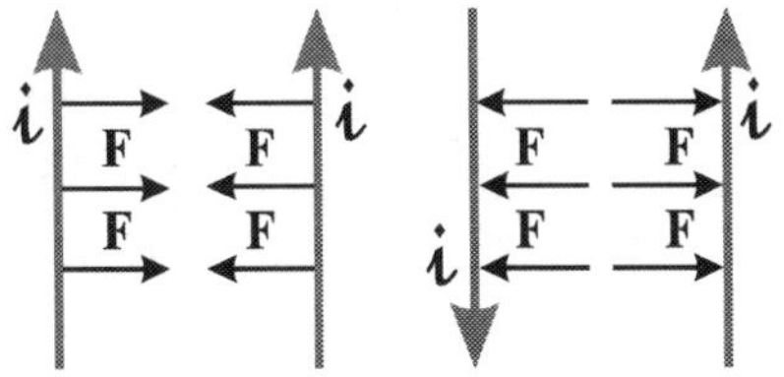

This force is a result of their magnetic field orientations at the interface.

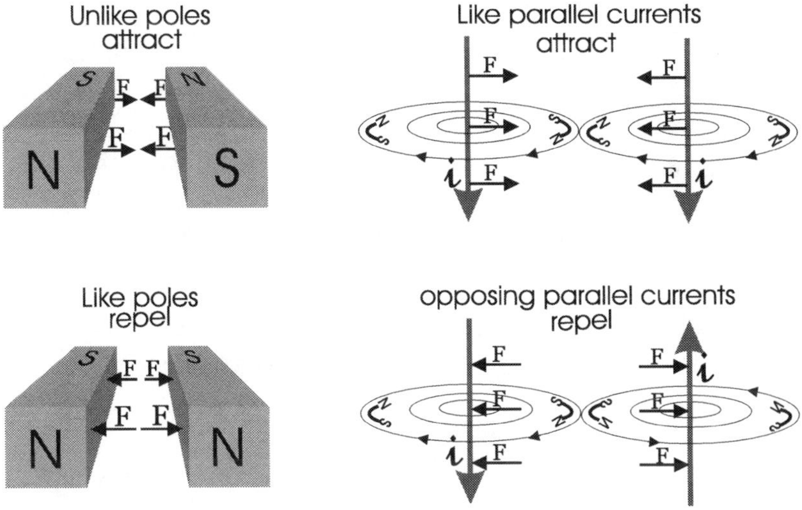

When the conductors are perpendicular to each other they have zero mutual attraction or repulsion.

When a current in a conductor flows through a magnetic field perpendicular to it, forces perpendicular to both occur. When either the current is reversed or the magnets are reversed, the forces are also reversed.

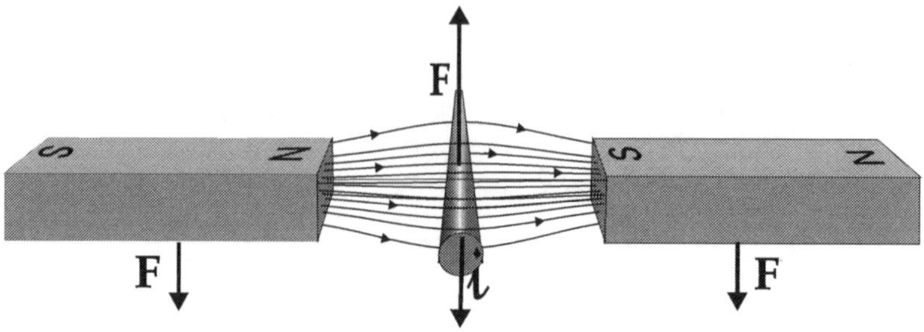

To determine whether the conductor is forced up or down, students are taught to make an x-y-z axis with the thumb, index finger and middle finger of the right hand. The thumb represents the force, the index finger the current direction, and the middle finger the direction of the magnetic field. Students often cheat their examiners by crib-writing *F, i* and *M* under their right thumbs and fingertips.

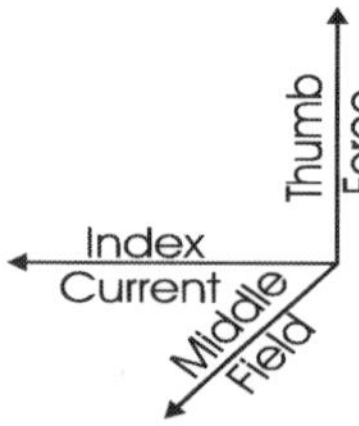

The direction of the magnetic field heading into the page is graphically represented by crosses symbolising the tail flights of arrows.

The direction of the magnetic field heading out of the page is graphically represented by dots symbolising the points of heads of arrows.

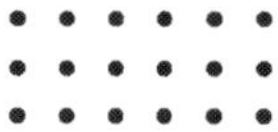

A magnet advancing towards a current loop induces a current in it whose direction opposes the change that produced it.

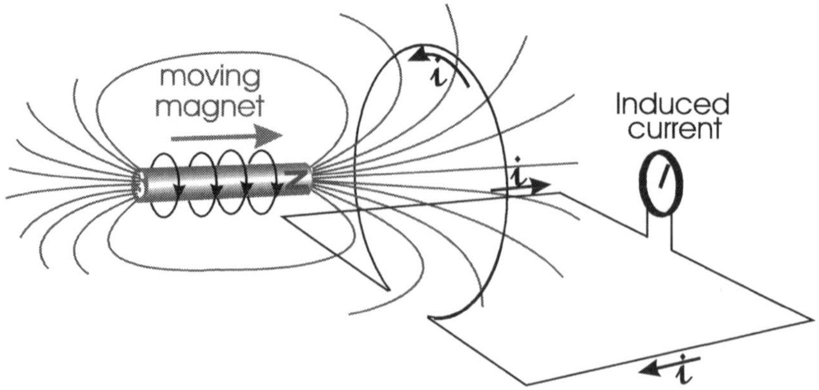

Electric Fields

The electric field, also called the electrostatic field or e-field was first observed as the effects of an invisible tentacle which reached out from materials like glass rods when rubbed with silk, to lure little bits of lint and paper. It occurs whenever there is an imbalance of charge between objects, which is when one object is more positively or negatively charged than the other. Since this can happen just as easily between moving charged objects as stationary ones, the static part of electrostatic is bit of a misnomer.

The difference in voltage between glass rods and lint is about a thousand volts. That between a synthetic jumper and one's hair is tens of thousands of volts.

The attraction between positive and negative charged objects implied that the invisible tentacle left from one object and entered the other. The choice was made to say the field left the positive object and flowed into the negative one.

Streamlines were drawn to graphically represent this flow which expanded out from, or contracted in, to charged objects and points. Where the streamlines were closer together the field was stronger.

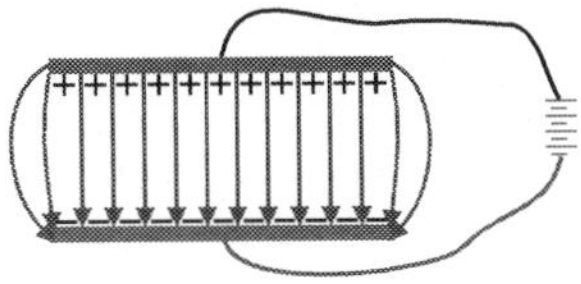

Where the electric streamlines were parallel, they represent a constant flow. Here an equal force acts on a body whether it is upstream or down stream.

Parallel lines of force in electric fields are characteristically found between two opposite charged conducting plates.

A negative charged particle, be it stationary or moving, is forced upstream in an electric field. A positive charged particle is forced down stream.

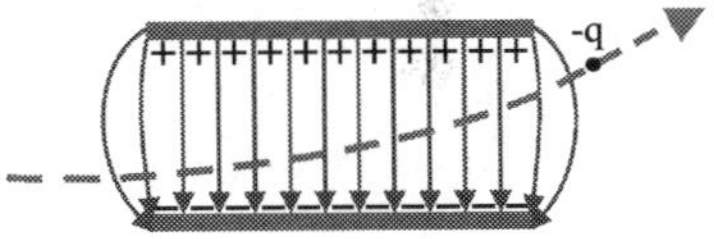

The force on the particle is given by: $F = Eq$.

where E = field strength, q = charge on the particle.

Magnetic and electric fields replaced the long standing 'action-at-a-distance' model which held that an inexplicable magical force accompanied electromagnetic *(and gravitational)* events. Field theory made it possible to measure these forces but they remained inexplicable, more so following the execution of the *ether*

Fields then are measurable actions-at-distances.

Eleven years after the execution of the *ether* English physicist Joseph J. Thomson (1856-1940) discovered the electron.

The Electron & J.J.Thomson

In the 1890s there were many arguments as to whether cathode rays were streams of particles or waves. If they were particles, Hertz thought that they would be deflected by electric fields as well as magnetic fields.

In 1992 he fired the rays through an electric field in the air of the laboratory and found no deflection. Thereafter he pronounced that cathode rays were only waves.

Five years later Thomson theorised that the air in Hertz's experiment would interfere with deflection, so he repeated the experiment in a good vacuum. The rays were clearly deflected and the particle was reborn. It had a negative charge.

Would that all boo-boos in science be so swiftly resolved.

Around the turn of the 19th century Thomson was the keystone of British and indeed world physics. He was the son of a bookseller who in 1876 earned a scholarship to Trinity College Cambridge where he remained for the rest of his life. In 1884 he became the head of physics at the Cavendish Laboratory.

He was renowned not only for his discoveries but also for his administration of the laboratory. Students came from all over the world to study under him and between 1895 and 1914 seven of these received the Nobel Prize.

Thompson himself received the Nobel Prize for Physics in 1906 and a knighthood in 1908. In 1918 he was made Master of Trinity, a post he held til his death at the age of 84.

The Thomson Experiment

In 1898 J.J. Thomson fired cathode rays through a magnetic field of strength 'β' and from its deflection determined the radius of its curved path 'r'. His aim was to determine the velocity 'v' of the ray.

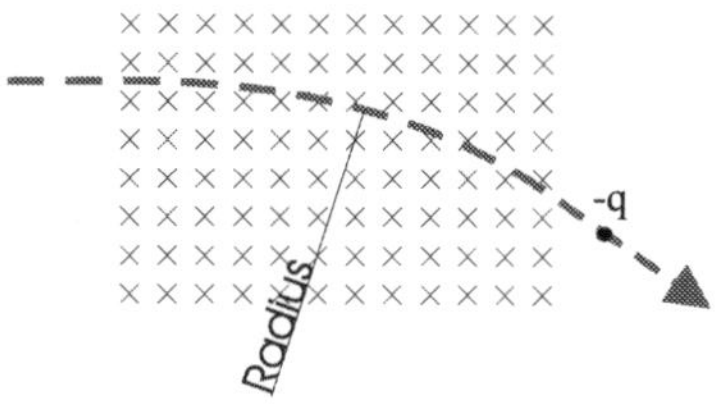

The centripetal force on the particle of mass m is: $\frac{mv^2}{r}$

and the force from the magnetic field β on a charge q is $qv\beta$

then: $$qv\beta = \frac{mv^2}{r} \qquad \frac{q}{m} = \frac{v}{\beta r}$$

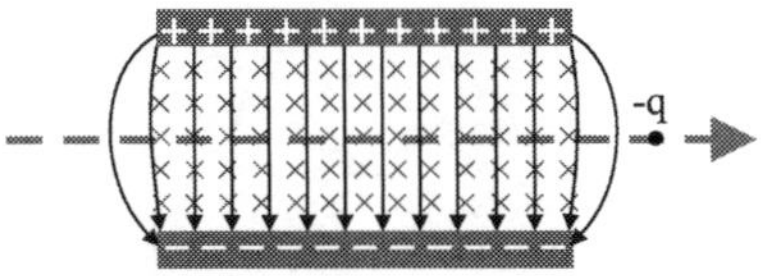

An electric field was adjusted so its force cancelled that of the magnetic field. Now the ray travelled in a straight line through both fields.

The electric field force now equalled the magnetic field force.

$$Eq = qv\beta \quad (E \text{ and } \beta \text{ are known})$$

$$v = \frac{E}{\beta} \qquad = 30{,}000 \text{ kilometres/second}$$

With the velocity discovered and the original radius and the magnetic field strength known, he was able to calculate the charge to mass ratio of the particles of the cathode ray.

$$\frac{q}{m} = \frac{v}{\beta r} = \frac{30{,}000}{\beta r} = 1.7 \times 10^{11} \text{ coulombs / kilogram}$$

This brilliant result is very close to the latest figure of:

1.75890×10^{11} coulombs/ kilogram.

The coulomb is the unit of charge the SI system of units, i.e. the International System of Units, (Système International d'Unités). It is the amount of charge transported by a current of one ampere in one second.

Thomson concluded that the rays were comprised of tiny particles he called corpuscles having a mass of about 1/2000th of a hydrogen atom travelling at close to 1/10th the speed of light. He thought these particles were an elemental part of all nature. Later these particles were called electrons which are said to date from his experiment.

From 1903 to 1907 JJ tried valiantly to support Lord William Thomson Kelvin's concept for atoms, which was that of a positive material with negative electrons scattered through it. Because of his investment and because JJ was head of the Cavendish Laboratories at Cambridge this representation became known as the Thomson's plum pudding model.

The 18th century ended with the *ether* buried in 'fields' of electromagnetism, and the birth of the electron.

THE CENTURY WITHOUT ETHER

CHAPTER 4

DISSECTING THE ATOM

Radiation

The period following the ether's execution was so fruitful that it is often referred to as the age of invention. Scientists were now liberated by the phantasmagorical 'fields' from the incongruous ether and its burdens, such as wondering why gravitational or electromagnetic events were immune to the friction of space.

This was also the age of the second communication revolution, the first being that of the printing press. Scientists could now exchange ideas and news via the telephone or radio. New discoveries were more and more closely linked to previous ones. Reclusive scientists were evermore an endangered species, unable to keep pace with the collective.

Europe and America and Asia were gripped by a scientific fever of discovery with each laboratory just a whisker in front of the other. When new eyes were discovered that could peer into atomic structure, the pace quickened.

X-ray Vision

The first of these new eyes was occasioned by that ubiquitous travelling companion of research called chance. In this instance chance was freakish.

On November 8th 1895 German physicist Wilhelm Conrad Röntgen (1845-1923) professor of physics at Würzburg University, was experimenting with large voltages crossing partially evacuated tubes similar to Crookes' tubes. If these had a metal foil window, phosphorescence was usually observed on luminous material outside the tube. Röntgen's tube had no windows but he wanted to see if there was phosphorescence without them. He covered the tube with black cardboard, turned off the lights and turned up the volts.

At first he saw nothing but then a tiny glow caught his eye coming from the desk near the experiment. Lights on showed that the glow emanated from a scrap of packaging paper which had part of its advertising print, the letter 'A', highlighted with a fluorescent solution. It was the 'A' that glowed and the unknown emission which caused it he called X-rays.

Further investigation showed that fluorescent screens also glowed even if the were in the next room. Now chance struck again! In a desk drawer beneath the experiment some photographic plates were stored in a black envelope weighed down by an iron key. When Röntgen developed the plates he found them darkened around the impression of the key.

By the end of the week he was showing doctors that by placing his hand holding a lead disc between the X-ray tube and a barium-platinocyanide treated screen, the screen showed an image of the bones of his fingers holding the disc.

Medicine, dentistry and industry were instantly and indelibly transformed.

Röntgen published three works on the nature of X-rays in 1895-6-7. He was awarded the worlds first Nobel Prize in Physics in 1901.

The röntgen or roentgen, symbol 'R', is a unit of radiation exposure for X-rays and gamma rays. It is equal to the quantity of radiation that produces ionization equal to one electrostatic unit of charge at 0° C and at standard atmospheric pressure.

Burns to the skin from overexposure to X-rays were soon painfully obvious.

Henry Antoine Becquerel

In 1896 Henry Becquerel (1852-1908) professor of physics at the Ecole Polytechnique in Paris discovered radioactivity in uranium.

He was a third generation French physicist, his father being Alexandre Becquerel, who studied light and phosphorescence and invented the phosphoroscope; and his father was Antoine César Becquerel, who discovered the photovoltaic effect, (the science behind solar panels).

Henry was working along his father's lines of phosphorescence when he heard of Röntgen's X-rays. He thought that X-rays might be the invisible element of sunlight that made certain materials glow in the dark. To test this on uranium crystals he placed them on a photographic plate, wrapped the lot with black paper then placed the package on the window sill. Disappointed with three days of cloud he repeated the experiment with a fresh plate hoping for sunny days.

Out of habit he developed the previous wrapping expecting nothing but the impressions of the crystals were there

Further experiments in the dark room confirmed that it was emission from the crystals that darkened the plate.

He soon discovered that the emission could also cause the leaves of an electroscope to collapse proving that the emission had an electric charge.

The becquerel, 'Bq', is the international unit of radioactivity.

Marie, and Pierre Curie

The Polish born Maria Sklodowska (1867-1934) moved to Paris in 1891 to study at the Sorbonne, from which she graduated with honours in 1894. The following year she married the renowned crystallographer Pierre Curie (1859-1906), who in 1880 with his brother Jacques had discovered the piezoelectric effect.

Looking for a subject for her thesis Marie among others chose to research the topical discoveries of Röntgen and Becquerel. Looking for other material that exhibited Becquerel's radiation she discovered that thorium was one.

German physicist Gerhard Carl Schmidt had discovered this two months earlier. Perhaps they both walked the corridors of the geology museum waving little electroscopes in front of rocks.

Thorium was discovered in 1828 by the Swedish chemist Baron Jöns Jakob Berzelius. It is found in the mineral monazite and occurs also in thorite and thorianite. Monazite is now known to be $(Ce,La,Y,Th)PO_4$ i.e. Cerium-lanthanum-yttrium-thorium-phosphate. (Monazite sands also hold helium.)

Marie then turned her attention to pitchblende, a black opaque pitchy mineral comprised mainly of the mineral uranite (uranium dioxide). She concentrated it by boiling and found that per volume, it darkened photographic plates 4 to 5 times faster than Becquerel's crystals did, therefore there had to be a stronger radioactive element it. The chase was on and now Pierre joined the hunt. (They coined the word 'radioactive'.)

The task of finding this element turned out to be gargantuan. At the Curies expense eight tonnes of pitchblende was purchased from a slag heap at Joachimsthal in Austria (now Jáchymov in Czechoslovakia) and transported to their back yard shed, then began the revolting work of rendering the goo.

Marie was the cook while Pierre analysed. He diligently studied the new radiations separating the ore into countless chemical fractions, while she spent long days and half nights stirring the pot in the stinking, stinking hot shed, with no chimney. At the same time she cared for her first daughter Irène, born in September 1897

In July 1898, using a chemical fraction with bismuth sulphide, a new radioactive element was isolated which they called polonium after her native Poland.

In December that year a radioactive barium chloride fraction was treated to isolate a very powerful new element which they called radium.

They had eventually garnered one gram of pure radium chloride from the eight tons of pitchblende.

The story goes that after dinner Pierre and Marie went into the laboratory turned out the lights and sat hand in hand for over an hour bathed in the golden glow of the speck of radium chloride on the desk. All sorts of objects in the lab were twinkling as if in candlelight.

Along the way they had discovered that beta rays were negative charged particles.

In 1903 Marie got her doctorate while both Curies got the Davy Medal of the Royal Society. The same year they shared the Nobel Prize in Physics with Becquerel.

Pierre was tragically killed in 1906 when run over by a horse and carriage. Marie was devastated.

In 1910 Mme Curie and André-Louis Debierne isolated 1/10th of a gram of the silver-white pure radium metal itself.

She worked on and received the Nobel Prize in Chemistry in 1911.

The Golden Fleece eventually killed her. In 1934 she died of Leukaemia.

In 1995 Marie Curie's ashes were enshrined in the Panthéon in Paris.

Hazards

Marie lent their friend Becquerel a few grains of radium for his experiments. This he kept in his vest pocket for a few days until he realised his skin itch was a burn just like the overdosed X-ray burns he'd seen. He alerted his medical associates who began treating skin diseases and some cancers with radium emissions.

*Radium made certain things glow in the dark and therefore it was ideal for the numbers on watch faces and the dials of instruments like pilot's altimeters and compasses. It was used for this purpose until the early 1950s.

During World War I it was applied by factory girls using fine paint brushes which they licked continuously to keep the points keen. Much was ingested and it concentrated in the bone marrow where it destroyed red corpuscle production resulting in leukaemia. Many suffered bone degeneration of their jaw. There are stories of prankster girls painting their teeth with radium so they glowed madly in the lights of the dance room.*

Rutherford and Alpha and Beta Rays

New Zealand born Scientist Ernest Rutherford (1871-1937) was the son of a wheelwright who had migrated from Britain. Ernest was the 4th born of 12 siblings who rode several scholarships on his way to Cambridge in 1895 where he began working under J.J.Thomson at the Cavendish laboratory of experimental physics.

He and J.J. began research on the new X-rays and soon discovered in early 1896 that they generated masses of ionised particles, both positive and negative in gases, which would thereafter reunite to form neutral gas.

Rutherford formulated a technique for finding the rate of reunification and published the results. His star was rising.

He then repeated the experiment using Becquerel's rays and by 1899 found that these also ionized air, but there was a difference. He found that they were comprised of two different rays which he named:

- Alpha rays: These produced lots of ionization but were easily absorbed, only penetrating a few 100th of a millimetre into aluminium foil.
- Beta rays: These could penetrate about a millimetre into alfoil but produced much less ionization.

The atom had always been thought of as the indivisible particle. Now Rutherford was showing that some atoms were decaying by producing radioactivity.

Neither Thomson nor Rutherford jived to the fact that the beta ray was a stream of electrons. It was Becquerel who made the connection.

In 1903 Rutherford detected another ray in atomic decay which he called the gamma ray.

All five rays i.e. cathode, alpha, beta, 'X' and gamma can ionise gases as indicated by a collapse of the leaves of electroscopes.

In 1907 Rutherford returned to England from a nine year stint as professor of physics at McGill University in Montréal Canada. He took up the post of professor of Physics at Manchester University, which he held for the next twelve years.

At that time, all over the world scientists were experimenting with the new radioactivity. By 1905-6, many believed the alpha ray was a stream of double charged helium ions. Rutherford proved this by placing an alpha emitter beside a thin glass window of a vacuum flask; the alpha could just penetrate in but couldn't get out. Within days the presence of helium was detected with a spectroscope.

Also in 1907, the German physicist Hans Wilhelm Geiger (1882-1945) became Rutherford's assistant at Manchester U. Within a year he had built the first Geiger counter

It consisted of a light metal tube with a cathode at one end and an anode at the other across which was a high potential difference. The tube contained a gas at low pressure. When an alpha particle penetrated, it ionised a gas molecule which travelled rapidly toward the cathode ionising other molecules along the way. Collectively these ions were sufficient to cause a tiny current which powered a click sound. The faster the clicking, the higher the level of radiation.

Like the a bat seeing with echoes, the Geiger counter could then track the paths of super high speed miniscule bullets, which either breezed through the gaps in matter or ricocheted off its internal kernels.

Its eyes were its ears

Discovering the Nucleus (1911)

While at McGill, Rutherford had experimentally observed some slight alpha ray diffusion when a straight beam passed through a thin sheet of mica. This suggested that enormous forces from within the mica were involved. He thought it worth a look to see if any could bounce backwards from heavier material, and asked Geiger and a student called Ernest Marsden to set up an investigating experiment.

They used polonium as an alpha source and a strip of gold foil 4×10^{-7} meters thin as the target. Most particles went straight through but to their amazement one in 10,000 bounced backwards through more than 90º, and one in a few million bounced right back towards the source.

Rutherford said; *"It was quite the most incredible event that ever happened to me in my life. It was almost as incredible as if you fired a 15 inch shell at a piece of tissue paper and it come back and hit you."*

They had expected to find support for Thomson's plum pudding atom, but the forces from that model were 100 million times too weak to create the rebound.

After some angst and much calculation Rutherford concluded that the atom had a tiny positive charged nucleus with electrons orbiting around it; Indeed a Saturnian model.

Hantaro Nagaoka

The plum pudding recipe had been challenged back in 1904 by Japanese physicist Hantaro Nagaoka who proposed a Saturnian atomic model with rings of electrons circling around a central nucleus. The Britts quickly shot this down with an argument which in 1913 Niels Bohr would show was flawed

*Japan wasn't very popular in 1904. On February 9th without warning it had attacked the Russian fleet in Port Arthur, now Lüshun China, torpedoing three battleships.

This kicked started the Russian-Japanese war of 1904-05 which culminated in the annihilation of the Russian Baltic Fleet in the Tsushima Strait in 1905. The superior armed and faster Japanese ships were commanded by the British naval schooled Admiral (Kōshaku = Marquis) Tōgō Heihachirō.*

The Millikan Experiment & the Mass of the Electron

Midst the research into rays, and as a direct result of their ionising properties, came the discovery of the mass of an electron. This was ingeniously achieved in 1909 by American physicist Robert Millikan (1868-1953).

Upper and lower chambers containing rarefied air were interconnected via

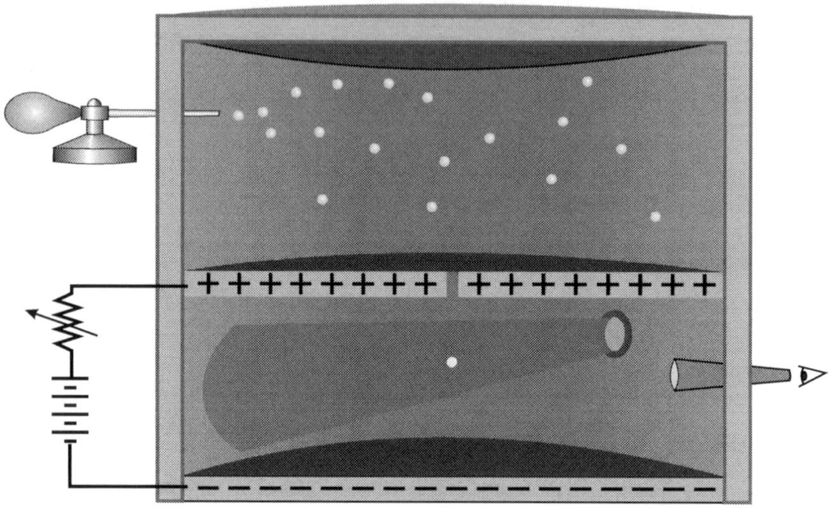

a small central hole. The floor and roof of the lower chamber were metal and insulated from the sides. It had a light in the wall. Oil drops were sprayed into the top chamber and some fell through the hole in its floor.

A burst of x-rays into the lower chamber ionised the air with some ions positive and some negative. The electric field between the metal plates was immediately activated. The positive ions fell rapidly to the floor while some electrons on some of the negative ions migrated to the oil drops whose fall could be arrested by adjusting the electric field. The density of the oil was known and the diameter of the shiny illuminated oil drop could be observed through a zoom lens.

The oil drops then held one or more electrons, but each required an electric field charged in multiples of 1.6×10^{-19} coulombs to arrest its fall, therefore this had to be the charge of an electron.

Combining this result with Thompson's $$\frac{q}{m} = 1.7 \times 10^{11} \text{ coulombs / kilogram}$$

yielded the mass of an electron as: $$m = \frac{1.6 \times 10^{-19}}{1.7 \times 10^{11}} = 9.19 \times 10^{-31} \text{ kilograms}$$

Discovering the Proton

In 1919 Rutherford fired helium nuclei, alpha rays (atomic number 2) emitted from radium, down a tube full of nitrogen (atomic number 7). The surprising result was the first observed act of artificial transmutation of one element into another, for afterwards the tube was found to contain some oxygen (atomic number 8) and a new positively charged particle, with the atomic number 1.

$$N_7 + He_2^{++} = O_8 + H_1$$

Rutherford named his particle the proton from the Greek prōton, a form of prōtos meaning first or elementary.

The approximate mass of the proton was 2000 times that of the electron, but the combined mass of the protons in the nuclei of elements could only explain about half the total mass. Rutherford suggested that there must be a third elementary particle to explain the difference.

He was awarded the Nobel Prize in chemistry in 1908, was knighted in 1914 and made '1st Baron Rutherford of Nelson and Cambridge' in 1931.

Discovering the Neutron

In 1930 the German physicists Walther Bothe and Herbert Becker bombarded beryllium with alpha particles and detected a radiation popping out the back which could pass through ten centimetres of lead.

Both would share the 1954 Nobel Prize in physics with Max Born.

In 1932 French physicists Irène and Frédéric Joliot-Curie showed that this new radiation could knock protons out of hydrogen atoms.

Also in 1932, Sir James Chadwick (1891-1974), first a student then a peer of Rutherford, showed that what knocked out the hydrogen atoms had a mass similar to that of the proton but it carried no charge. This confirmed the prediction of a neutral particle and it was sensibly named the neutron.

The neutron completed the elementary picture of the neutral atom. It has a nucleus of concentrated protons and neutrons, (with the exception of hydrogen which has only one proton), surrounded by a cloud of orbiting electrons which are exactly as numerous as the protons in the nucleus

The neutron is comprised of an electron and a proton into which, if free of the nucleus, it fragments in an average time of 15 minutes. It is slightly magnetic so it is affected by the electromagnetic force.

Electrons of the Atom

Between the discovery of the nucleus and that of the neutron there were remarkable discoveries concerning the cloud of orbiting electrons.

Answers were needed and found for:

- How the electron orbits were arranged?
- What was the relationship between orbits and Mendeleyev's periodic table?
- Why don't electrons run out of puff and fall into the nucleus?
- What were the spectrum lines or Fraunhofer lines which were unique to each element?

One of the earliest contributors to unscrambling some of the above was Niels Bohr and he used a key discovered by Max Planck who used a key found by Boltzmann.

Ludwig Boltzmann

In 1870 the brilliant Austrian Physicist Ludwig Boltzmann (1844-1906), had shown among other works that in huge systems, the second law of thermodynamics (there's no such thing as a free lunch), could be explained by statistically analysing its microcosm. This idea fragmented or particle-ized nature.

The Photoelectric Effect and Einstein

Just when it seemed that the light war was over and Maxwell's vision of light being electromagnetic radiation would prevail the war erupted yet again.

In the very year that M&M claimed they executed the ether, scientists playing with light beams, (as they do), discovered the photoelectric effect.

Light shining onto a negative charged metal disc caused a current to flow across a gap into a surrounding positively charged ring.
Aspects of this created turmoil in the wave-works.

After 1897 it was found the light rays excited the newly discovered electrons with enough energy to jump across the divide to the positive ring. The electrons carried the current.

Wave theory could not explain why:

- Increased intensity of light increases the amount of electron emission but does not increase their energy.
- Each surface has a characteristic cut off frequency below which the photo-electric effect disappears irrespective of the intensity of the beam.

- Theoretically electromagnetic light, feeble or strong, would need to be soaked up by an area of the surface equivalent to more than a few atomic areas, thereby collecting enough energy to excite the electron to escape. This should take some finite amount of time; however there is no detectable time lag. This is exemplified when the photoelectric material is a gas. Here the energy for the electron must be sucked out of the beam by its single mother atom but this is insufficient.

Max Planck and the Birth of Quanta

In 1900 understanding cavity radiation was the outstanding scientific puzzle of the time. The white hot glow in the depths of a coal fire is one example of this radiation.

Various blocks of different metals with internal cavities of any shape and size had a small hole drilled into each of them.

When they were heated to the same temperature, say 2000º C, the various surfaces radiated heat at different rates, however, the cavities of each block radiated heat through the drill hole at precisely the same rate.

The German physicist Wilhelm Wien (1864-1928) got very close to the solution with the formula:

$$R_\lambda = \frac{c_1}{\lambda^5} \frac{1}{e^{C_2/\lambda T}}$$

Where c_1 and c_2 are constants determined by fitting theoretical formula to experimental data, and 'e' is the emissivity which depends on the material and the temperature.

Wien was awarded the Nobel Prize in 1911.

German physicist Max Planck (1858-1947) tweaked Wien's formula and nailed it with:

$$R_\lambda = \frac{c_1}{\lambda^5} \frac{1}{e^{C_2/\lambda T} - 1}$$

The cavity heat was a form of electromagnetic radiation cause by excited electrons jumping up and down.

To explain the common radiation of the cavities Planck postulated that it had to be quantized. To this end, and to fit experimental results, he daringly lashed out with the famous assumptions:

1. An oscillator can only have energy given by:

   $E = nhv$ (where E = energy, n = an integral number, h = Planck's constant and v = frequency.)

2. The oscillators do not radiate energy continuously, but only in jumps or quanta. Therefore if the quantum number 'n' changes by 1 then:

   $$\Delta E = \Delta nhv = hv$$

3. An oscillator in one of its quantum states or so called 'stationary states' neither emits nor absorbs energy.

In 1916 Robert Millikan made the first good measurement of Planck's constant. Today it sits at:

$$h = 6.625 \times 10^{-34} \text{ joule-sec.}$$

This very small number, though significant at the atomic level, pales to insignificance in the macro world. For example, a 1 kilogram weight bobbing up and down when suspended on a spring, can only have single jumps of quanta energy of the order of 10^{-30} joule, which is so miniscule it is undetectable.

Also in 1900, Planck resolved one of the difficulties associated with radiant energy posed by a 'black body'. This theoretical body absorbs all the radiation that falls onto it, then after reaching some equilibrium temperature, proceeds to re-emit energy as fast as it absorbs it. He concluded that these radiations were emitted in discrete amounts called quanta or theoretical photons each with an energy equal to Planck's constant 'h' multiplied by the frequency of radiation.

$$E_2 - E_2 = hvz$$

Planck was awarded the 1918 Nobel Prize in Physics.

The Zeeman Effect

The Dutch physicist Hendrik Antoon Lorentz, (1853-1928) had hypothesised that the oscillations of electrons as they jaunt from orbit to orbit inside an atom produce light which a magnetic field would affect.

In 1896 Dutch Physicist Pieter Zeeman (1865-1943) discovered that spectral lines of a light given off when elements and compounds burn could be split into several components by a very strong magnetic field. Some levels would gain energy while others either lost it or remained the same.

This Zeeman Effect confirmed the electromagnetic nature of light and its explanation is the foundation stone of quantum theory.

Lorentz and Zeeman shared the 1902 Nobel Prize.

In 1925 the Dutch physicists Samuel Goudsmit and George Uhlenbeck proposed that the Zeeman Effect was caused by electrons spinning in opposite directions.

The Photoelectric Effect and Albert Einstein

In 1887 Hertz had shown that the presence of ultraviolet light could expand the minimum gap size a spark could jump across.

In 1900 the German Physicist Philipp Lenard showed that strong light could produce an electric discharge from the surface of metals and these were the particles of the cathode ray called electrons.

In 1905 Albert Einstein, (1879-1955) re-ignited the smouldering; "Is light a wave or a particle?" argument. In the second paper of his doctorate thesis at the University of Zürich, he theorised that light could at times be thought of, not as a continuous stream but as a stream of particles or quanta whose individual energy was given by:

$$E = hv$$

He argued that when these theoretical photons smashed into electrons in the metal they could knock them out at a fantastic speed. The metal would naturally try to prevent this by holding them back, so restraining work would be done on the electron on the way out. The kinetic energy of the totally liberated electron would be given by:

$$E_k = hv - \omega \qquad \text{where } \omega \text{ is the work function}$$

The third paper of his thesis on relativity was put into the reject basket by most scientists. This rejection cast a belittling four year shadow over the second paper on light; however when the relativity theory became accepted, so also was his interpretation of the photoelectric effect. Thereafter he was feted.

During this time in the boondocks his principle supporter was not surprisingly Max Planck. Planck may have thought; "Das formula ist ja selbstverständlich Albert", but kept his council. At that time Einstein was one of Planck's few supporters on this issue.

Niels Bohr

Danish Scientist Niels Henrik David Bohr (1885-1972) finished his doctorate at Copenhagen University in 1911, popped briefly into the Cavendish, then moved to the Rutherford in Manchester. Between 1913 and 1915 he cooked Rutherford's atomic model with some Max Planck sauce and dished up what is known as 'Bohr's atomic model'.

He theorised that the hydrogen atom exists in certain stationary states in which it does not radiate. Heating the gas could cause electrons in lower orbits to absorb specific quanta of energy and jump to higher states However when it drops from a higher state to a lower one it radiated energy in the form of a theoretical photon which is given by:

$$hv = E_k - E_j$$

The waves have specific frequencies for each jump between different orbits and their presence and energy is revealed by the Fraunhofer lines in the gas's spectrum.

He went further and formulated that the angular momentum 'L' of each orbiting electron was given by the quantum equation:

$$L = n\frac{h}{2\pi} \qquad n = 1,2,3 \ldots\ldots$$

Since $L = mvr$ *(where m = mass, v = velocity, r = radius)* then the radius, energy and frequency of each electron is known.

The radius of orbit r is proportional to n^2, while the energy $E \propto \frac{1}{n^2}$

Bohr was awarded the 1922 Nobel Prize in Physics.

The Stark Effect

Since Zeeman scientists had been trying to similarly split the spectral lines in luminous gases, however their high electrical conductivity defeated them.

In 1913 Johannes Stark (1874-1957) German physicist and lecturer at the University of Göttingen observed the hydrogen spectrum emitted just behind the perforated cathode in a positive-ray tube. He arranged a second electrode only millimetres from the cathode and applied a huge electric field with an intensity of over 100,000 volts/cm causing the spectral lines to split into what is called Balmer lines. Some of these were linearly polarised and some were perpendicularly polarised.

In 1916 quantum theory was able to produce a reasonable reason for the Stark Effect. This gave quantum theory great kudos.

Stark received the Nobel Prize in physics in 1919.

As a rabid anti-Semitic Nazi he became president of the Reich Physical Technical Institute from 1933 to 1939. In 1947 he was sentenced to four years jail for his Nazi-ness

Lorentz–FitzGerald Contraction and the Lorentz Transformations

The Irish physicist Sir Joseph Larmor (1847-1942), a diehard ether theorist, was the first to explain the Zeeman effect; the splitting of spectrum lines by a magnetic field.

He was the first person to publish transformations which deal with the relationship between events in one co-ordinate system and another which is moving in relation to the first. His dimensions were 3-dimensional.

The M&M experiment in the 1880s showed that the speed of light was the same for every direction the Earth was moving. It was a constant.

Along with the experiments came probably the first appearance of the fraction:

$$\frac{1}{\sqrt{1-\frac{v^2}{c^2}}}$$

This formula is central to relativity theory.

In 1892 the Irish physicist George Francis FitzGerald (1851-1901) postulated that a moving body is shorter along the line of motion than when it is at rest.

The shortening factor was called *the Beta factor:* $\beta = \sqrt{1 - v^2/\,c^2}$

In 1895 Hendrik Antoon Lorentz independently came to the same conclusion and extensively elaborated it with what is known as the Lorentz Transformations. His dimensions were 4-dimensional; *x,y,z,* and *t* for time.

In 1905 the great French mathematician Jules Henri Poincaré, (1854-1912), tweaked the algebra and then attributed the transformations solely to Lorentz.

Sadly, Larmor didn't get a gong.

Both Lorentz and Fitzgerald had clung to a Newtonian approach for the contraction. They held that since the measuring devices like M&M's table shrunk with motion, then the speed of light might also have slowed. This however was undetectable because the ruler as well was shorter.

It is possibly or even probable that this conservatism blocked Lorenz from discovering the famous $E = mc^2$ formula which literally falls out of his equations.

Minkowski

In 1908 the German mathematician Hermann Minkowski developed Space time diagrams for the motion of particles in accordance with the Lorentz transformations.

Einstein & Relativity

In his 'third paper' in 1905 Einstein stated that the speed of light is invariant, therefore a universal constant in any reference frame, be it moving or stationary.

At the time M&M may have thought *"We know that Albert. We proved it in 1887."*

The beta contraction was now a property of space and time and was independent of physical phenomena like compression or temperature.

All particles or objects in the universe must have what Einstein called a 'world line' which requires them to have both space and time co-ordinates simultaneously. When two time lines intersect there is an event

- Absolute rest then becomes an impossibility since it requires both place and time to stop.

The beta factor was now applied to mass and time.

- The faster something moves the heavier it gets. At light speed its mass would be infinite which is impossible.
- The faster something goes the slower its clock goes because its seconds gets longer. At light speed its time would be infinitely long which means it stands still which is impossible.

$$m = \frac{m_0}{\sqrt{1 - v^2/c^2}} \qquad t = \frac{t_0}{\sqrt{1 - v^2/c^2}}$$

$$\text{As } v \longrightarrow c\text{, then } \sqrt{1 - v^2/c^2} \longrightarrow \sqrt{1-1} = 0 \text{ and } 1/0 = \infty$$

This time lengthening is the basis of fictional stories about Space-folk travelling close to the speed of light for a few decades, then returning to Earth to find their grandchildren now years older than themselves.

Einstein went on to link mass to gravity.

He saw that if one were in a box and were being pulled down to the floor then one couldn't distinguish whether one was on a planet or in a Spaceship which was accelerating in free Space.

If mass were linked to gravity then so was time and therefore the Space-time continuum was linked to gravity.

Consequently he argued that the continuum was curved in the neighbourhood of masses, and particularly curved near large masses like suns. He predicted that light passing these masses would follow a curved path.

In 1919 Einstein achieved international fame when this prediction was verified by the British astronomer Sir Arthur Eddington. He had photographed the position of stars both before and after and during a total eclipse of the Sun witnessed on the 29th of May at Principe Island in the Gulf of Guinea. Einstein's prediction that light from a star which passed close by the Sun would be bent through 1.745 seconds of arc was spot on.

The general theory of relativity predicts 'frame dragging'. Here when massive bodies are rotating they drag the Space time around with them. This may have been detected around a spinning neutron star in 1997. There is possibly some frame dragging being detected in our own satellites. They appear to speed up as a result of being dragged on a bit by the Earth's rotation.

The general theory also predicted the existence of black holes which appear to be substantiated.

One vital aspect of the theory which remains elusive is the prediction that gravity operates via gravity waves. So far no Cuban cigar here.

Einstein was awarded the 1921 Nobel Prize in physics for his photoelectric work.

Quantum Theory and the Atom

In the 1920s Bohr's ideas on quantum steamed ahead.

In 1923 the young French physicist and aristocrat Prince Louis de Broglie theorised via his probability formulae, that electrons, which were thought of as particles, should also be radiations with the properties of waves.

German physicist Werner Heisenberg (1901-1976) using his matrix mechanics reached the same conclusion. In 1927 he published his 'uncertainty principle'; "the more accurately one measures one quality of a particle the more uncertain other qualities become."

The wave properties were proven experimentally in 1927 when American physicists Clinton Davisson and Lester Germer fired an electron beam through a crystal and observed diffraction.

This duality of the electron led to the creation of the electron microscope. The higher the energy is, the shorter the wavelength, therefore the smaller the object one can peer into.

In 1925 Austrian physicist Wolfgang Pauli (1900-1958) published his Exclusion Principle. 'No two electrons have the same properties or locations.'

In 1926 both the British physicist Paul Dirac (1902-1984) and Italian physicist Enrico Fermi (1901-1954) developed equations for the electron which confirmed Pauli's exclusion principle.

In 1927 Dirac predicted that electrons had a property called spin.

Also around 1927, Austrian physicist Erwin Schrödinger (1887-1961) extended Luis de Broglie's work (who had extended Einstein's theories) on wave mechanics.

Luis de Broglie (1892-1987) was awarded the 1929 Nobel Prize in physics.

Dirac shared the 1933 the Nobel Prize with Schrödinger.

Heisenberg got the 1932 Nobel Prize for physics.

Pauli was awarded the Nobel Prize in physics in 1945.

Some of the other greats (too many to name) who contributed to the progress of quantum theory include Arthur Compton, Samuel Abraham Goudsmit and George Eugene Uhlenbeck, Roberet Mulliken, Friedrich Hund, Gilbert N. Lewis, etc.

They discovered that:

- Electrons orbited a nucleus in shells.
- The dynamics of each electron (symbolised by quantum numbers) is unique.
- All matter is both a particle, and if it moves is also a wave whose wavelength 'λ' is given by the de Broglie formula:

$$\lambda = h/p = h/mv$$

where h = Planck's constant, p = momentum, m = mass, v = velocity

- The length one orbit of an electron's path is a multiple of half its wavelength (The first quantum number). Because of this electrons don't have permission to fall into the nucleus. This half wavelength property also determines the shape of a shell.
- Shells are comprised of subshells whose electrons have the same angular momentum. They have either of two possible orbitals, each having opposite spins.
- It is impossible to predict the precise location of an electron. One can only predict the level of probability that it will be somewhere.
- Electrons progressively filled each shell and its subshells and their orbitals.
- Electrons in higher orbits have higher energy.
- The number of electrons in the outer shell (the valence shell) of an atom is what determines its chemical properties. There are between 1 and 8 electrons here.
- Sometimes electrons in an inner orbit can have higher energy than the valence electrons and these therefore dominate the atoms chemical properties.

- More than 8 electrons in the outer orbit ist verboten!
- Atoms with 1 to 3 electrons in the outer shell are conductors and those with 5 to 7 are nonconductors.
- Atoms with 4 electrons in their outer shell are AC/DC; two of them, silicon and germanium are well known semi-conductors.
- Atoms with 8 electrons in their outer shell are the inert elements or the noble gases. Helium is also an inert noble gas but it has only one shell filled by 2 electrons.

Quantum Classification

Each electron in an atom has its own unique barcode or set of quantum numbers:

n The principle energy state or shell.
l The angular momentum.
m The magnetic orientation in space in the plane of the orbiting electron.
m_s The electron spin.

Principle quantum number n	Secondary quantum number l	Magnetic 3rd quantum number m	Spin 4th quantum number m_s		Electrons per shell
1	s	0	½	-½	2
2	s	0	½	-½	8
	p	-1	½	-½	
		0	½	-½	
		1	½	-½	
3	s	0	½	-½	18
	p	-1	½	-½	
		0	½	-½	
		1	½	-½	
	d	-2	½	-½	
		-1	½	-½	
		0	½	-½	
		1	½	-½	
		2	½	-½	
	p	-1	½	-½	
		0	½	-½	
		1	½	-½	

Mendeleyev

When Mendeleev attended a conference of heavyeeights in london, nobody recognised who the little man was. He introduced himself about by saying "ich bin Mendeleev".

By 1930 Mendeleev's periodic table was comprehensively filled.

Today it is all but complete.

- The tail end heavy elements are totally unstable, existing only for a flicker of time *(Only 4 of the last 8 have been created so far in the laboratory.)*
- Elements in the columns of the table have similar properties
- Also in general the elements of the transition metals have similar properties.

Periodic Table

= TRANSITION METALS

SUB-SHELLS	1 IA	2 IIA	3 IIIB	4 IVB	5 VB	6 VIB	7 VIIB	8 VIIIB	9 VIIIB	10 VIIIB	11 IB	12 IIB	13 IIIA	14 IVA	15 VA	16 VIA	17 VIIA	18 VIIIA
1s	1 H																	2 He
2s,2p	3 Li	4 Be											5 B	6 C	7 N	8 O	9 F	10 Ne
3s,3p	11 Na	12 Mg											13 Al	14 Si	15 P	16 S	17 Cl	18 Ar
4s 3d,4p	19 K	20 Ca	21 Sc	22 Ti	23 V	24 Cr	25 Mn	26 Fe	27 Co	28 Ni	29 Cu	30 Zn	31 Ga	32 Ge	33 As	34 Se	35 Br	36 Kr
5s 4d,5p	37 Rb	38 Sr	39 Y	40 Zr	41 Nb	42 Mo	43 Tc	44 Ru	45 Rh	46 Pd	47 Ag	48 Cd	49 In	50 Sn	51 Sb	52 Te	53 I	54 Xe
6s,6f 5d,6p	55 Cs	56 Ba	71 Lu	72 Hf	73 Ta	74 W	75 Re	76 Os	77 Ir	78 Pt	79 Au	80 Hg	81 Ti	82 Pb	83 Bl	84 Po	85 At	86 Rn
7s 6d,5f	87 Fr	88 Ra	103 Lr	104 Rf	105 Db	106 Sg	107 Bh	108 Hs	109 Mt	110 Uun	111 Uuu	112 Uub		114 Uuq		116 Uuh		

4f,5d	57 La	58 Ce	59 Pr	60 Nd	61 Pm	62 Sm	63 Eu	64 Gd	65 Tb	66 Dy	67 Ho	68 Er	69 Tm	70 Yb
5f,6d	89 Ac	90 Th	91 Pa	92 U	93 Np	94 Pu	95 Am	96 Cm	97 Bk	98 Cf	99 Mt	100 Fm	101 Md	102 No

Matter

- Matter is an assembly of 114 elemental atoms locked or linked together by electric forces.
- The lattice and/or crystal structure and/or the molecular structure of the material, and/or certain rules of quantum mechanics keep the sources of electric charge separated.
- Protons in the nuclei of the atoms are responsible for its positive charge while electrons are responsible for the negative charge.
- Electrons surround the nucleus in shells.
- 66 elements have two electrons the outer shell, about 18 have 1 electron and the remaining 32 and perhaps a few more highly unstable ones have between 3 and 8. These outer shell electrons are the 'valence electrons' which determine chemical properties. They are loosely bound and may be shared by other atoms or may wander about from atom to atom.
- Within the above constraints electrons progressively fill their lower shells with specific maximum numbers of electrons. The sequence is 2,8,18,32,32,18,(8).
- Quantum rules prevent an atom's electrons from plummeting into the nucleus.
- *Externally fired high energy electrons can penetrate the nucleus.*

Conductors

Metals are good conductors of heat and electric current. Copper, silver, platinum and gold are some of the best. They have a valence of one, meaning there is one electron in their outer shell which is very loosely bound. These flit from atom to atom, either briefly occupying outer orbits if vacated, or bouncing off densely negative inner shells.

In copper for example electrons spend 85% of their time away from home

Electron shell configuration of copper is 2-8-18-1.
Silver is 2-8-18-18-1. Gold is 2-8-18-32-18-1.
Platinum is 2-8-18-32-17-1. In this latter variation the atom finds greater stability in popping an electron into the sixth shell prior to filling the fifth.

When an atom has a valence electron absent it is said to have a hole.

The Noble Gases

Helium, Neon, Argon, Krypton, Xenon and Radon

Shells:		K	L	M	N	O	P	Q
He	2	2						
Ne	10	2	8					
Ar	18	2	8	8				
Kr	36	2	8	18	8			
Xe	54	2	8	18	18	8		
Rn	86	2	8	18	32	18	8	
??	118	2	8	18	32	32	18	8

The outer shells of the noble gases contain the maximum of eight electrons and therefore they are very stable though not totally inert. The heavier noble gases have their outer shell electrons somewhat shielded from the protons in the nucleus by the myriad of inner electrons. This allows them to bond with fluorine to form compounds such as xenon, krypton or radon fluoride.

Diagram and Chart for the First Three Orbitals

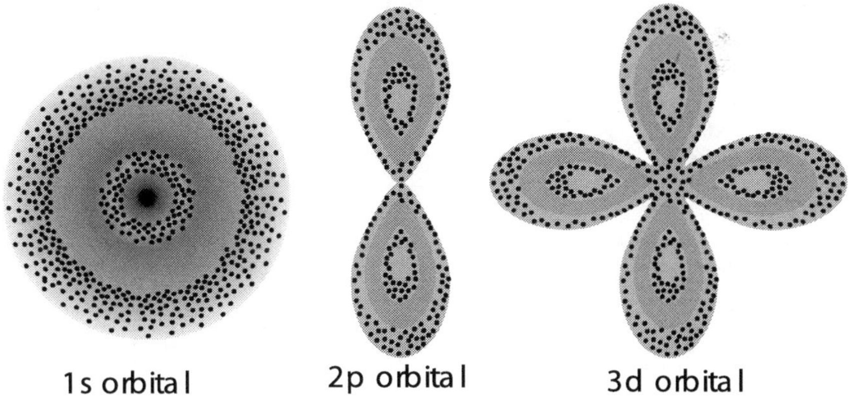

Shell Shapes

The inner s-shell with only two electrons (only one for hydrogen or deuterium) is spherical. Devices like tunnelling microscopes supported predictions that larger atoms have rather strange looking orbitals.

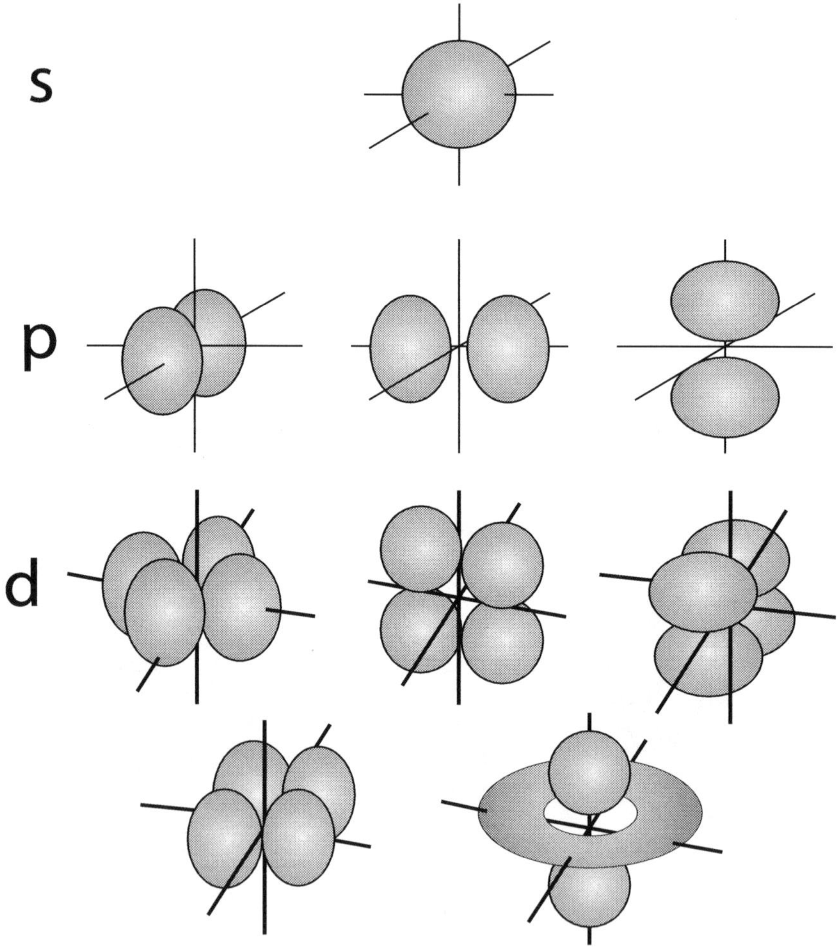

Atom Sizes

As each shell fills with electrons there are corresponding increases in the number of protons in the nucleus making it more positively charged and therefore a greater attractor of electrons. This draws the electrons into closer orbits and therefore the size of the atom shrinks. However when a shell is filled, subsequent electrons must enter orbit in a new higher energy shell, so there is a jump in the size of the atom.

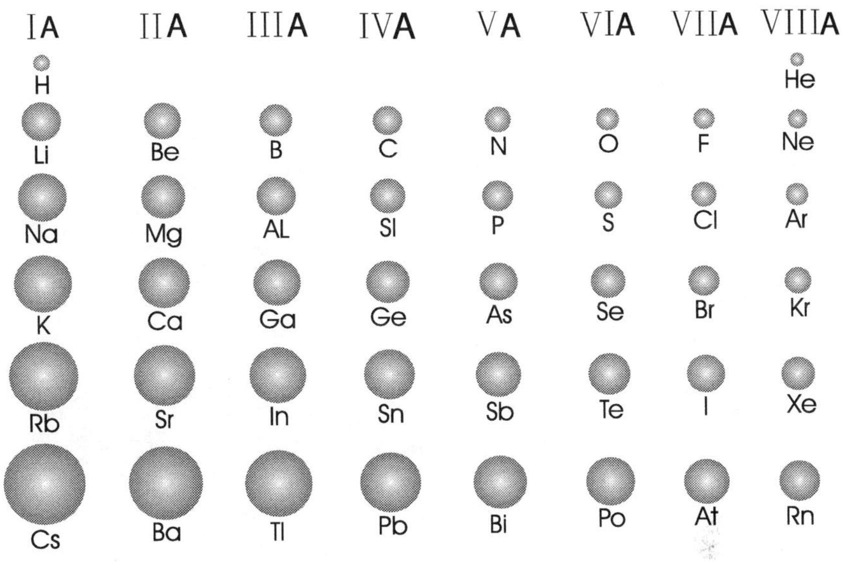

The diameter of the nucleus varies between 1 & 7 x 10^{-15} meters.
The diameter of the atom varies between 1 & 3 x 10^{-10} meters.

Electrons occupy orbitals. Similar orbitals occupy shells whose electrons have similar energy *(both kinetic and potential).*

As a general rule electrons progressively fill their shells (energy level by energy level). However with medium to large atoms, when the outer shell is almost full, one or two electrons can find it more energy efficient to occupy a further orbit before totally filling the one inside it.

Bonding

In 1916 the American physicist Gilbert N. Lewis laid much of the foundations for atomic chemical bonding with his octet rule.

The maximum number of electrons permitted in the outer valence shell is either 2 or 8. After this extra electrons must jump into a larger and higher energy orbit. There is a fair amount of stability in having the outer orbits filled, i.e. having 2 or 8 electrons. The noble gases do this and are so stable they are all but inert.

Other elements must juggle combinations of their valence electrons to achieve the 2 or 8 stability, and they do this basically in three ways.

Covalent Bonds

These are strong bonds where elements share their valence electrons so that their combination adds up to 2 or 8.

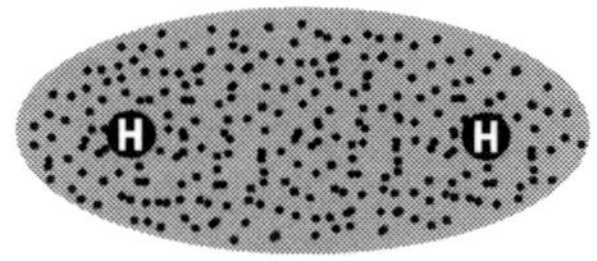

Hydrogen H_1 has only 1 proton and 1 orbiting electron. By combining with another hydrogen atom and sharing electrons they form the stable H_2 molecule found in nature.

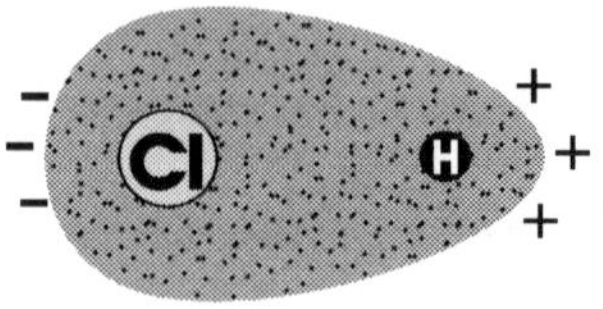

H_1 will combine with chlorine, $Cl_{valence7}$ to create the stable hydrochloric acid HCl sharing 8 electrons in the outer shell. Chlorine has 17 electrons arranged 2-8-7.

Since the electrons spend more time near the chlorine side of the molecule it becomes slightly negatively charged while that of the hydrogen side becomes proportionally positively charged. The molecule becomes an electric dipole.

Ionic Bonds

Sodium 'Na' has 11 electrons arranged 2-8-1 therefore a valence of one.

Sodium chloride NaCl (table salt) is created differently from HCl. Here the sodium atom totally gives up its outer electron and donates it to the chlorine atom. This electron is not shared.

The sodium atom now has more protons than electrons and is positively charged (a cation), while the chlorine now has one electron too many and is therefore negatively charged (an anion).

The two atoms are bonded into a dipole which juxtaposes itself upside down with a neighbouring NaCl. This repeats over and over to form a lattice or crystal of salt.

Cations which have lost an electron from their valence shell are smaller than their parent atoms, sometimes by as much as 50%.

Conversely anions which have gained an electron (usually in the valence shell) inflate the atom.

Metallic Bonds

Because the atoms of metals are closely packed, their outer valence shells overlap. This enables those electrons to wander from one atom to another. This means that the metal is comprised of positively charged cations surrounded by a current of randomly flowing electrons. It is the attraction between cations and roving electrons which compacts the atoms into metallic crystals.

Van der Waals Bonds

The shape of a neutral molecule might create a weak dipole. This can induce a reciprocal dipole in a neighbouring molecule which sets up a process which can domino through a gas or through many organic liquids and solids. Since one dipole is the obverse of its neighbour, there is mutual attraction.

NUCLEAR REACTIONS

AZ Masses

Nuclei are clusters of protons and neutrons closely bound together by what is called the strong force. The simplest nucleus is that of hydrogen which in its simplest form has only one proton and no neutrons. All the other nuclei contain some neutrons. The larger the nucleus is, the greater the ratio of neutrons to protons. For example uranium-238 has 146 neutrons and 92 protons.

Convention uses the letter 'A' to represent the number of neutrons plus the number of protons in nuclei *(the mass number),* and the letter 'Z' (the atomic number) represents the number of protons. Thus element XX would be represented as

$$^{A}_{Z}XX$$

Atomic Mass Unit, amu

Masses of atoms nuclei and their components are measured in atomic mass units or amu's. One amu = $1/12^{th}$ the mass of the carbon-12 atom which has 6 protons and 6 neutrons.

The stand alone mass of a proton is 1.00728 amu.

The stand alone mass of a neutron is 1.00866 amu.

The combined stand alone mass of 6 neutrons + 6 protons is 12.09564 amu, whereas the mass of the carbon atom is 12 amu.

Like building a car out of spare parts, building a carbon atom from spare particles is mass expensive.

In fact the mass of every nucleus is less than the sum of the stand alone masses of the protons and neutrons which make it up. *(The masses of the electrons are negligible.)*

Mass and Energy

In his 1905 paper Einstein showed that mass and energy were inseparable, being governed by the formula $E = mc^2$.

The mass of the nucleus is a measure of its energy.

When particles combine to form a nucleus the energy released is called the binding energy ($E = mc^2$).

Scientists measure the mass of atomic and sub-atomic particles in 'million electron volts' (MeV) rather than in kilograms.

$$1 \text{ MeV} = 1.78 \times 10^{-30}\text{kg} \times c^2$$

1 electron volt is the energy a particle acquires when it is accelerated across a potential difference of 1 volt = 1.602×10^{-19} joule.

An electron has a rest energy 0.511 MeV

A proton has a rest energy 938.3 MeV

A neutron has a rest energy 939.6 MeV

Isotopes

In most elements the number of neutrons can vary slightly. Some of these variations exist in their natural state and some can be created artificially. These varieties of each other are called isotopes. Most elements in nature are comprised of two or more isotopes. There are about 280 naturally occurring stable non radioactive nuclei.

Natural chlorine for example is comprised of isotope Chlorine -35, i.e. $^{35}_{17}\text{Cl}$ @ 76%, and isotope chlorine -37 i.e. $^{37}_{17}\text{Cl}$ @ 24%. Chlorine has the atomic weight of 35.457.

The number of protons in an element determines the same number of electrons in orbit which determine the chemical properties of the element. Thus isotopes of the same element have pretty much the same chemistry.

Since one isotope is heavier or lighter than its sister, each can be identified using a mass spectrometer developed by British scientist Francis William Aston in 1919 .

The isotope is vaporised then ionised (stripped of an electron) , then deflected one way by an electric for ce, then the other by a magnetic force and finally photographically analysed, which determines its mass and velocity. It is a sophisticated version of the Thompson experiment.

In 1912 J.J. Thomson deflected neon ions ($Z = 10$) with magnetic and electric for ces and found some were deflected further than others therefore these had to have less mass than the others. He found the heavier ones to have the mass number $A = 22$ i.e. $^{22}_{10}\text{Ne}$, and the lighter ones $A = 20$ i.e. $^{20}_{10}\text{Ne}$.

There is also a neon -21.

Hydrogen Isotopes

Everyday hydrogen which has only one proton and no neutrons is also known as the isotope protium, $P = {}^{1}_{1}H$

In 1931 the American physicist Harold Clayton Urey (1893-1981) published his discovery of the isotope deuterium, $D = {}^{2}_{1}H$ *(1 proton +1 neutron)*

and its offspring, the heavy water molecule D_2O.

The nucleus called deuteron, being twice as heavy as a proton makes a powerful missile for inducing nuclear reactions.

Urey was awarded 1934 Nobel Prize in Chemistry.

In 1934 Rutherford, Mark Oliphant, and Paul Harteck, bombarded deuterium with deuterons and created the isotope tritium (1 proton + 2 neutrons)

$${}^{2}_{1}H + {}^{2}_{1}H \rightarrow {}^{3}_{1}H + {}^{1}_{1}H$$

Hydrogen Fusion

Nature is always trying to organise its elements to operate in the simplest and therefore the most energy efficient way. Consequently there is a tendency for groups of particles to modify themselves into groups of lesser energy.

If conditions are favourable small atoms will fuse to form larger ones which have less mass. The missing mass appears as energy.

All three isotopes of hydrogen can be involved in the fusion process which transmutes them into helium. The above tritium equation can under certain conditions become:

two deuterium atoms fusing to form

a helium-3 isotope + 1 neutron + energy.

$${}^{2}_{1}H + {}^{2}_{1}H \rightarrow {}^{3}_{2}He + {}^{1}_{0}n + 3.2 \text{ MeV} \quad (\text{or } 5.0 \times 10^{-13} \text{ joule})$$

Thereafter two helium-3 atoms can then fuse into

one helium-4 atom + two hydrogen-1 atoms (protiums) + energy.

$${}^{3}_{2}He + {}^{3}_{2}He \rightarrow {}^{4}_{2}He + 2({}^{1}_{1}H) + 12.82 \text{ MeV} \quad (\text{or } 2 \times 10^{-12} \text{ joule})$$

Both the above reactions release energy because of the larger helium atom has less mass than its spare part assembly.

Decay

At the heavy end of the elements list another phenomenon occurs.

Here there is a tendency for larger particles to modify themselves into groups of smaller particles which have less energy than the parent. Here too the missing mass is converted into energy by $E = mc^2$. This is what drives radio-active decay.

Atoms decay either by ejecting from their nuclei:

- an alpha particle *(α-particle or* ${}^{4}_{2}He$ *α-ray)* which is the helium nucleus,

- or a beta particle *(β-particle or β-ray)* which is an electron.

The α-particle is spat out at about 1/20th the speed of light while the much faster β-particle is fired out at close to the speed of light.

Essentially a neutron is comprised of a proton merged with an electron, thus by ejecting an electron a neutron is transformed into a proton.

A large atom ejecting an α-particle will become an element with two less protons in its nucleus and be found two steps lower in the periodic table.

However an atom ejecting a β-particle from its nucleus will have one extra proton in its nucleus *(even though it is a smidgen lighter than its parent).*

Heavy elements like uranium with 92 protons can and do progressively decay, going down 2 protons *(by shedding an alpha)* up 1 proton *(by shedding a beta)*, down 2 up 1......until they arrive at an isotope of lead which will have 82 protons. Through the sequence of decay, isotopes of all the elements having between 92 and 82 protons, make an appearance.

All elements with atomic numbers higher than 82 *(lead)* are radioactive and consequently decay.

Some lighter elements are radioactive. Potassium Z=19 has three isotopes $^{39}_{19}K$ (abundant), $^{40}_{19}K$ and $^{41}_{19}K$. The $^{40}_{19}K$ is radioactive with a half-life of 1.28 billion years.

A half-life is the amount of time it takes for a radioactive atom to reach half its radioactive intensity.

Fission

Fission occurs when larger atoms split into a few smaller atoms and a few elementary fragmented particles. The combined mass of the flotsam and jetsam is less than that of the parent atom and energy is liberated.

In the mid 1930s fission reactions were conducted in many countries though what was occurring was not fully understood. It is fortunate for the world that this also applied to Nazi Germany who could, but for a thought, have had a five year head start on the nuclear bomb.

While there are many fissionable elements only uranium and plutonium are capable of having chain reactions.

Uranium & Fission

Uranium is found naturally in oxides or complex salts like pitchblende. It has three natural isotopes:

$^{238}_{92}U$ @ 99%, $^{235}_{92}U$ @ 0.7% and traces of $^{234}_{92}U$.

(Artificial isotopes $^{233}_{92}U$, $^{237}_{92}U$, $^{239}_{92}U$ can be generated.)

It can be split into (usually two) different elements, resulting in about 40 different combinations creating about 200 radioactive isotopes. The usual ratio is about mass number A = 95 to 140 though some are from A = 70 to 165. Some such as cesium-137 and strontium-90 with half-lives of 53 days can be taken up by the food chain and are lethal. The latter blends in with calcium and gets deposited in bones. Some products have half-lives of decades and therefore persist for centuries.

U-235 is the fissionable isotope which is responsible for the chain reaction. Reactor fuel needs to be at least 3% U-235 so the uranium has to be enriched from its natural 0.7% value which isn't easy.

The rapier which splits the U-235 atom is the neutron.

Nuclei occupy only about 1/100,000th of an atom's diameter, and the neutron being neutral and impervious to the pushes or pulls of protons or electrons can sail straight through them unchallenged. Inevitably however, if a material is dense and thick enough, the neutron will strike a nucleus.

In natural uranium there are some neutrons zipping about which in the main bounce off the nuclei. Occasionally one will get captured by one of the rare (0.7%) U-235 nuclei and cause it to split and simultaneously create an extra 2 or 3 neutrons which zip around or zip off into Space.

The operative word here is 'capture'. A too fast neutron will often bounce off nuclei, whereas a not so fast neutron (called a slow neutron) is more likely to be captured by the nucleus it strikes. Nuclear reactors use materials like heavy water to slow neutrons down thereby making collisions more numerous.

When a neutron is added to the U-235 nucleus, instead of creating U-236, nature dictates that the preferred state is to have two smaller atoms (here cesium and rubidium) and a surfeit of 3 neutrons plus energy.

$$ {}_0^1\mathrm{n} + {}_{92}^{235}\mathrm{U} \rightarrow {}_{55}^{140}\mathrm{Cs} + {}_{37}^{93}\mathrm{Rb} + 3\,{}_0^1\mathrm{n} + 200\ \mathrm{MeV} \qquad (\text{or } 3.2\times10^{-11}\ \text{joule}) $$

Plutonium

Plutonium was discovered in 1940 as a by-product in uranium reactors. The most important of the 15 isotopes of plutonium is plutonium-239 which is chain reaction fissionable. ${}_{94}^{239}\mathrm{Pu}$

To make plutonium the uranium in the reactor core is flooded with slow neutrons. Some of these are captured by the U-238 nuclei becoming U-239. This is unstable and after 23.5 minutes it transforms itself by β-emission into neptunium Np-239, then after 2.35 days by another β-emission into plutonium Pu-239.

$$ {}_{92}^{238}\mathrm{U} + {}_0^1\mathrm{n} \rightarrow {}_{92}^{239}\mathrm{U} \xrightarrow{-\beta} {}_{93}^{239}\mathrm{Np} \xrightarrow{-\beta} {}_{94}^{239}\mathrm{Pu} $$

The plutonium, having 94 electrons in orbit has different chemical properties to those of uranium so it can be extracted chemically from the mix.

This is much easier than the isotopic separation of the U-235 from the uranium mix, because isotopes have the same chemical properties.

Critical Mass

Critical mass can be likened to spontaneous combustion. Vegetable matter releases heat as it decomposes. Damp conditions and heat from sunlight accelerates the decomposition. The middle of a small pile of damp lawn clippings in sunlight will get quite hot but not hot enough to ignite the pile because the heat from decomposition escapes through the surface grass. If the pile is large enough, much of this heat is contained and the temperature in the core rises toward ignition point.

In fission it is the neutrons that cause the chain reactions that heat the pile. If the pile is below a certain size most the neutrons escape before they can split other U-235 atoms.

If the pile is large enough, escape is cut off by the depth of the U-235 atoms which when split create evermore neutrons that can't escape. The size or mass of the pile where this chain reaction starts is called the critical mass.

In modern nuclear reactors which have core volumes in the order of 80 metric tons of uranium oxide, fission is controlled by pushing neutron absorbing rods into the pile.

The Bomb

Atomic bombs need weapons grade fissionable material. This is either highly concentrated U-235 or Pu-239. The more concentrated the material is the less of it is needed to achieve critical mass.

In the early 1940s a team of scientists from England, Canada and the United States gathered in the U.S. under the umbrella name 'The Manhattan Project' in a desperate race to beat Germany to the bomb. By 1945 it was clear that Germany could not build the bomb as most of its pertinent infrastructure had been destroyed, but the desperate race continued in the hope that they could nuke the joint before it capitulated.

Two locations in the U.S. were trying hard to supply the fissionable material in time. Oak Ridge in Tennessee was laboriously producing the U-235, and as well took over production of stage 1 of transmuting U-238 into plutonium.

Stage 2 was conducted at the Hanford Engineer Works in Pasco Washington.

At Oak Ridge uranium was enriched by gaseous diffusion.

Firstly it was converted to the gas uranium hexafluoride UF_6, and then forced through filters. The lighter U-235 fluoride moved faster than the U-238 fluoride arriving earliest at the end of the funnel where it was captured. Later arrivals were recycled.

*More recently the fluorides enter tall gas centrifuges wherein the U-235 rises to the top.
Another modern cheaper method excites the gas with lasers of a particular frequency peculiar to U-235. This ionises it enabling it to be drawn off with electro-magnetic forces.*

During the fuel production race some boffins went from Los Alamos to Oak Ridge to see the progress and were horrified to find partially refined stockpiles of material approaching critical mass. The riot act was read and A-team went home. A short time later they sent their youngest boffin Richard Feynman over to check on progress. He found that the piles were now of moderate size but stored in separate neighbouring rooms but they were now stacked up, to his horror against the common stud wall!

On 16th July 1945, first atom bomb (which had cost $2,000,000,000) was exploded at a site on the Alamogordo air base which is about 300 km south of Los Alamos.

Three weeks later on August the 6th 1945, the United States dropped the U-235 atomic bomb 'Little Boy' on the Japanese city of Hiroshima. The bomb was detonated when a golf ball sized lump of U-235 was fired into a waiting depression in a baseball sized lump of the same material and turned the mass critical. In less than a second the equivalent of 20,000 kilotons of TNT exploded killing 70,000 people instantly. 140,000 had died by the end of the year.

Three days after that, the plutonium bomb 'Fat Man' was exploded over Nagasaki killing 40,000 people. The pilot on that flight spent ten minutes over Kokura (now part of Kitakyūshū) but couldn't find his target point so he flew on to the second target Nagasaki.

Since then in the name of world peace there have been over 2,200 nukes exploded on Earth of which over 1200 were American. The footage of the surface explosions shows all manner of buildings, bunkers, cars, trains etc. being blown to smithereens. Sometimes bunkers survive. Included in the ground zero zones are the ubiquitous vaults or safes. These appear post explosions in various degrees of obliteration and one can't help thinking that the real purpose of the polluting test was to create bomb proof vaults to preserve diamonds and documents.

Fusion and Tokamaks

Since the discovery of hydrogen fusion the world has had high hopes that a device could be built which could maintain nuclear fusion and replicate the energy produced in stars and our Sun. This would be clean limitless energy.

The Sun contains its nuclear fusion with its huge gravity. In the Sun's core the temperature is around 15,000,000° C. Every second 700 million metric tons of hydrogen is transmuted into 695 million metric tons of helium. About 5 million metric tons are converted to energy.

Two methods to replicate the Sun's fusion have been tried, both with some success.

One method involves firing and converging very high energy pulsed laser beams onto a tiny glass sphere containing hydrogen. Some say better success has been achieved by pulse firing ion beams like heavy barium ion beams instead of lasers at the sphere.

The second method involves super-heating the hydrogen plasma while it is contained in a magnetic trap called a tokamak.

Plasma describes a gas which has had its electrons stripped off, leaving only the nucleus. The plasma of hydrogen (protium) is the proton and of deuterium is deuteron.

A tokamak is a big doughnut or torus shaped metal vacuum chamber surrounded by massive coils that induce magnetic 'fields' of around 50,000 gauss. These excite and therefore heat the plasma which fuses in the centre of the doughnut where temperatures are as high as 40 million degrees. This heat is manageable since the fusion occurs in the centre of a vacuum chamber half a meter from its walls. The plasma atmosphere there is so thin that if a pin fell into it with its billions of cold atoms, it would put the fire out.

In 1992 Joint European Torus (JET) produced 1.7 million watts from controlled fusion. In 1993 the tokamak at Princeton University produced 5.6 million watts. Both of these produced less energy than they consumed but where there is smoke there is fire.

A new player in the field is the 'stellarator' which can operate continuously whereas the tokamak operates with pulses.

SUBATOMIC PARTICLES

There are only 4 subatomic stable particles that can exist in free space. These are protons, electrons, electron neutrinos and muon neutrinos. Even the mighty neutron decays into some of the above after a mean life of 887 seconds; the positron can briefly be electromagnetically stored, however the the remaining hundreds or so subatomic particles exist for a mean life span of less than a millionth of a second.

Cosmic Rays

Before the advent of particle accelerators scientists experimented using natures own high speed particles, cosmic rays.

In 1911-12 the Austrian physicist Victor Franz Hess (1883-1964), flew in balloon baskets to perilous heights of between 5,000 and 9,000 meters, accompanied by his trusty electroscope. He measured the levels of ionisation of the atmosphere at various heights. Since the ionisation was there it had to be ionised by rays of some sort which were called the cosmic rays.

Cosmic rays travel at fantastic speeds close to the speed of light. They originate in suns. About 3,000 hit every square meter of the top of Earth's atmosphere every second.

Particle Accelerators

In the 1920s the linear accelerator 'Linac" was conceived. It accelerated charged particles through a series of mesh grids which were oppositely charged to attract the oncoming particle. Most of the particles being a trillionth the size of the mesh holes zip through. As the voltage of the grids gets progressively higher, so does the velocity of the particle.

The latest linac is at Sanford University. It is 3.3 kilometres long and accelerates electrons to around 50 GiV or 50 billion electron volts.

A particle gains energy of 1 electron volt when it is accelerated through a potential difference of 1 volt. $1\text{eV} = 1.602 \times 10^{-19}$ Joule.

In the early 1930s the American physicist Ernest Orlando Lawrence (1901-1958), built the first cyclotron. It consisted of two back to back D shaped chambers enclosed in a vacuum. These were sandwiched between strong (north above south below) electromagnets which forced the moving charged particles into tight circles. Every time the charged particles pass through the gap between the Dees, they are accelerated.

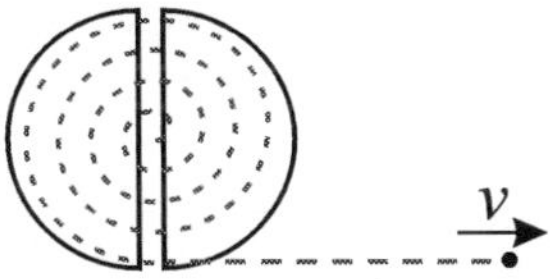

A variation of the cyclotron is the synchrocyclotron which copes with the increase in mass of particles as they get faster. The Russian one has a diameter of 6 meters and has 7,200 metric tons of magnets.

The latest and greatest developments are the synchrotrons. Of these, Fermilab's Tevatron produces collisions of about 2 TeV or 2 trillion electron volts. Here a 1 TeV charged particle going clockwise collides with another going anticlockwise. There are superconducting magnets spread around its 4 kilometer circumference to hold particles in orbit.

The CERN synchrotron has a circumference of 27 kilometers with energies reaching 100 GeV, i.e.1 TeV or 1 trillion electron volts. There are also some smaller yet huge storage rings.

Early New Particles

In the 1920s the proton-neutron nucleus with orbiting spinning electrons seemed to complete the atomic picture but anomalies appeared, both mathematically and experimentally. Energies measured before and after events weren't quite equivalent. This led to speculation that there might be other particles as yet unidentifiable.

One of the earliest speculators was Pauli who predicted the existence of il neutrino, the little neutron. He recognised that some phantom like particle was carrying energy away during β-decay. Even a 125mm lead shield around the radioactive source couldn't stop it.

This non-charged, probably massless entity was not detected until 1956, two years before Pauli's death. A cubic meter volume of fluid, (scintillating fluid), surrounded a radioactive pile emitting 10^{18} neutrinos per second. When they were stopped by some atoms in the fluid they produced recordable flashes of light. About 2 or 3 per hour were recorded so around 10^{23} got through. Only about 1 in 10^{12} neutrinos gets stopped when zipping right through the Earth.

In 1928 Dirac theorised that a positive electron, (positron) was possible and that both it and an electron could be created out of nothing, provided enough energy was directed at one point.

In 1932 the American physicist Carl Anderson (1905-1991) discovered it. It has the mass of the electron but has opposite charge.

Positrons are often produced when gamma rays pass through matter.

In 1934 the great Japanese theoretical physicist Yukawa Hideki (1907-1981) predicted that heavy quanta particles existed with masses of about 300 times the mass an electron. They would carry either a positive or a negative charge equivalent to that of the proton. In 1935 he deduced the existence of the meson particle and predicted the existence of the pi-meson or pion.

This was discovered in 1947 by British physicist Cecil F. Powell using photographic plates he sent up in balloons.

In 1937 American scientists Carl Anderson and Seth Neddermeyer believed they had found the pion but it didn't quite fit the predictions; it was in fact a muon.

Pions are rare at sea level but plentiful in the stratosphere being created when cosmic rays collide with atoms. They can be manufactured abundantly in particle accelerators.

- Pions are 273 times heavier than electrons but live for about 10^{-6} seconds before they break up into a neutrino and a muon or mu-meson.
- Muons are 207 times heavier than an electron but have the same quantity of negative charge.
- They can live about 80 times longer than pions.
- Whereas pions have difficulty reaching Earth, muons with their longer life span don't.
- About 40 of them strike each square metre of the planet per second.
- Muons can go straight through atomic nuclei.

Anderson and the Austrian cosmic ray physicist Victor Hess shared the 1936 Nobel Prize in physics.

Yukawa Hideki was awarded the 1949 Nobel Prize in physics.

Powell was awarded the 1950 Nobel Prize in physics.

From the late 1950s onward until today, is the era of the particle accelerators which energise and thereafter smash particles to smithereens. These smithereens were populated with new particles and a new model of the atom emerged.

Also floating amongst the debris were Nobel Prizes for those who controlled the colliders.

By the dictum "Might is Nobel", Russian scientists should have reaped their share of fame in the sub-atomic discovery race, for it was reported that the Soviets had also built a super-collider called UNC which didn't work very often, because when it crashed it required spare parts which were only attainable according to the five year plan rules. (Anything including spare parts had to be requisitioned five years in advance.)

Visiting scientists were flabbergasted by requests from the UNC team like: "When you come over can you bring some transistors, resistors and capacitors?" The lack of $5 worth of parts often paralysed the trillion rouble UNC.

Protons and Neutrons Mapped

In 1950 American physicist Robert Hofstadter (1915-1990), using a scattering machine he designed, probed the proton and the neutron with great precision. He found each to consist of a positively charged dense point-like core surrounded by two intermingling layers of meson clouds.

He shared the 1961 Nobel Prize in physics with Rudolf Ludwig Mössbauer.

Antimatter

- Every particle has its anti-particle. They are all-but mirror clones of each other. In charged particles the all-but difference is the charge itself.
- The positron or positive charged electron has the same mass, spin and magnetic moment as the electron.
- In 1955 antiprotons and antineutrons were found in the collision debris in particle accelerators.
- The antiproton or negative proton has the same mass, spin and partial decay as the proton.
- The antineutron has the opposite magnetic moment to that of the neutron when they are spinning in the same direction.
- Pions and muons exist with either a positive or a negative charge and one is the antiparticle of the other.
- A neutral pion exists but lives only 10^{-12} seconds then disintegrates into two theoretical photons.
- The theoretical photon is an anomaly. It is theoretically its own anti-particle since if all its magnetic moments are reversed, it is still the same.

- Antimatter particles are, unless somehow kept in isolation, very short lived. As soon as they strike their mirror matter they both disintegrate releasing gamma radiation.
- Some antimatter particles like positrons can be contained and temporarily stored by electromagnetic forces in storage rings. Here they can be accumulated as they circulate before being directed into collision chambers.
- Intense gamma radiation observed emerging from large clouds in the Milky Way indicates that the cloud contains antimatter. These clouds might be created in catastrophic astronomical events.

Fundamental Particles

Fundamental particles are particles that cannot be split into anything smaller. They fall into two groups called fermions and bosons.

Fermions, named after the great Italian physicist Enrico Fermi (1901-1954), are particles that make up matter.

- They obey the exclusion principle.
- They have non-integral spin. like ½ 1½, 2½,

½ is short for: $$\frac{1}{2}\left(\frac{h}{2\pi}\right)$$ h = Planck's constant

They are further divided into:

a) Hadrons, which are subject to strong nuclear forces.

b) Leptons like electrons muons and neutrinos which are only subject to weak forces.

Bosons are the conventionally theoretical particles that make up the four forces:

Gravity
Electromagnetism
Strong force (which holds nuclei together)
Weak force (which preserve atoms from decay)

- They don't obey the exclusion principal.
- They have integral spins like 0, 1,
- Theoretical photons are also bosons as theoretically they have zero spin.
- The theoretical Higgs boson is believed to give matter and antimatter mass.

Bosons are named after the Indian mathematician and theoretical physicist Satyendra Bose (1894-1974). In his 1924 paper 'Planck's Law and the Hypothesis of Light Quanta', he derived Planck's equation purely from quantum theory. This he sent to Einstein who collaborated with him to produce the Bose-Einstein statistics.

Rules of Addition for Fermions and Bosons

- Adding an even number of fermions together produces a boson.
- Adding an odd number of fermions together produces a fermion.
- Adding any number of bosons together produces a boson.

Quarks

By the early 1960s there were literally hundreds of debris particles listed and in 1964 the American physicists Murray Gell-Mann and George Zweig independently theorised that those particles, which interacted via the strong force, were composed of elementary particles carrying fractional electric charges. Gell-Mann called them quarks, (derived from Joyce's Finnegans Wake; "Three quarks for Muster Mark.") Zweig less artistically called his particles 'Aces'

Gell-Mann was awarded the 1969 Nobel Prize in Physics.
Zweig didn't get a Gong.

Quarks were confirmed experimentally in 1969 by American physicists Henry Kendall and Jerome Friedman, and Canadian physicist Richard Taylor. They shared the 1990 Nobel Prize in physics for this discovery.

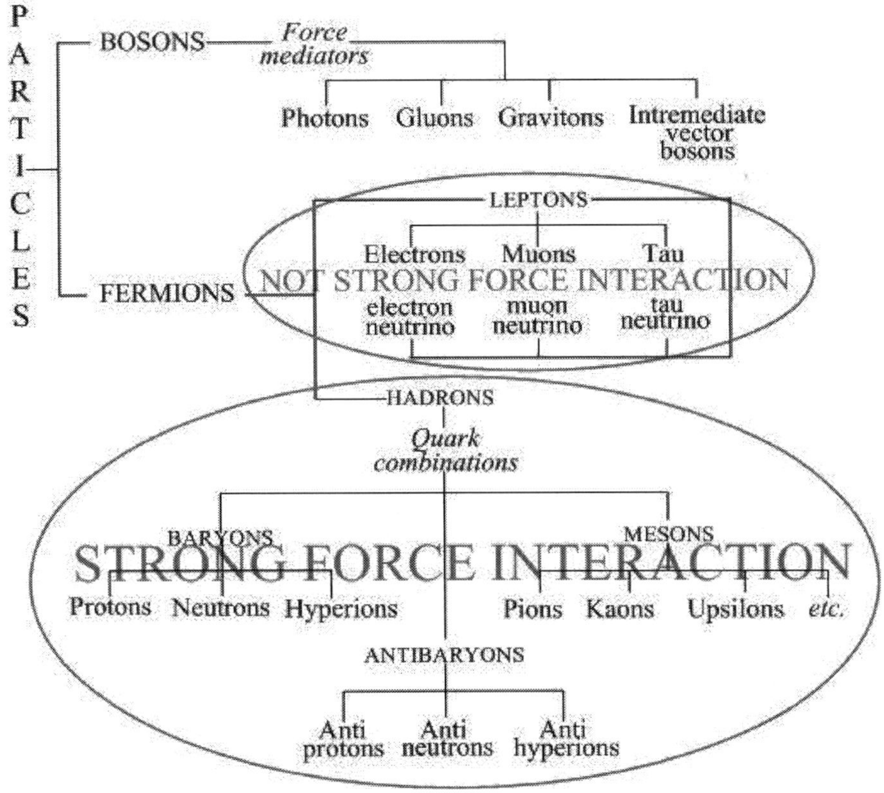

Quark Properties

Quarks are symbolised with bold type.

THE SIX TYPES or FLAVOURS of QUARKS		
up & down u = +⅔ d = - ⅓	charm & strange c = + ⅔ s = - ⅓	top & bottom t = + ⅔ b = - ⅓
The lightest quarks found in atoms in nature. These combine in triplets to form protons and neutrons	The middleweights. Most are found in particle accelerators	The heavyweights Found in particle accelerators. (Also thought to exist just after the big bang)

- The mass of a particle is directly proportional to the energy of the cue ball that breaks it out of the pack.
- The mass of the top quark is 30,000 times heavier than the up-quark which is $1/200^{th}$ the mass of the proton. (A proton has a mass of 1.)
- No one knows why the mass of the top quark (12 GeV) is heavier than the bottom quark (4.8 GeV).
- Quarks have charges of +2/3 or -1/3
- Like mice, there is never just one quark by itself. They cannot stand alone so an isolated 1/3 or 2/3 charge will never be observed.
- Antiquarks combine to form antimatter.
- Antiquarks have most of the properties as their corresponding quarks but some properties are the opposite.
- Antiquarks have charges of - 2/3 or +1/3.
- Antiquarks and antiparticles are symbolised by a bar over them.

Hadrons

- Literally hundreds of hadrons have been observed in the debris in particle accelerators.
- Hadrons are comprised of quarks and fall into three groups; baryons, anti-baryons and mesons.
- Baryons include protons neutrons and a large group called hyperons all of which have three quarks.

- Hyperons all contain the strange quark. They are only observed in particle accelerators.
- Mesons have only two quarks one of which must be an antiquark.

Baryons quarks		symbol	Mesons quarks		symbol
uud	proton	p	$\mathbf{u\bar{d}}$	positive pion	p^+
udd	neutron	n	$\mathbf{d\bar{u}}$	negative pion	p^+
Hyperons			$\mathbf{u\bar{u}+d\bar{d}}$	neutral pion	p^0
uds	lambda	1	$\mathbf{u\bar{s}}$	positive kaon	K^+
uus	sigma plus	S^+	$\mathbf{\bar{s}u}$	negative kaon	K^-
dds	sigma minus	S^-	$\mathbf{d\bar{s}}$	neutral kaon	K^0
uds	sigma zero	S^0	$\mathbf{\bar{d}s}$	neutral antikaon	$\bar{K}^0$
dss	xi minus	Xi^-			
uss	xi zero	Xi^0			

Hyperons break down very quickly:

A sigma decays into a lambda and a pion.

A lambda decays into a proton and a pion etc.

Quark additions:

A proton has 2 up-quarks and 1 down-quark:

uud with charges $+⅔ +⅔ -⅓ = +1$.

A neutron has 1 up-quark and 2 down-quarks:

udd with charges $+⅔ -⅓ -⅓ = 0$

A positive pion has 1 up-quark and 1 antidown-quark:

$u\bar{d}$ with charges $+⅔ +⅓ = +1$

Leptons

Leptons are fundamental particles which are not affected by the strong forces which operate on the quarks.

There are six leptons:

Three are negatively charged:

electrons - muons - tau particles

Three have no charge and possibly no mass:

electron neutrinos - muon neutrinos - tau neutrinos.

LEPTONS		
Electrons e	muon m	tau τ
electron neutrino ν_e	muon neutrino ν_m	tau neutrino ν_τ

Leptons are included in groups called generations:

GENERATIONS		
FIRST	SECOND	THIRD
up & down quarks	charm & strange quarks	top& bottom quarks
electrons	muons	taus
electron neutrinos	muon neutrinos	tau neutrinos
antiparticles of above	antiparticles of above	antiparticles of above
Found in nature	Found in particle accelerators and theoretically in the big bang	

Neutrinos

- Are not affected by the strong or electromagnetic forces.
- If they have mass it is incredibly minute.
- They are affected by the weak force.
- They are detectable in retrospect. If an electron say appears in certain debris then it owed its existence to the presence of an electron neutrino before or during the collision.

Forces

FORCE CARRIERS			
GAUGE BOSONS whose SPINS = 1 (i.e.1x$h/2\pi$)			
8 gluons	photons	weakons $W^+ W^- Z^0$	gravitons
Mediate the strong nuclear force	Mediate the electromagnetic force	Mediate the weak nuclear force	Believed to mediate the gravitational force

Strong Force

- The strong force is the short-range force that holds atomic nuclei together.
- Its range is about 2 fm (fermi) or about 2×10^{-15} m.
- Since the diameters of small nuclei are 5 or 6 fm, nucleons on opposite sides only feel the strong force from their close neighbours.
- If particles are forced closer to each other than 2 fm the strong force turns repulsive. Particles seem held at a 2 fermi range by a spring.
- Gluons are said to be the elementary particles of energy which carry the strong nuclear force between particles of matter.
- Gluons have energy but no mass.
- The greater the number of gluons exchanged between quarks, the stronger the binding force is.
- The strong force gets weaker as quarks close on each other. At a distance of 10^{-15} meters (the diameter of the proton) they behave as if they were free. (this is called asymptotic freedom)
- The strong force gets stronger as quarks move away from each other because some of the energy from departing quarks is said to create more gluons.
- If more energy is directed at knocking out a quark, this can then create new quarks and antiquarks which then combine with the target quark.
- The strong force does not affect leptons.

The Weak Force and Neutrinos

In 1956 the theoretical specks called the neutrinos, which were predicted to escape from nuclei during transmutations, were eventually detected.

They were so small, believed massless, and so fast, light speed, that billions of them pass through the Earth each second without hitting anything. And there are billions of them that come our way every second since one is produced with every transmutation in the Sun when 4 protons fuse into one helium-4 atom, plus 4 neutrinos and 6 theoretical photons. The probability a neutrino from the Sun interacting with any matter in the Earth as it zips through is 1 in 5×10^{13}.

These little neutral particles are untouched by electromagnetic forces because of their neutrality but also they are untouched by the strong forces of the nucleons they escaped from.

It was argued that there had to be some other type of force responsible for their containment in the nucleons and this force was christened 'the weak force'.

This weak force has an unbelievably short range of only 0.1 fermions or 1/10th the radius of a proton.

The Weak Force Carriers; the Vector Bosons

In conventional physics every force must be the comprised of particles called gauge bosons. The electromagnetic force is the movement of theoretical photons, the strong force is the movement of gluons, the force of gravity is attributed to the yet undiscovered movement of gravitons, and the weak force is the movement of vector bosons.

In 1983 using the CERN accelerator, Italian physicist Carlo Rubbia and Dutch physicist Simon van der Meer discovered the W^+, $W^-$and Z force particles. These decayed nearly instantly into a neutrino or an antineutrino and other particles.

Weak Force Control

- The weak interaction is mediated by the heavy W^+, W^- and Z bosons.
- It controls beta decay.
- The weak force can change electrons, muons, or tau particles into their corresponding neutrinos and back again.
- However it cannot cross over generations. Therefore for example, muons will not change into tau neutrinos or electron neutrinos.

The weak force governs why:

- Stable matter contains only first generation leptons and quarks.
- Fermions must come in pairs.
- Each particle must have its antiparticle.
- Leptons interact weakly with one another and with other particles.
- Reactions that absorb neutrinos do not absorb antineutrinos and vice versa.

Spontaneous Transformations

Some nuclei can transform into other nuclei with the same mass number by:

1. Negative beta decay, which transforms a neutron into a proton ejecting an electron with a speck of leftover called a neutrino.
2. Positive beta decay, which transforms a proton into a neutron ejecting a positron and a neutrino.
3. Electron capture, where a nucleus drags in an electron (usually from the inner K shell) which transforms a proton into a neutron ejecting a neutrino.

Negative Beta Decay (β- decay)

The neutron decays into a proton, an electron and an electron antineutrino which turns its atom into an atom with one more proton.

$$n \longrightarrow p + e^- + \bar{\nu}_e$$

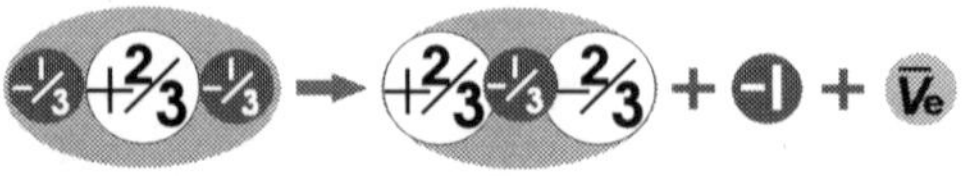

This via a W^- boson: $n \rightarrow W^- + p \rightarrow p + e^- + \bar{\nu}_e$

where either:

a) One of the neutron's down-sink-quarks is transformed into an up-pump-quark and a W^- boson, and then the W^- boson immediately decays into an electron and an antineutrino.

or

b) Both the neutron's down-sink-quarks turn into up-pump-quarks while its only up-pump-quark turns into a down-sink-quark . The switcheroo simultaneously releases a W^- boson which immediately decays into an electron and an antineutrino

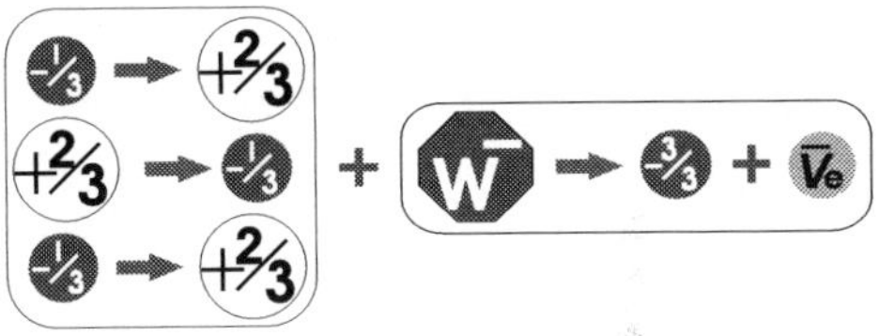

Positive Beta Decay (β^+ decay)

Here the proton decays into a neutron, a positron and an electron neutrino which turns its atom into another atom with 1 less proton.

$$p \longrightarrow n + e^+ + \nu_e$$

This is via the W^+ particle: $p \longrightarrow n + W^+ \longrightarrow n + e^+ + \nu_e$

where either:

a) One of the proton's up-pump-quarks is transformed into an down-sink-quark and a W^+ boson, then the W^+ boson immediately decays into an positron and a neutrino.

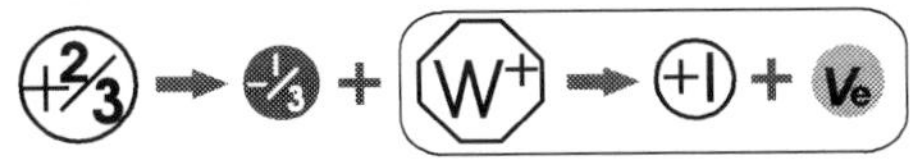

or

Both the proton's up-pump-quarks turn into down-sink-quarks while its only down-sink-quark turns into a up-pump-quark. The switcheroo simultaneously releases a W^+ boson which immediately decays into an positron and a neutrino.

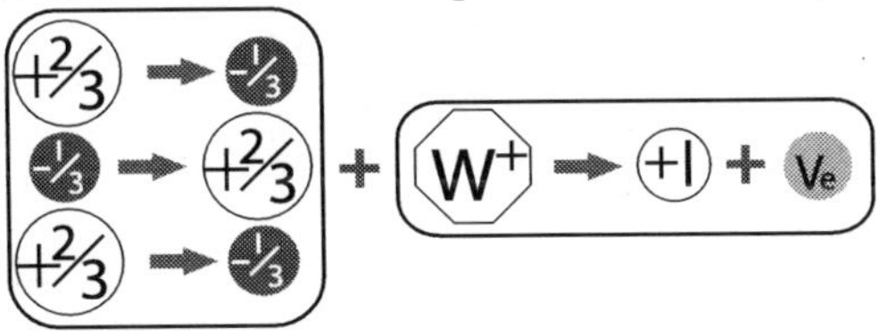

Electron Capture

This was detected because when an electron is removed from the high energy K-shell, then another electron from its electron cloud jumps into its place emitting a detectable X-ray.

Quantum Chromodynamics

The colour here is symbolic. It doesn't mean that the particles are actually coloured.

COLOURS			
quarks	red	blue	green
Antiquarks	antired cyan	antiblue yellow	antigreen magenta
6 gluons	are a combo of a colour and some other colour's anticolour		
2 more gluons	are complex mixture of colours		

Some Colour Rules

- Baryons are made from three quarks and one will be red, another will be blue and the third will be green. Their combination yields white or neutral.
- Mesons are made from a quark and an antiquark. One will always be the anticolour of the other. Their combination will be white or neutral. For instance a neutral kaon will have a green down quark and a magenta anti-strange up-quark.
- Unlike theoretical photons, gluons carry a colour charge therefore they can attract one another. Possible combos are red anti-blue, green anti-red and blue anti-green.

UNIFIED FIELD THEORIES

In 1967 American physicist Steven Weinberg and Pakistani physicist Abdus Salam independently developed a mathematical theory that unified the weak force and electromagnetism by linking an intermediate vector boson with a theoretical photon. The weak force is shown to be comparable to the electromagnetic force when particle energies are 70-100 GeV.

In 1968 American physicist Sheldon Glashow and Italian physicist Luciano Maiani showed how the model could be applied to hadrons. Theoretically they would meld at 10^{15} GeV.

Salam, Weinberg and Glashow shared the 1979 Nobel Prize in physics.

Gauge Theories

- Gauge theories evolved the 1970s to explain the behaviour of gluons which theoretically mediate the strong force.
- They expanded to unite, or try to unite, any of the four forces of nature; strong, weak, electromagnetic and gravitational.
- They revolve around symmetrical interaction between two kinds of particles. Objects and particles are symmetrical if they are unchanged when they become their mirror images.
- Baryons and leptons exhibit certain symmetries. For example if a neutron turns into a proton or vice versa the strong force is unchanged.
- Electromagnetic phenomena are symmetrical since changes in the combination of electric and magnetic potentials have no effect on the results.
- Gauge theories hold that these symmetries are only approximate symmetries. There are violations. It incorporates these violations (such as parity violations) into what it calls gauge symmetries where internal symmetries can be different at each point in space time.

In 1957 parity violation was proposed by Chinese physicists Chen Ning Yang and Tsung-Dau Lee to explain the "tau-theta puzzle." They shared the 1957 Nobel Prize in physics for this alarming and soon proven thorn to symmetry.

Grand Unification Theories – GUTs

These try to reduce all symmetries to gauge symmetries. They try to unite the strong and weak forces.

Theories of Everything TOEs

In the early 1960s British physicist Peter Higgs proposed the existence of Higgs fields to accommodate huge bosons without destroying symmetry. In 1964 he reported that one Higgs field called for the existence what became known as the Higgs boson which has no electric charge or intrinsic spin. It ostensibly carries the force of gravity.

- Theories of everything cannot theoretically be realised unless the Higgs boson can be proven to exist. In 2012 a particle of mass 125-127 GeV/c^2 was found at Cern which seems to fit expectations.
- Supersymmetry theories SUSYs, invent twins with opposing spins for fermions and bosons.
- The SUSY 5 (SU5) theory holds that charge is quantized into multiples of 1/3.
- Superstring theory suggests all particles are dimensionless objects that extend in one dimension to lengths of 10^{-35} meters.
- Etc...

The New Particle Addendum

The discovery of a new particle was always celebrated. I recall listening in on one broadcast on the radio in about the middle of 1975. It was over thirty years ago and I'm a bit hazy but this is pretty close to what was broadcast, albeit garnished with artistic license.

A journalist was interviewing a spokesman (possibly the leader) of an American research team while in the background champagne corks popped midst an excited whoopee hubbub. It sounded like they were celebrating the birth of a royal child.

Q. And what sex is it?
A. Negative; have a cigar!
Q. And how much did it weigh?
A. It's absolutely huge. Have a drink!
Q. And what made you look for something that big?
A. Oh it was Bill's fault (or Fred's or Sam's, I can't remember the name).
He left the deflector switch to maximum by mistake
and the whopper was dragged into the detector.
Q. I see. Is there any theory that predicted its existence?
A. Not that I know of; Of course we'll have to look back
through all of them and check;
Have a cocktail!

The particle in question was the tau-lepton anticipated by Antonino Zichichi. and independently by Yung-su Tsai.

In 1975 a team headed by Martin Perl was studying electron/positron collisions and observed in the debris anomalous events which included two inexplicable particles. These were deduced to be the Tau+ and the Tau− pair. They were 3,500 times heavier than an electron.

In the documentation describing its discovery, red pencilled out is any reference to Bill's fault or to chance. Never the less it was a team effort which delivered a spectacular result.

Perl shared the 1995 Nobel Prize for physics.

The Bugorski Addendum

In 1978 Russian scientist Anatoli Petrovich Bugorski was checking a malfunction in Synchrotron U-70 when unbeknownst to him, its fail safe device had also failed. Unfortunately he stuck his head into the proton beam which passed straight through it.

He saw lights "brighter than a thousand suns", but curiously felt no pain. Soon his face swelled beyond recognition, and after subsidence the left half remained paralysed. He was deaf in one ear and his head was always filled with a buzzing noise.

Miraculously his mind stayed alert and he completed his Ph.D. He was a bit of a celebrity .

ELECTROMAGNETIC RADIATION

Radiation is the transfer of energy through space or matter by particles or waves.

Particle Rays

Mechanical radiation like sound waves is transmitted through matter.

Conventional theory holds that electromagnetic radiation has a split personality. It can take be observed as a stream of particles or a wave of energy.

Streams of particles or theoretical photons are called rays.

- Alpha rays, beta rays and nearly all cosmic rays are particle rays.
- About 87% of cosmic rays are protons (hydrogen nuclei), about 12 percent are alpha particles (helium nuclei) and 1% are electrons, muons pions. These all travel close to the speed of light.
- A fraction of 1% of cosmic rays are gamma waves which travel at the speed of light.

The speeds of the rays are:

Cathode ray of electrons	@ 30,000 kps
Beta rays of electrons	@ almost light speed (c = 300,000 kps)
Alpha rays of helium nuclei	@ 16,000 kps = 1/20th c
X ray	@ light speed
Gamma ray	@ light speed

The alpha and beta speeds are from standing starts. Even at alpha's 'slow' speed the ray in a vacuum would circle the Earth in 2 seconds.

All the above can ionise certain atoms especially gas atoms. Gamma rays and X-rays do this by knocking out electrons from atoms. (A burst of X-rays ionised the air in Millikan's oil drop experiment.)

- Alpha rays may only be able to penetrate a few sheets of paper before they are stopped or slowed down by the electrons in the paper's atoms.
- Beta rays can penetrate over 100 sheets of paper.
- Gamma rays and X-rays, depending on their energy can penetrate several centimetres into lead.

The apparent weakness of alpha rays belies their hazardous nature.

When alpha emitting radioactive particles (like those in fallout dust) are ingested, they fire off high energy helium nuclei like canon balls which smash and maim neighbouring cells.

The immune system ignores these damaged cells since they are the host's cells. Some mutations are still able to divide and multiply and some of these become cancers.

Various radioactive isotopes emit their alpha beta or gamma rays at speeds or energies peculiar to each and therefore these speeds can reveal the identity of their launch pad.

- For example polonium-210's alpha particles penetrate 3.8 centimetres of air before being stopped, whereas those of polonium-212 penetrate 8.5 centimetres.
- Beta rays can be bent by strong magnets. The average and maximum angles of deflection are proportional to their velocity and therefore are signatures of their launch pads.
- Gamma radiation nearly always occurs simultaneously with alpha or beta decay. This more or less cleans up any left over energy. Each radioactive isotope has characteristic gamma-ray energy. They have about the same energy as high energy X-rays.

The Electromagnetic Spectrum

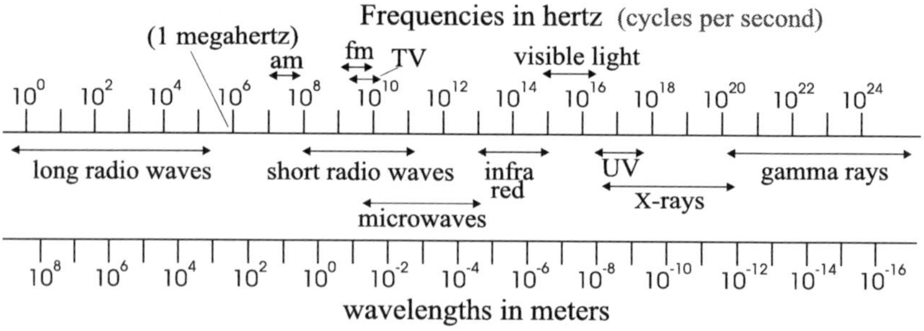

Electromagnetic radiation is generated by the movements of electric charges. This radiation is in the form of electromagnetic waves. Some of these waves (such as light waves) theoretically behave as particles and are described as the movement of theoretical photons.

- All electromagnetic radiation travels through free Space in a straight line at the speed of light, however when it travels through matter its speed, direction and intensity are affected. For example light travels through water at ¾ of its air speed.
- Any wave in the electromagnetic spectrum can be diffracted or undergo interference.

High speed electrons either from Space, or generated in particle accelerators, or those known as beta rays, also behave as waves. As they zip by atoms or molecules they are diffracted or bent into several beams which spread outwards

These then interfere with each other creating constructive and destructive zones similar to those seen in slit experiments using light rays.

The Momentum of theoretical photons

Momentum is defined as the product of mass and velocity.

Einstein defined light as being comprised of massless theoretical photons moving at the 3×10^8 mps. Only massless particles can travel at the speed of light.

Despite having no mass, light is able to impart momentum to certain objects it hits. At a suitable frequency an electron in an outer orbit around a nucleus is thought to absorb a theoretical photon causing it to jump to a higher energy outer orbit. Moments later this electron sheds the theoretical photon and falls back to its original orbit. Because the rejected theoretical photon flies off in any direction and the incident light beam is coming from only one direction, momentum in the direction of the beam is imparted to the electron and thereby to its atom.

Because of this light pressure or radiation pressure, massless theoretical photons are said to possess momentum. A beam of light directed to one side of a paddle wheel causes it to rotate.

X-rays

A lot of heat as well as X-rays are produced by the abrupt deceleration of high speed electrons. X-ray producing tubes are essentially cathode ray tubes with hard metal anodes set at an angle to the electron beam which deflect and direct the X-rays out of the tube.

As previously stated, very fast electrons are able to knock other electrons out of the inner orbits of metal atoms. The vacancy is then immediately filled by an electron from an outer shell which falls into the slot. Since this electron is moving from a high energy orbit to a low one it sheds energy in the form of a high energy theoretical photon.

When these knockouts occur in the outer orbits the theoretical photons tend to be of the frequency of visible light. When knockouts occur in the inner orbits, which require higher energy electron cue balls to penetrate the atoms electron cloud, theoretical photons are produced in the X-ray frequency range.

X-rays can also be produced when high energy electrons are deflected or retarded by large electric 'fields'. This is called bremsstrahlung (braking) radiation

X-rays are absorbed by heavy objects like bones but pass through light ones like skin.

Micro-Micro-scopes

To see into something we need a radiation whose wavelength is smaller than it, since waves with larger wavelengths bounce off the object in all directions and can't define its shape. Consequently the shorter the wavelength is the smaller the object it can see.

Visible light has wave lengths around 10^{-7} meters which is too large to see into atoms which have diameters of around 10^{-9} meters and nuclei of around 10^{-14} meters.

Very high energy X-rays (hard X-rays) with wavelength around

0.01 Å = 0.01 Angstroms = 10^{-12} meters, see right into atoms.

1 Angstrom (unit of wavelength) = 10^{-10} meters

Even shorter wavelengths can be created by accelerating electrons to enormous energies in the gigavolts range. These can delve into protons and neutrons.

The Mössbauer Effect & Gamma Rays

In 1955 German physicist Rudolph Ludwig Mössbauer discovered that when gamma rays were emitted from certain radioactive substances with tight crystalline structures those atoms did not recoil. Here the frequencies of the gamma rays correspond to the exact difference between the nuclear ground-state energy and the excited-state energy. Now if these rays strike their tight crystalline neighbours they emit gamma rays of the same frequency.

This effect is used to precisely measure electrical, magnetic and gravitational effects.

Mössbauer shared the 1961 Nobel Prize in physics with American physicist Robert Hofstadter.

The Ionosphere & Radio Waves

In 1902 British physicist Oliver Heaviside and American engineer Arthur Kennelly postulated the existence of a layer of high level ionised atmosphere which would bounce radio waves back to Earth from which they could bounce back up and so on. Eventually they could bounce around the globe. It is called the Kennelly-Heaviside layer or more often just the Heaviside layer.

The ozone layer floats between 19 and 48 kilometres above the Earth's surface. Above this and extending for a thousand kilometres up is the ionosphere. It is made up of several layers of atmosphere which have been ionised by ultraviolet rays and X-rays from the Sun. The lower layer builds up ionisation in sunlight which fades away into the night.

These layers both partially absorb and partially reflect electromagnetic waves in the radio wave spectrum. Longer wavelength waves of over 10 meters are generously reflected; however shorter wavelengths below 10 meters are either absorbed or they pass through.

Very shortwave radio waves with wavelengths of around 0.2 meters are absorbed by water droplets or clouds. Below this wavelength transmission is absorbed by atmospheric water vapour.

Van Allen Belts

The upper regions of the ionosphere blend into the magnetosphere. It is a radiation belt consisting of high energy protons and electrons from cosmic rays or solar winds trapped by the Earth's magnetic field

It stretches from about 200 to 64,000 kilometres above the Earth. The protons have energies around 10 MeV and tend to zip about near the 3,200 kilometre level while the electrons prefer the outer regions. Both spiral back and forth between the magnetic poles whose stronger magnetism reflects them.

Back in the early 19th century the mighty Gauss had suggested the existence of such a phenomenon. It was discovered in 1958 by a Geiger counter inside the first U.S. satellite. James Van Allen headed the team that was responsible for the detector experiment and the radiation belt bears his name.

Microwaves

Microwaves have wavelengths of between 1 and 300 millimetres. These are associated with devices called travelling wave tubes, klystrons and magnetrons.

Travelling Wave Tubes

These were the brainchildren of British scientist, R. Kompfer of Oxford University's Clarendon Laboratories.

- They are essentially modified cathode ray tubes. There is the gun or cathode at one end and the collector or anode at the other.
- Electrons are fired from the convex cathode and immediately focused with an electric grid lens into parallel motion, forming a beam which travels in a straight line to the collector.
- Around this beam is a helix in which a high frequency current is oscillating. Generally currents in this range are said to be radio frequency currents and are said to generate radio frequency or RF fields.
- The system is arranged so that the current in the spiral going the long way round matches the velocity of the electron beam taking the short cut up the centre of the helix.
- Now small amperage oscillating currents in the helix cause very large bunching and thinning or compressions and depressions of electrons in the beam and every change in charge density has its companion magnetic fluctuation.
- The RF current frequency is in the range of microwaves. An electric potential across the electron tube of the order of around 10,000 volts will amplify it by around 50 decibels.

Klystrons

Klystrons are vacuum electron tubes or cathode ray tubes with two pairs of electric grid gates in them. These speed up then slow down the velocity of the high energy electrons as they race toward the tubes anode. Here fluctuations or variations in electron speeds create magnetic disturbances or electromagnetic waves which are collected in and redirected from chambers called cavities which surround the grids.

Magnetrons

Magnetrons don't have to amplify out-sourced high frequency oscillations since the device itself, by dint of its design and its magnetised block creates copious amounts of microwaves.

The block containing the cavities looks like a revolver's cylinder but has corridors connecting each cartridge chamber to a hollow core. In the centre of the hollow core is the cathode.

Electrons leave the cathode and travel in a circular direction passing corridors which lead to the chambers called cavities. As they pass the corridors their speed slows and then accelerates between doorways. These fluctuations generate electromagnetic waves in the microwave range.

Magnetrons are the drivers in microwave ovens. They produce oscillating electromagnetic waves whose orientation alters millions of times a second. These escape through the stirrer, a fan whose spinning metal blades scatter the waves throughout the oven. (It is the fan noise one hears when it is cooking.) The microwave frequency is tuned to water molecules which are jerked every which way so rapidly that they heat up the food.

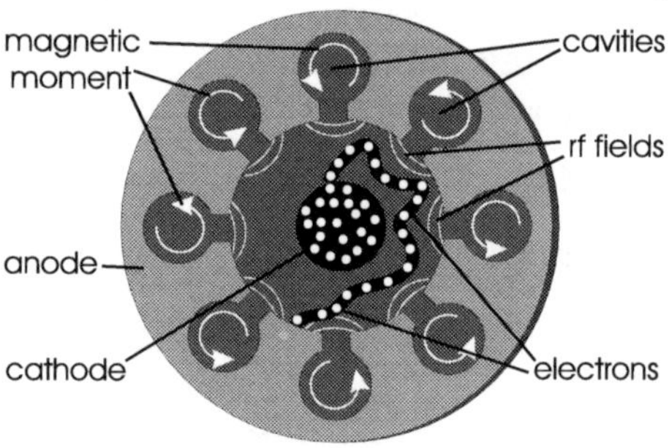

Unprotected magnetrons are absolutely lethal. Do not open up to investigate or try to fix your microwave oven while its on.

Synchrotron Radiation

Any change in velocity of a charge changes its companion magnetism. When a charged electron is travelling in a circle in a cyclotron for example, even though its velocity is constant, it has an acceleration towards the centre.

This is a change and therefore it generates electromagnetic waves. These are very constant waves if the velocity of the electron is constant.

Synchrotron Radiation is the name given to the electromagnetic radiation created when high energy electron beams are bent away from straight lines.

This radiation is ever present around circular particle accelerators.

Some scientists believe the radio emissions coming from deep Space sources like the crab nebula are caused by very high energy electrons being forced into spiral paths by strong magnetic fields.

LOW TEMPERATURE PHYSICS

Both Humphry Davy and Michael Faraday by liquefying gases were pioneers of cryogenics. The process involved first heating certain gases at the big end of a funnel then chilling them as they rose up through a jacket of salt ice mixture surrounding the neck which was closed at the top. As the gases rose, their pressure increased as they were forced into the neck and they liquefied. When the neck was opened the gases boiled off.

By evaporating solid carbon dioxide mixed with the chemical gas ether $(C_2H_5)_2O$ at low pressure Faraday was able drop temperature as far as -110° C.

The Inversion Temperature

Warmish gases heat up as they expand through a valve however cold gases which are initially below what is called the 'inversion temperature' atypically get colder as they expand. This is called the Joule-Thomson effect. (This Thomson is Willie Thomson aka Lord Kelvin.)

By a series of gentle compression then fast release and expansion, one can cascade gas temperatures progressively lower.

In 1877 the French physicist Louis Cailletet and the Swiss scientist Raoul Pictet cascaded out droplets of liquid oxygen, boiling point BP -182.96° C and liquid nitrogen, BP -195.79° C.

Now gases previously thought to only exist as gases were shown to also exist as liquids at very low temperatures. A new branch of science was born called 'cryogenics' and this had to have new rules.

Cryogenics is a term applied to all temperatures below -150° C.

Absolute Zero

The first recorded physicist to ask how cold can things possibly get was French physicist Guillaume Amontons in 1702. He asked that if a temperature fall of x degrees caused the height to fall y centimetres in a long straight thermometer, what temperature would it be if the column fell to zero height? His answer was -240° C which is not far off.

In a treatise on heat in 1779, German mathematician Johann Lambert, (who showed that π is irrational), set zero temperature at -270° C, almost spot on.

Somehow in their 1780 treatise, Pierre-Simon Laplace and Antoine Lavoisier plunged it to between -1500° C and -3000° C

A little later in 1808 British chemist John Dalton in his 'Chemical Philosophy' also put it at -3000° C.

In 1848 Lord Kelvin silenced this excessive temperature madness. He devised a scale of absolute temperature based solely on the laws of thermodynamics using J.P. Joule's mechanical equivalent of heat. This scale set absolutes zero at -273.15° C.

Today it rests at -273.16° C which is called 0° Kelvin or zero K

Johannes van der Waals

In 1880 the Dutch physicist Johannes van der Waals (1837-1923) published a theory which gave continuity between fluid and gas states.

Boyle's law and Charles law relate to theoretically perfect ideal gases. The ideal gas equation is:

$$PV = nRT$$

where P = pressure, V = volume, n = number of moles,
R = a constant for all ideal gases, T = Kelvin temperature.

If the volume is measured in litres and the pressure is measured in atmospheres then R = 0.0821 litre atmospheres per degree Kelvin per mole.

The ideal gas equation implies that there is always positive pressure between gas atoms and molecules so why were liquefied gases viscous?

Van der Waals then amended the gas law to: $\left(p + \frac{a}{V^2}\right)(V - b) = nRT$

Where (a and b) are parameters determined from experimental measurements carried out on actual gases. The term (V – b) indicates the permissible proximity between atoms or molecules.

Van der Waals forces, also known as London forces, occur when atoms or molecules temporarily develop polarization, which causes inter-attraction resulting in phenomena like viscosity and surface tension.

He was awarded the 1910 Nobel Prize in physics

The only way to test van der Waals's ideas was to measure gaseous behaviour at extremely low temperatures. Here gases with extremely low boiling points would behave very much like ideal gases whereas gases with higher boiling points like carbon dioxide (dry ice) BP = -78.5° C don't.

Sir James Dewar and Liquid Hydrogen

A new science race now began in the laboratories of Europe to liquefy hydrogen and consequently approach the Kelvin limit of absolute zero.

One of the two protagonists in this race was the British chemist and physicist Sir James Dewar (1842-1923). In 1898 he used a jacket of liquefied air to cool pressurised hydrogen which then liquefied as it escaped through a valve.

He also invented the Dewar flask now commonly called the vacuum flask as well as the smokeless gunpowder called cordite.

He got hydrogen to liquefy at –252.8° C and freeze at –259.14° C beating the other protagonist Heike Onnes, but already the next great challenge was afoot.

In 1868 the astronomer Pierre Janssen had discovered a new gas in the spectrum of the corona of the Sun during an eclipse. It was soon identified as an element by the British chemist Sir Edward Frankland and the British astronomer Sir Joseph Norman Lockyer and named helium after the Greek word 'helios' for Sun.

In 1895 William Ramsay had discovered helium on Earth and the atoms of this gas were found to exert extremely weak forces on each other resulting in it having a boiling point even lower than that of hydrogen.

Although it weighs twice as much as hydrogen Helium has 92 percent of its lifting power.

Liquid helium now became the holy grail of cryogenics and the protagonists raced on. This time the champion was Onnes who thereafter became decorated in totally unexpected laurels.

Heike Kamerlingh Onnes and Liquid Helium

The Dutch physicist Heike Kamerlingh Onnes (1853-1926) was professor of physics at the Leiden University from 1882 until his death. His mentor from his doctorate days was Van der Waals who set him on the road to cryogenics. Onnes established a cryogenics laboratory at Leiden employing the finest craftsmen and instrument makers in Europe.

In 1894 he set up the first liquid air plant in using the cascade principle. (In 1895 the great German refrigeration engineer Carl von Linde set up a large-scale plant for its production.)

The helium gas was difficult to procure. Onnes's brother however, as a director of the Office of Commercial Intelligence in Amsterdam, was able to procure tons of monazite sand from North Carolina. From this Onnes extracted about 300 litres of helium.

Now with volumes of relatively cheap liquid air, reasonably cheap liquid hydrogen could be manufactured which in turn was used to cascade his stock of helium gas into liquid form.

It took a great deal of experimentation but eventually on July 10th 1908 sixty cubic centimetres of liquid helium was produced. It had a boiling point of 4.2° Kelvin.

Liquid helium was an extremely difficult material to handle, but by 1911 Onnes had developed a helium cryostat capable of maintaining it reasonably well at those very low temperatures.

Now with manageable liquid helium further experimentation could proceed. The very first experiments using electricity revealed totally unpredicted literally mind boggling phenomena.

Superconductivity

There were lots of theories as to what would happen to matter when the temperature fell to absolute zero. Most of them proposed (theoretically) that all motion would stop, the nucleus would cease to vibrate or spin, electrons would either freeze or fall into the nucleus, etc.

These theories were purely theoretical since everyone knew the Heisenberg principle forbade the electron to stop; because this would fix it both somewhere and somewhen simultaneously.

On the other hand there was nothing to stop scientists legitimately theorising about atomic motion in the super-cool zone just above absolute zero, however it's fair to say no one was even close to imagining what lay ahead.

In 1911 Onnes was the sole proprietor of liquid helium in the world. His first electric experiment was conducted using mercury which had been purified through many distillations to eliminate any contamination from other elements. Impurities in conductors cause their resistance to rise, often quite considerably.

The mercury was then poured into a U-tube then immersed in the liquid helium which was sitting at about 1.7° K. (Mercury becomes a solid at -39° C.) When a current passed through the mercury wire, extensive testing could not discover any resistance in the circuit. Everyone thought that there must be a short circuit somewhere but exhaustive checking failed to find one. When a W-tube was substituted for the U-tube the zero resistance persisted.

Repeated testing over the next few days shed no light on the enigma until a student charged with minding the cooling mechanisms nodded off. The temperature now accidently rose and suddenly at 4.2° K resistance appeared in the circuit. Superconductivity was discovered.

Soon it was detected in other metals like tin and lead.

In 1914 Onnes induced a current in a supercooled lead ring, and then removed the inductor. The current continued to flow for days evidenced by the directions of tiny weak compass needles. To physicists looking on this freakish experience was perhaps akin to that of observing Robert Boyle's silent bell clanging away in a glass vacuum chamber.

The event rocked quantum theory to its roots. Here was a ring of lead though which electrons circulated in perpetual motion as if they were in the outer orbit of a gigantic atom.

For an all too brief euphoric period Onnes and his team believed that superconductivity was a phenomenon that heralded an electromagnetic Utopia; its potential appeared limitless. So it was a great let down for them when they discovered that a weak magnet would cancel the zero resistance.

To be of any practical use, superconductivity would have to support large electric currents whose magnetism would obviously be totally extinguished by external magnetism or induced magnetism.

Hopes were dashed then, yet 40 years later this obstacle to large current would begin to crumble though not without resistance. Many intermediate discoveries paved the way.

Indeed superconductivity turned out to be (and still is) full of surprises. Many of the important discoveries were totally unpredictable and consequently were stumbled upon by chance.

Onnes was awarded the 1913 Nobel Prize in physics.

The Meissner Ochsenfeld Effect

In 1933 Karl W. Meissner and R. Ochsenfeld of Germany discovered strong diamagnetism in superconductors.

Diamagnetism is the property of matter in which the magnetic domains are repelled by magnetic poles. In other words diamagnetic material wants to be halfway about the waist of a magnet. Weak diamagnetism occurs in many of the elements and substances like water or paraffin.

Many of us are familiar with the picture of a small magnet floating or levitating above a superconductor amidst vapours resulting from liquid nitrogen boiling off.

Here, as the magnet is lowered, a current which can flow forever is induced in the superconductor which has no electrical resistance. This current has its own companion magnetism which repels that of the descended magnet. Not only will the magnet float above the superconductor but it has been shown capable of carrying a weight of around 300 grams per square centimetre. This phenomenon only occurred with weak magnets whose magnetism couldn't penetrate and destroy superconductivity.

In 1988 Palmer N. Peters of the U.S. Space Science Laboratory found that a small piece of a ceramic superconductor could be suspended in space under a magnet until it warmed up.

Superfluidity of Helium

In 1937 Russian physicist Peter Leonidovich Kapitza (1894-1984), and independently in 1938 the British physicist John Frank Allen, showed that when liquid helium, which at normal pressure becomes liquid at 4.2° K, is cooled below 2.17° K, it changes from essentially a normal liquid into a totally eccentric one.

This metamorphosis takes place at what is called the lambda point because of the shape of the specific heat graph here.

The lambda point for helium is 2.17° K.
Between 4.2° K and 2.17° K liquid helium is called helium I.
Between 2.17° K and $(0+\Delta)^{\circ}$ K liquid helium is called helium II.

Helium II is a mix of normal atoms and superfluid atoms. The colder it gets the greater the percentage of superfluid ones. These are ground state atoms and don't carry any thermal energy.

Helium II cannot be solidified. It has a viscosity of zero.

Helium II flows spontaneously from cold to not-so-cold places while helium I flows from cold to colder ones, consequently helium II has a thermal conductivity 3 million times higher than helium I because in the mix the helium I atoms carry thermal energy down while the helium II atoms carry it up where it then reverts or converts into helium I.

Helium II can flow through microscopic holes that helium I can't. It even climb up the walls of its container against gravity resulting in a film of it on the walls 100 atoms thick .This Houdini easily escapes through a container's tight lid.

Kapitza showed that at about 1.7° K, helium II is a better conductor than copper.

At temperatures below 0.00093° K the stable isotope helium III is also a superfluid.

It is impossible to solidify helium at atmospheric pressures.

Kapitza shared the 1978 Nobel Prize in physics with American radio astronomers Arno Penzias and Robert W. Wilson, (the discoverers of cosmic background microwaves radiation.)

BCS Theory

In 1957 the American physicists John Bardeen, Leon N. Cooper, and John R. Schrieffer proposed that superconductive electrons move as pairs, (Cooper pairs), with opposite spins.

They shared the 1972 Nobel Prize in physics. (It was Bardeen's second)

The Lattice Key

In 1958 H. Froehlich, working at the University of Liverpool, and J. Bardeen of the University of Illinois provided the ground breaking step to understanding how the electrons travel without resistance through some supercooled essentially rigid lattices and not through others.

They viewed the electrons as an integral element of the vibrating lattice, not as vehicles on a highway.

The energies of a superconducting electron and a normal one differ by only one part in a million.

Josephson Effect

In 1962 (based on BCS theory) Cambridge graduate student Brian D. Josephson proposed that an electric current (the Josephson current), (a current of paired electrons), would tunnel through (Josephson tunnelling) an ultra-thin insulating layer (<10 angstroms thick) separating two superconductors. (there are 10^7 Angstroms per millimetre)

This soon substantiated phenomenon is called the 'DC Josephson Effect'. The current only flows when the conductors are not connected to a DC source.

Contrarily, when a DC current is connected across the superconductors, the current oscillates between them whereby no net charge crosses over. This is called the 'AC Josephson Effect'.

Here the energy of an electron pair increases by frequency 2eV as it crosses over, and since the energy

$$E=hf$$

$$f = 2eV/h,$$

where e = electron charge,

V = the voltage, h = Planck's constant.

The oscillation is fairly easy to measure and consequently this can give a very precise measurement of the DC voltage. Indeed the defining standard of the volt is now taken as that potential difference which generates an AC oscillation of 483597.9 GHz across a 'Josephson junction'

SQUIDS

Superconducting quantum interference devices, convert DC current to AC via Josephson junctions extremely quickly.

The junction is so sensitive that it has led to revising the value of Planck's constant from 6.62559×10^{-34} to 6.626196×10^{-34} joule-second.

The presence of a magnetic 'field' however weak, affects the AC frequency across the junction which here becomes an ultra-sensitive magnetic measuring device.

The Josephson tunnelling was confirmed experimentally by the Japanese physicist Leo Esaki with his tunnelling diode (the Esaki diode). It was also confirmed by the Norwegian physicist Ivar Giaever via semi-conductors.

Josephson, Esaki and Giaever shared the 1973 Nobel Prize in physics.

Quantum Hall Effect

In 1980, a century later than the discovery of the Hall Effect, the German physicist Klaus-Olaf von Klitzing experimenting with very strong magnets on conductors at near absolute zero temperatures, discovered that the Hall voltages increased in discrete steps which he called the Integer Effect or the Quantized Hall Effect.

As the magnetic force increases, the electrons begin collecting on one side of the conductor which is revealed with a voltmeter. After a gradual build up, the voltage levels off, then start building up again at a higher level. The difference between the voltages at each level shows that at each level is made up of whole numbers of electrons.

This discovery of the quantized variations in resistance is responsible for achieving the high accuracy required in the structure of electronic devices. Our computers depend on it.

Fractional Quantum Hall Effect

In 1982 German born American physicist Horst L. Stormer and Chinese born American physicist Daniel C. Tsui, repeated Klitzing's experiment at even colder temperatures using greater magnetic force. They were expecting similar results but discovered that the number steps between levels were too numerous. Charge had apparently been split into quantities smaller that integers, therefore smaller than the unit electron charge which theoretically is inviolable.

In 1983 American physicist Robert B. Laughlin arrived with a new theory; groups of electrons could act together as units he called quasiparticles. Each electron can belong to more than one quasiparticle and divide its charge amongst them. The effect is known as Fractional Quantum Hall Effect.

Von Klitzing was awarded the 1985 Nobel Prize in Physics.

Stormer, Tsui and Laughlin shared the 1998 Nobel Prize in physics.

Bose-Einstein Condensate

In 1995, 70 years after its conception by Satyendra Nath Bose (the father of the boson) and his corroborator Albert Einstein, American scientists succeeded in creating a Bose-Einstein Condensate.

Here at temperatures very close to absolute zero where atomic activity almost stops, atoms become indefinite and merge into each other forming a fuzzy cloud of what approximates as one huge atom.

In 1999 scientists fired two beams of light through a condensate. The first beam combined atoms into a hugely refractive volume. This slowed down the second light beam passing through it at right angles about 20 million times to 61 kph.

Superconductors Today

- All known superconductors are solids.
- All conduct finite quantities of direct current DC (not alternating current) without resistance.
- All have near infinite thermal conductivity.

They

- repel magnetic 'fields'.
- have enhanced thermal conductivity.
- have higher optical reflectivity.
- have reduced surface friction.

Since the mid 1950s superconductivity has been divided into two categories:

1. Earlier Type I low temperature superconductors.
2. Type II conductors:

    a) Comprised of metal alloys.

    b) More recently discovered high temperature superconductors which are complex oxides.

Type I Superconductivity has been observed in more than 21 metallic elements. These have transition temperatures (symbol TC) between 0.5° K and 11° K above which they regain their electrical resistance; many of these are very poor conductors at room temperature.

Type I's also regain their electric resistance in the presence of weak magnetism when its strength increases above the so called critical 'magnetic field' level, (symbol H_C).

As the temperature of superconductive elements falls toward absolute zero its H_C rises, making the element able to withstand stronger magnetic attack.

Metal	T_C° K	H_C ousted
Zinc	0.88	53
Aluminium	1.20	99
Indium	3.41	282
Tin	3.72	306
Mercury	4.15	411
Lead	7.19	803

Type II

The element that heralded the birth of Type II superconductors was the transition metal niobium Ni:

Atomic number 41, configuration 2-8-18- (T_C = 9.15° K).

In 1954 George Yntema built a superconducting magnet from niobium wire which at 4.2° K generated a magnetic 'field' of 0.71 teslas.

1 tesla = 10^4 Gauss. The Earth's magnetic flux is about 1 gauss.

In 1961 Kunzler discovered the extremely useful properties of Niobium-three-tin (Nb_3Sn) (T_C of 18.5° K) which pushed the critical limit much higher.

Then came Niobium-three-germanium (Nb_3 Ge), ($T_C = 23.2°$ K) which held the T_C record until 1986.

High Temperature Type IIs

In 1986 the West German physicist J. Georg Bednorz and superconductor Swiss physicist K. Alex Müller were systematically testing supercooled oxides containing nickel or copper. They discovered that barium lanthanum copper oxide became superconductive at a scientifically astonishing temperature of 100° K.

This oxide was a very viable product since it would easily go superconductive in liquid nitrogen which boils at 77° K and is itself fairly user friendly.

The new oxide held the promise of better things to come, and in the of the laboratories world, particularly in semiconductor laboratories, research quickly rose to a fervour pitch. Soon the transition bar went higher still.

These new compounds (which are ceramics) are mostly cuprates.

$Bi_2Sr_2Ca_2Cu_3O_{10}$	110° K.
$Tl_2Ba_2Ca_2Cu_3O_{10}$	127° K.
$Hg_2Ba_2Ca_2Cu_3O_{10}$	134° K.

And the race goes on.

Bednorz and Müller shared the 1987 Nobel Prize in physics.

High Temperature Type IIs changed the face of superconductivity. While their properties are essentially the same as earlier super conductors, the ranges of those properties are off the scale in comparison.

The most radical variation apart from the high level transition temperature is the very high critical magnetic 'field' levels. Some compounds are capable of maintaining their superconductivity in huge magnetic 'fields'. These materials also get braver as their temperature is further lowered, and some when immersed in liquid helium are capable of withstanding magnetic 'fields' of an enormous one million gauss!!

In Type I superconductors, the disturbing external effects of electromagnetism penetrate to depths of several thousand Angstroms (gradually lessening) which explains their vulnerability.

The distance is called the superconducting coherence length or Ginzburg-Landau coherence length (symbol ξ).

In Type II superconductors the superconducting coherence length is a mere 1 to 10 Angstroms, hence they are fairly magnet-proof.

High temperature Type IIs are constructed from planes or sometimes chains of copper and oxygen crystals. They are very anistropic, meaning their properties vary with their orientation to current density and/or invasive magnetism.

Type IIs can be further categorised:

- 3D Most.
- 2D Some, like surface doped sodium tungsten trioxide ($NaWO_3$) or a few organic ones.
- 1D Rare, di-lithium copper dioxide (Li_2CuO_3), or single walled carbon nano-tubes.

Superconducting Electromagnets

Individual loops of conducting coils generate magnetism which affects neighbouring loops. Taken as a whole, the magnetism generated by a superconducting coil attacks its superconductivity. If this magnetism equals or exceeds the critical magnetic 'field' limit H_C, the superconductivity breaks down; (it is quenched). Quenching also occurs if the critical current 'J_C' is exceeded.

The design and construction of superconducting coils is not easy. They are constructing from wires or thin tapes to overcome the brittleness of the material. These may themselves be made from filaments of around 20 microns thick. Most are made of niobium-titanium Nb_3Ti ($T_C = 10°$ K) which stays superconductive up to about 15 teslas. A magnetically better but more user unfriendly and more expensive material is niobium-tin ($T_C = 18.5°$ K) which can be cranked up to 30 teslas. When demanded a combo Nb_3Ti - Nb_3Sn is used.

High temperature superconductors crank it up higher still if demanded. They are also used as connecting leads between the room temperature current source and the super cold super magnets as they limit heat leakage.

The filaments are usually embedded in a copper matrix which:

a) *Provides an emergency low resistance path if quenching occurs.*
b) *Adds stability.*

The coil must be braced against its own Lorentz forces generated between huge co-parallel currents in its windings. These are so strong they tend to crush or squeeze out the insulation between them.

The filaments need to be tiny since in these types of superconductors the current only flows skin deep, so the finer the filaments the greater the total surface area becomes.

Impurity Problems

Impurities have a tendency to migrate to the surface of materials. In superconductivity this interferes substantially with their performance since they increase resistivity, lower the J_C and raise the H_C.

Occasionally an impurity causes a hot spot which spreads in a matter of seconds throughout the windings cancelling superconductivity as it goes. The conductor becomes resistant, heating rapidly and boiling the surrounding cooling liquid. The vapours are hazardous. A loud bang usually accompanies the breakdown.

Most superconducting electro-magnets have safety devices built in to protect them from H_C or J_C overload. Often this is a quench wire which short-circuits the coils.

Magnetic Superconductors

In 1997 the first discovery of magnetic superconductors occurred. This happened when an alloy of gold and indium was cooled to a few millionth of a degree above absolute zero.

In 1999 New Zealand physicist Dr. Jeffery Tallon and his German colleague Dr. Christian Bernhard discovered belatedly that a superconductor of ruthenium and copper, [$RuSr_2(Gd,Eu,Sm)Cu_2O_8$] they had cooked up was magnetic. It had a $T_C = 45°$ K.

Industry had been trying to achieve this result by laminating magnetic material onto superconductors with little success. The sleeping compound contained this layering in its crystal structure.

Forty years earlier the Russian scientist Valery Ginsberg had proved this crystal structure event was impossible so nobody looked for it!

By 2000 they had improved the material to have a $T_C = 58°$ K. They are predicting much higher results.

Super Insulators

The obverse of a superconductor is a super-insulator. Certain materials like thin films of titanium chloride when cooled far below their superconductivity state become super-insulators.

It is thought that the cooper pairs in highly squashed material tend to collect in puddles, interconnected by Josephson junctions. When these become over crowded a current cannot flow therefore the resistance is infinite.

Applications

Applications of superconductivity range from the ultra-sensitive, such as measuring devices, electronic switches, memory storage and other computer accessories, through to max-force mechanisms such as frictionless bearings, large generators, transformers, transmission lines, electromagnets used in maglev trains, medical magnetic resonance imaging etc……

Racetrack Super-Electromagnets

The enormous energies involved in ultra-high-velocity particle accelerators demand huge super magnetic 'fields' to force particles to follow circular paths of around 27 kilometres (CERN) or guide particles down the centre of 60 kilometre straight tubes.

To do this the superconductive electromagnets have been modified to do their most effective work when they are formed into what is called racetrack shapes which stretch along the accelerator tubes.

Some of them are over 10 meters long.

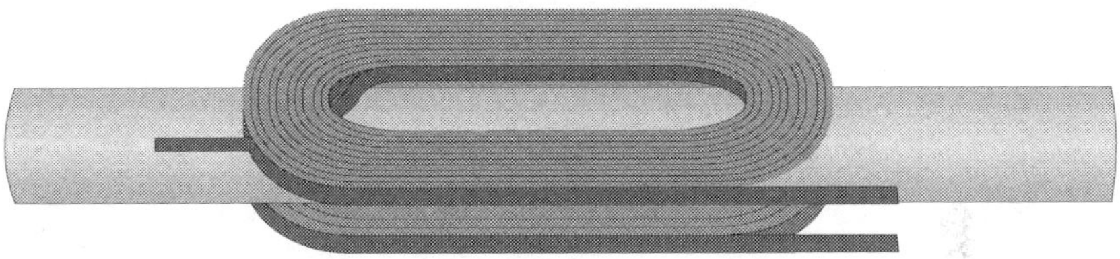

Part II

THE ODYSSEY FROM MAGNETS TO GRAVITY

MAGNETISM

CHAPTER 5

Illusions

Voltaire said; "Those who can make you believe absurdities can make you commit atrocities."

Nature is peppered with illusions and first impressions are their architects. Investigation and time will eventually unravel them, but the beguiled often dogmatically and desperately soldier on, many committing atrocities.

The greatest natural illusion ever, has to be that of the Sun apparently circling the Earth, but the 'magnetism illusion' is just as formidable.

It took 100 years before the establishment completely accepted the Copernican heliocentric solar system.

The Magnetism illusion

The magnetic illusion is a directional one. It has been conjuring for over two and a half millennia.

Early investigations gave the clear impression that something left the front end of a magnetite or lodestone rod and looped around to its rear end.

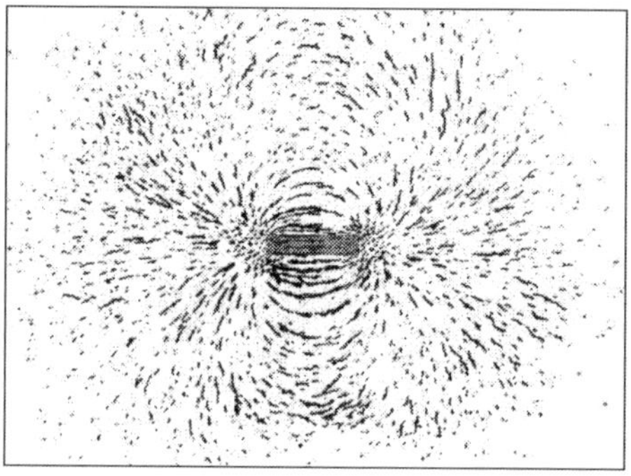

The patterns resulting from iron dust or filings sprinkled on a page stationed above the rod reinforced this impression.

Faraday saw these as evidence of his 'lines of force' which paralleled them. They were in fact only theoretical lines of force which science appropriated to underpin the direction of its 'magnetic flux'.

Because one end of a lodestone rod aligned itself towards the north star, a convention was adopted which decreed that the flux flowed from a magnet's north pole to its south pole.

After centuries of repetition this pseudo direction became gospel.

The Oersted Fork

In 1820, Oersted was preparing for a lecture on electric currents thru a length of wire when he noticed a desk compass deviating when the current was applied.

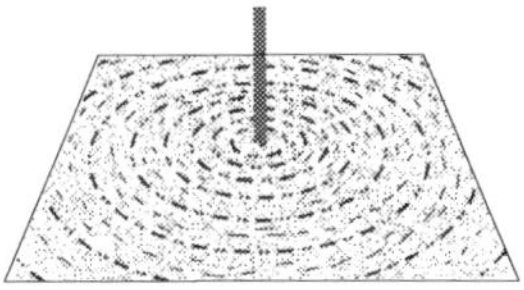

Eureka he cried!; As he manoeuvred it around the conductor and it pointed transverse to the current.

Had there not been an incidental compass on the desk science would have forked into a more logical cosmos; at least for a little while.

B4 long, perhaps he would have tired of tinkering with one conductor and added another one parallel to it.

Eureka!; He would have seen them bend in when the current was co-parallel and spread out when anti-parallel.

Aware of Bernoulli, he would have deduced that the current drags along an ether wind with it. Like winds attract, unlike winds repel.

But alas there was a conjuring compass, and Eureka!; it pointed around the conductor and this became the limb that physics followed, where cause vectors, having perpendicular effect vectors, prevailed over common sense.

Perpendicularitis

Perpendicularitis is the term I coined for a virus that has been infecting science for two centuries.

It is caused by an irrational interpretation of the perpendicular forces operating when moving charges collide with transverse magnetism.

Its symptoms are visible when scientists or students:

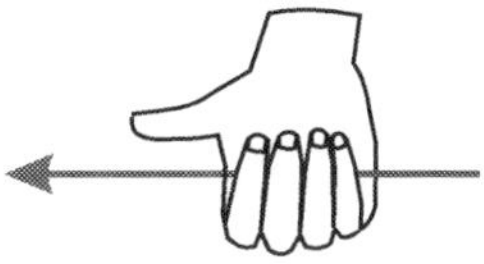

1) Pretend to grasp a wire with their right hand fingers and thumb extended.
   W
2) Contort their right hands with their digits strangely extended.

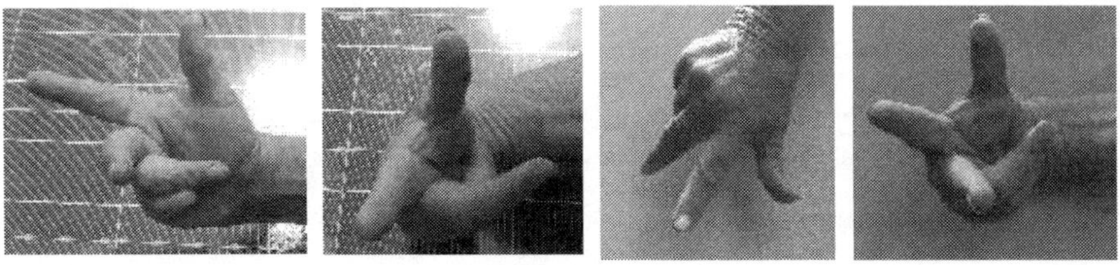

- The index finger points in the direction of the current.
- The middle finger the direction of magnets.
- The thumb the direction of the force

Here science illogically accepts that the intersection of two perpendicular events, namely magnetism and an electric current, generates a force perpendicular to both of them is hard to justify. Moreover if one event reverses its direction the perpendicular force reverses.

Comparable bizarreness is:

- Fly a plane through a crosswind and it gets lifted. Turn through 180° and fly back through the same crosswind and it plummets. Ridiculous!

We know that an arrow does not have a vortex wind circling its shaft, yet ruled by the historic tenets of magnetism, science blindly accepts that a shaft of electricity has a vortex called magnetism circling it. ***Science calls it the 'curl'.***

Fields & Conventional Magnetism

Before 1887 science believed that the *ether*, aka *aether,* was the medium through which magnetic and electrostatic forces and light operated.

In 1887 however Michelson & Morley deduced that Space was a void; there was no *ether*.

Now Space was gutted from empirical *'ether fields'* leaving behind magnetic and electrostatic fields operating without any support or reason.

Consequently:

- Not only was the conventional direction of magnetism circling its generating current mechanically illogical.

but!

- 'Magnetic fields' became abstract and incorporeal. They might as well be called magic spells or Merlins.
- Magnets then, cast abracadabra spells of attraction or repulsion at each other
- The bigger the spell the greater the force.

Obedience to the magnetism illusion and its incorporeal magnetic fields encouraged the acceptance of incorporeal electric fields, thereby impeded inquiry into the intrinsic nature of electrostatic forces.

Science built a sky-scraper on an illusion.

The Fluid Dynamics of Magnetism

- This work holds that magnetic energy flows logically co-parallel to a moving positive charge and anti-parallel to a moving negative charge.
- Co-parallel flows attract, because the magnetism between them requires less work (reducing magnetic pressure).than beyond them.
- Anti-parallel flows repel because the magnetism between them requires more work (increasing magnetic pressure).than beyond them.

Flow Types and Terms

Just as there are differences in conventional 'fields', so too are there differences in the types of electromagnetic flow.

- The conventional magnetic field is redefined as magnetic-energy-flow abbreviated to me-flow.
- The conventional electric field (i.e. the electrostatic field) is redefined as electric-energy-flow abbreviated to ee-flow.

A Geometric verses a Mathematical Universe

Post Keppler orbits, scientific practice has concentrated on to interpreting events mathematically, often loosing sight of the geometric significance of formulae. This work will try to rectify this impiety wherever possible.

For example in a mathematical universe, the strength of companion magnetism at any point is inversely proportional to the distance from its host current. *(a one dimensional image)*

In a geometric universe strength of the companion magnetism at a distance *'d'* from its host current is inversely proportionally to the area it spreads over, divided by the distance. *(A two dimensional image.)*

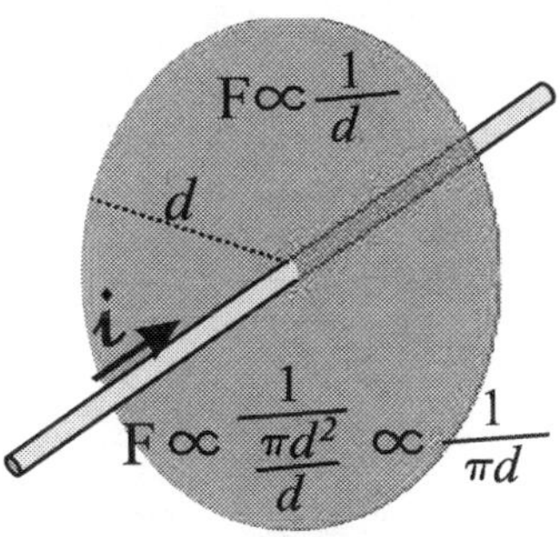

This perceives Space as having a property. It acquires embodiment.

Me-flow Streamlines

Streamlines in the following diagrams are used to graphically represent me-flow.
Crowding or thinning means respectively higher or lower energy flow.

Single Conductor

An electric current, which is by convention flowing from a positive source to a negative one, moves co-parallel to its omnipresent magnetic energy flow.

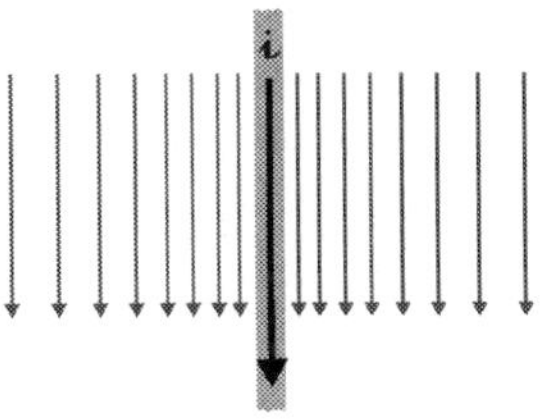

Two Conductors

When electric currents are co-parallel (in parallel conductors) their respective me-flows combine increasing the attraction between and around them.

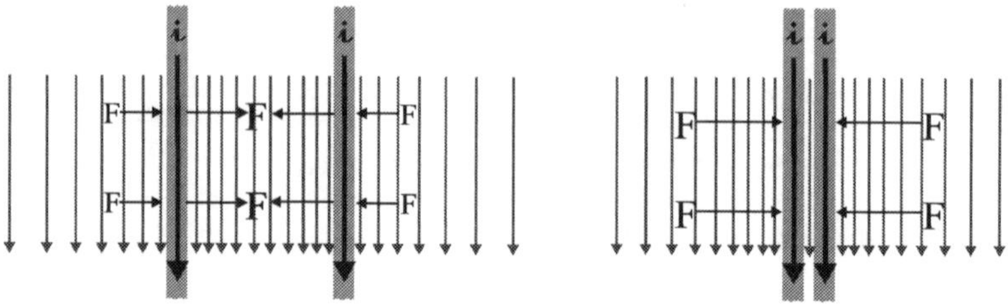

The closer the conductors get the greater the squeezing force becomes. This necessarily implies that if the conductors are touching, the external me-flow keeps pushing.

Therefore the electric current and its xcompanion me-flow are always transversely pressurised.

When electric currents are anti-parallel, (in parallel conductors), their respective me-flows repel each other.

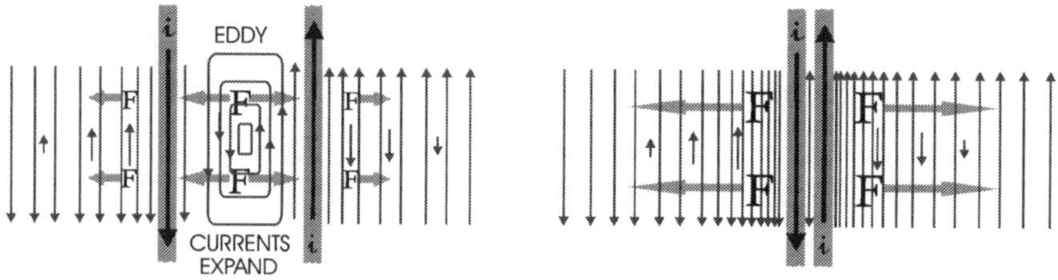

The conductors are internally pushed and externally pulled apart. The closer the conductors get, the greater the repulsive force becomes.

This necessarily implies that if the conductors are touching, the external anti-parallel me-flow keeps trying to rip them apart.

2D magnetism

The force between oblique straight conductors is:

$$F \propto d\ i\ \cos\theta \text{ ....when } \theta = 90°,\ F = 0$$

Bi-verse theory explores the phenomena that perpendicular conductors don't effect each other, *(which science can't explain).*

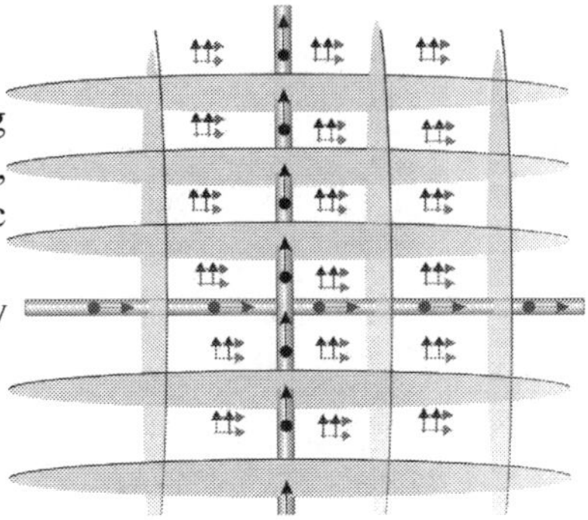

It suggests that radiation, after being generated by a conductor current (aerial), proceeds thereafter as purely 2D magnetic phenomenon.

Perpendicular 2D energies are mutually transparent.

Circular Conductors

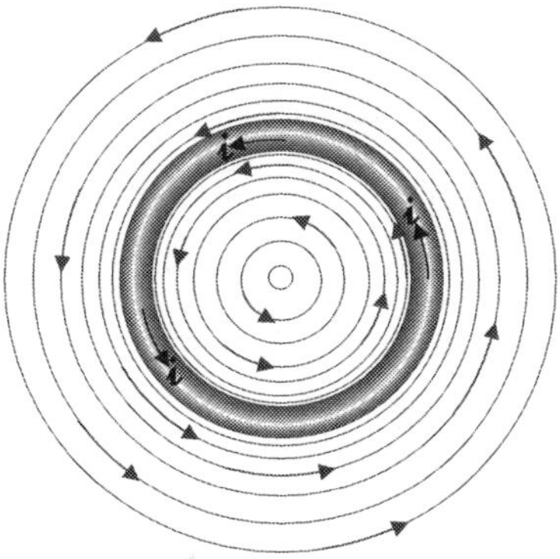

When a current flows around a ring the me-flow rotates the same way generating a vortex which circles around the axis of the ring.

The strength/energy of the internal and external vortex declines with distance from the ring.

Cylindrical current coils have stronger internal vortexes and weaker external vortexes surrounding them.

Go to Appendix 1: for detailed me-flow analyses

Renaming the Poles

- The face of the north pole of a cylindrical coil presents an anti-clockwise flow of current and an anti-clockwise me-flow, both flowing in the opposite direction to electrons in the conductor.
- A bar magnet's north pole has anti-clockwise me-flow face.

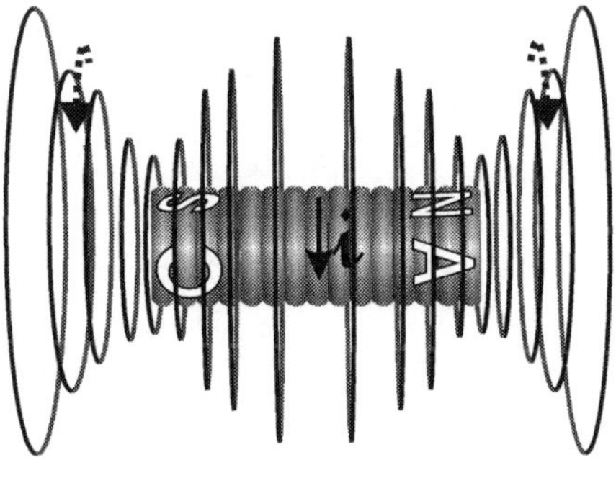

N-pole face ⟶ Anti-clockwise magnetic-energy-pole ⟶ A-pole
Anti-clockwise current -pole
S-pole face ⟶ Clockwise magnetic-energy-pole ⟶ C-pole
Clockwise current -pole

Redefined magnetism shows that poles with:

- Like spinning vortexes attract.

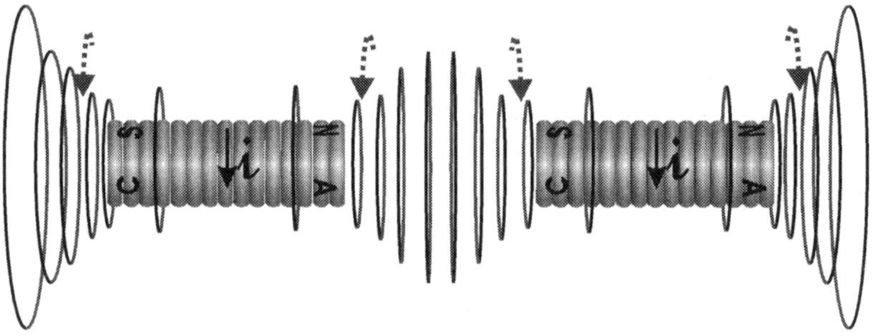

- Opposite spinning vortexes repel.

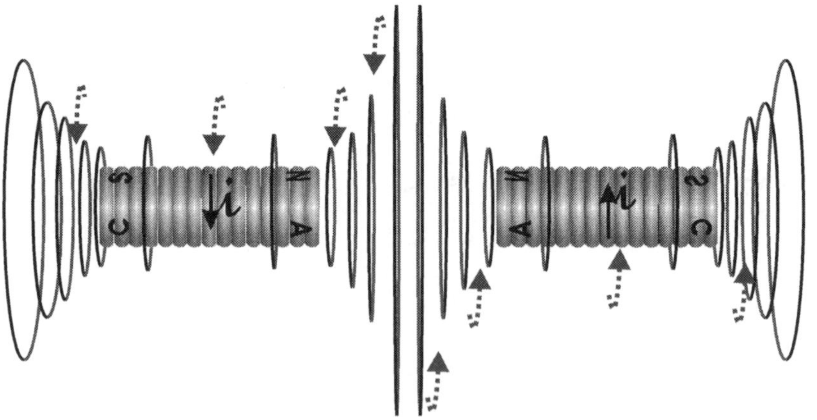

Co-parallel Cylindrical Coils

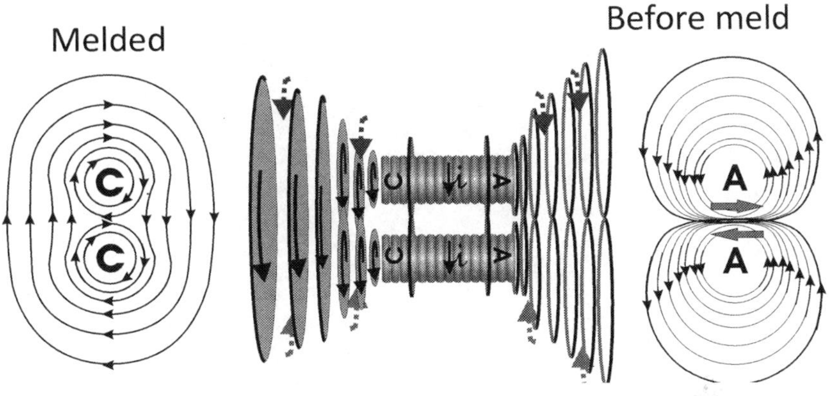

The vortexes emerging from the poles of co-parallel cylindrical coils Meld completely after a short distance

The filling map of two anti-parallel magnets shows me-flow vortexes meld after a short distance.
melding progressively

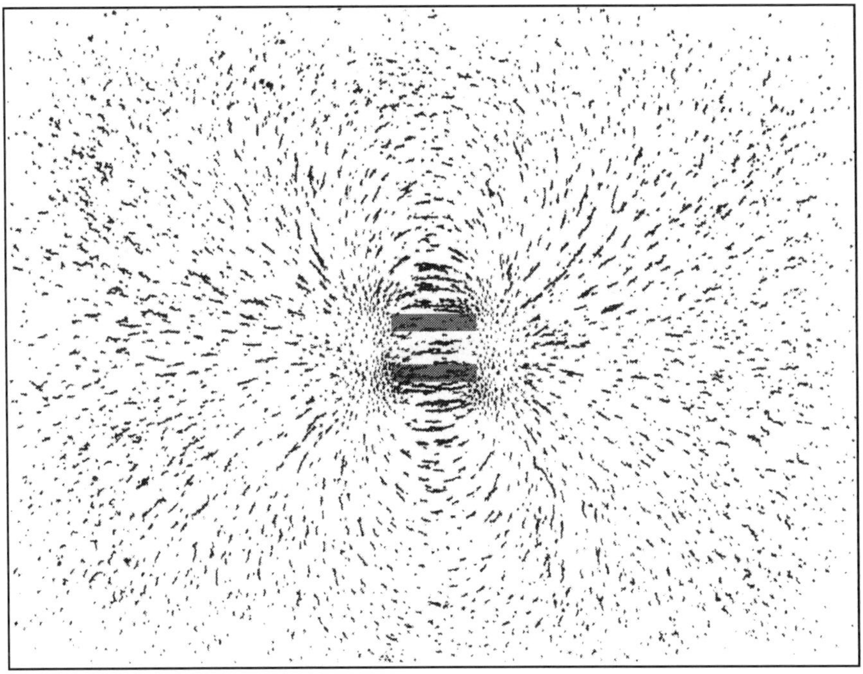

Melding process

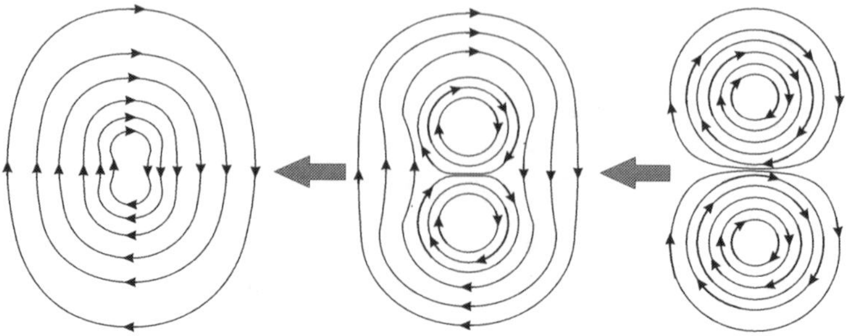

Magnets

- The orbits of electrons in the atoms and domains of magnetised material drive magnetic vortexes in the reverse direction.
- Those with aligned axes combine to form wobbly columns something like tyre stacks.

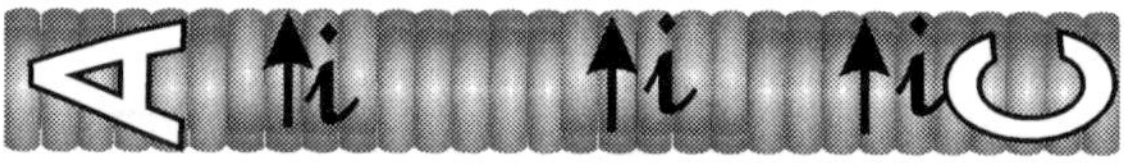

Magnets are comprised of a multitude of independent columns of mini- vortexes or vortlets.

- They behave like multi-co-parallel cylindrical coil vortexes.
- Each vortlet repels its neighbour because their respective domain me-flows oppose at their tangent contact zones.
- Within the magnet these vortlet columns are constrained by the atomic structure (coulomb forces) of the metal lattice.

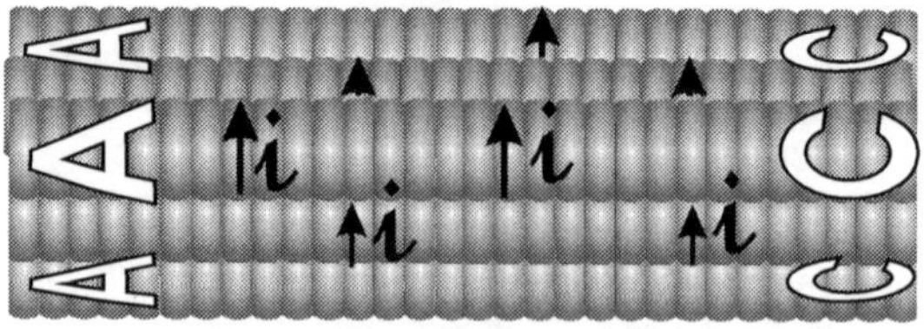

- Thousands of these vortlets emerge from the poles.
- Once free of their host's constraints they inflate rapidly and meld into a large vortex some of which feeds back outside the magnet and merging in the middle (which necessarily implies that space is viscous.)

Building Fining Patterns

When an filing are drops onto a page above a magnet it gets magnetised as it fall becoming mini magnets itself.

- It now has its own mini vortlet turbulence which blends into and augments the major vortex turbulence.

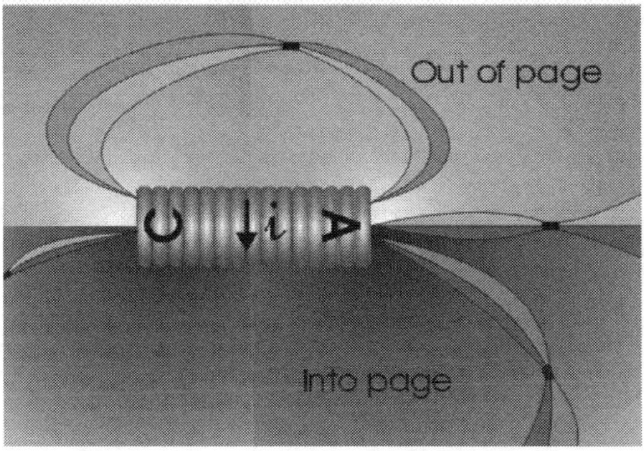

- The next filing either forms another individual vortlet ,or drops into the turbulence of the first filing vortlet starting a filing chain
- Chains generate vortlet streams
- Neighboring vortlet streams *(same spin)* repel each other
- Aligned iron filings form in the eyes of vortlet tunnels like Helmholtz filaments.
- ***Note there are no filament tunnels or lines of force without the presence of iron filings***

Vortex experimenton the above diagram.

- Generate a major impeller vortex in the coil/tube.
- Substitute battery mini-impellers for in situ iron filings

When the major vortex wind, turns them slowly;

- It motion chip-starts the battery power to the impeller's mini motors.
- These now generate their own mini-vortexes or vortlets;
- Colored smoke reveals all.

Iron fillings align along the eyes of the vortlets they create.

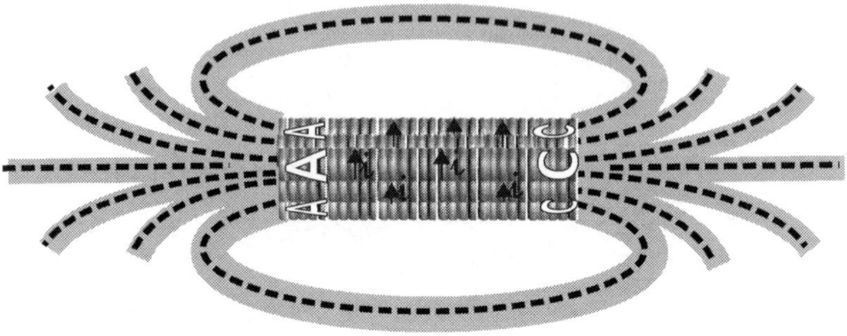

Pairs of reversed magnets interlink clockwise and anticlockwise poles with magnetic slinkies

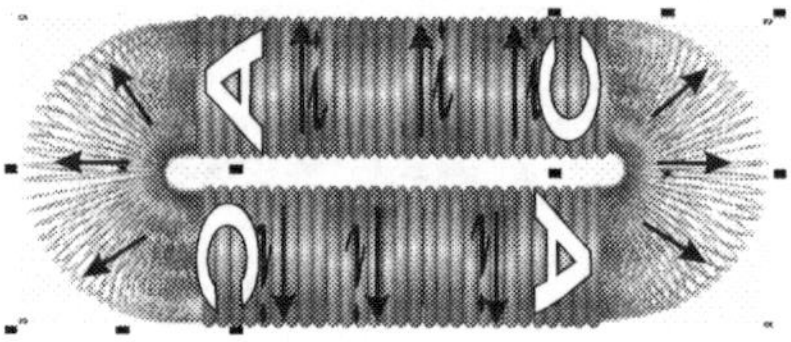

This chapter has shown that magnetism can be interpreted as a simple form of Bernoulli fluid dynamic

Though perpendicularitis is spooky, science has worked with it over two or three centuries and created many ingenious inventions.

Now, cured of perpendicularitis the next logical step is to discover.
What are the vortexes/vortlets spinning???

Go to Appendix 2: for more filing maps analyses.

COULOMB FORCES

CHAPTER 6

Conventional Electric Attraction and Repulsion

In the late 19th century science substituted the word 'fields' for the old expression 'action at a distance' to explain paranormal interaction between electric charges.

With the arrival of the theoretical photon scientists found suppositional explanation and, in the absence of the *ether,* expounded a truly masterpiece of absurdity.

Their analogy of repulsion between like charges, has one person standing on ice throwing a heavy ball to another person, also on ice, who catches the ball. Both are forced apart.

They 'reason' that an electron repels another electron by firing a theoretical photon at it. The shooter recoils while the target gets pushed back.

One wonders:

- Who fires the first shot?
- Do they take it in turns?
- Do they have machine guns?
- What guidance system do they use that enables them to lock on to their respective targets when they are each traveling at say 90% the speed of light in different directions?
- Do the theoretical photons from each collide mid way or there abouts?
- Do the theoretical photons pass through each other en route?
- According to Heisenberg each electron in an atom can only have a probable location at a specific time. So how can the gunner sergeants on the electrons be such dead-eye-dicks? Are they omnipotent guessers?

**The mind boggles under the raft of improbabilities.*

The conventional explanation for the nuclear force of attraction between a proton and a neutron is even more bizarre. Here the nucleons play pion yo-yo.

- It is believed that one particle throws a pion to the other but the attached string pulls the thrower after it.
- If the yo-yoing is fast enough it would appear that the particles are being pulled toward one another. "brilliant!?!?"

The First Law Of Thermodynamics

The first law of thermodynamics first proposed by Gabrielle Émilie Le Tonnelier de Breteuil, Marquise du Châtelet in the early 18th century. is 'energy can be changed from one form to another, but it cannot be created or destroyed'.

To say Charges have intrinsic Coulomb energy **per se** violates this law.

- ***Charges are not divine.***
- Coulomb energy does not begin at a zero point and perish at another, it passes thru them.
- Since it can't be intrinsic anywhere else en route, it must somehow circulate.
- It travels from positive to negative then somehow/way returns to positive thru some other set of mirror dimensions ;
- Either thru another set of Spatial dimension I call 'the antiverse' where time is mutual.
  or thru the same dimensions in a different time zone; 'the anti-time-verse' or thru both/other, like Brane theories' predicted fourth dimension.

Science doesn't say what <u>form</u> Coulomb or electrostatic energy was in **before** it becomes the so called 'electric fields', but whatever it is/was, it must travel thru the charge, which defines charges as being inter-dimensional energy portals.

Each 'mirror-verse' is the domain of anti-charge not antimatter.

This work chooses to work with the antiverse option for the mirror-verse; Here:

- Coulomb energy circulates thru two sets of three spatial dimensions.
- A particle or event in the "here-verse" is its anti-particle or anti-event in the antiverse
- An electron 'here' is a positron 'there' and vice versa.
- Particles/events are linked at the inter-dimensional equator.
- They are joined/entangled at a timeless hip.
- ***Coulomb bi-verse kinetic energy is zero.***

This work speculates that since coulomb energy circulates, then all kinetic energy circulates.

"Energy can neither be created nor destroyed; it always circulates."

SWER

The single wire earth return analogy lends weight to the inter-dimensional possibility of coulomb energy.

SWER was pioneered in New Zealand in the 1920's and then employed extensively (110,000km) in outback Australia where deep earth rods are used to bypass the very dry non conductive surface strata. After work the current tunnels back to its source via the ground or more directly via water-ways.

Detectable electrostatic energy must somehow theoretically perform similarly, even though this is 'invisible' to all and incomprehensible to conventional Alice in quantum-land.

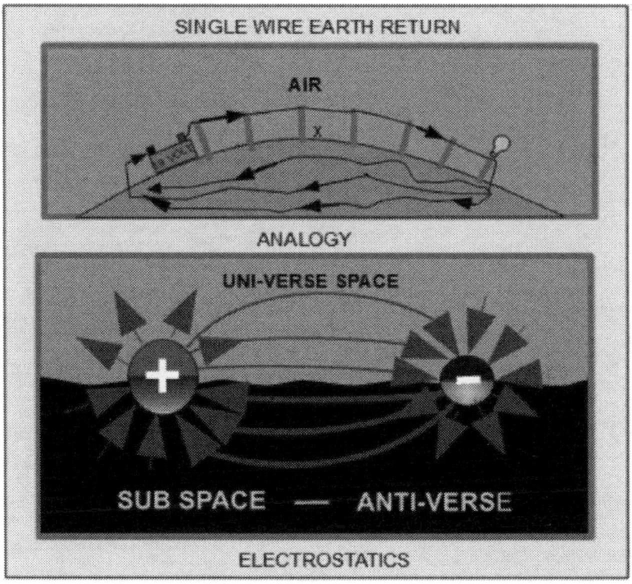

Fluid Maps and Ee-flow

A fluid map is made when water flows over a flat plaster surface previously impregnated with dye which changes colour when wetted. A heavy sheet of glass rests on the plaster.

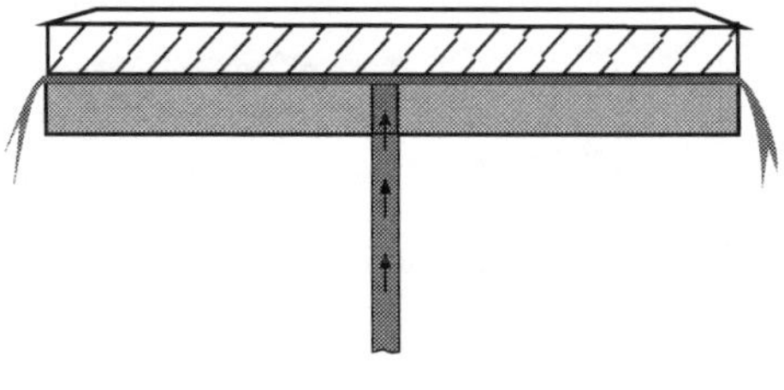

When water is pumped through the central inlet it flows outwards through the sandwich. It flows along a wetted surface much easier than it does across a dry one, so as soon as these wetted avenues get started, the water bursts forth as clear expanding spokes of radial flow, separated by expanding spokes of dry no flow.

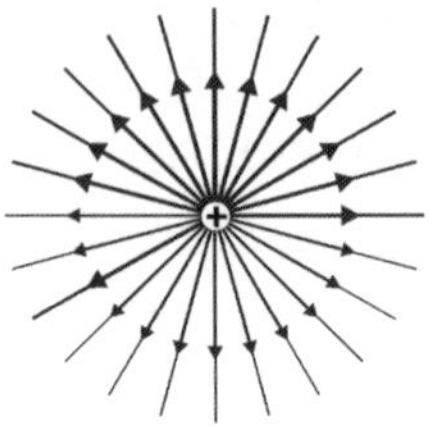

2D representation of a positive pole

Stylised graphic of a pump fluid map

For a pump-sink map the perimeter of the experiment is submerged in a holding tank. Water is pumped through the inlet and sucked out with equal force through the sink hole a short distance away.

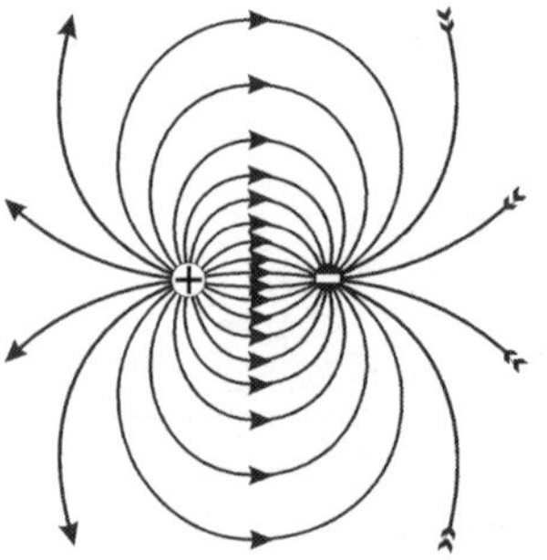

2D representation of electric dipole

Fluid map pressure inlet and vacuum outlet

Coulomb energy cycles continually thru a mirror-verse e.g. the antiverse

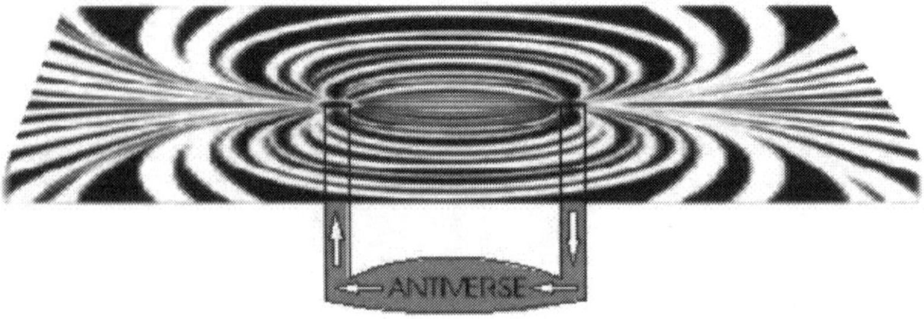

Geysers and Sinks

- An ee-flow geyser like a proton, radiates a measured amount of coulomb energy per nanosecond.
- *Because its charge is constant, irrespective of external events or its velocity, the proton must expel this and only this amount per nanosecond. It is its raison d'être.*

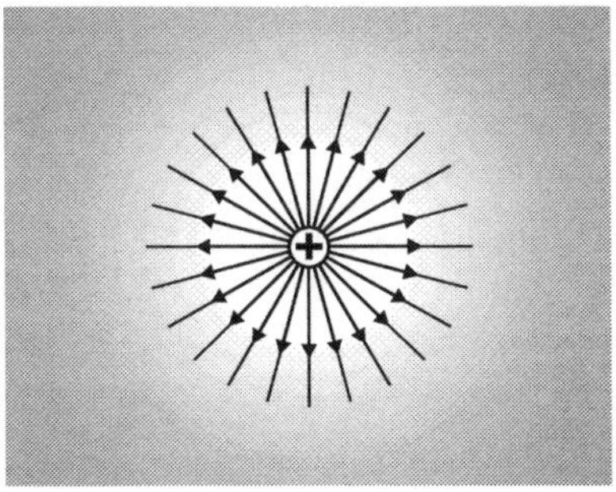

- An ee-flow sink like an electron, voids the equivalent amount per nanosecond
- Because its charge is constant, irrespective of external events or its velocity, the electron must ingest this and only this amount per nanosecond. It is its raison d'être.

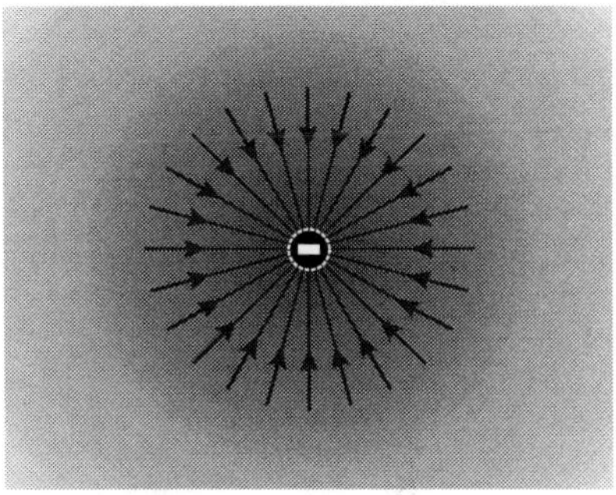

- The electron is so minute that it is thought to be a point charge. Thus to ingest Space-energy, Space-energy itself must be pure energy with no spatial dimensions.

Attraction and Repulsion of Geysers and Sinks

The statement 'like charges, repel one another and unlike charges attract one another' needs clarification

'ee-flow geysers' repel each other because the increased pressure between them is greater than the far-side pressure.

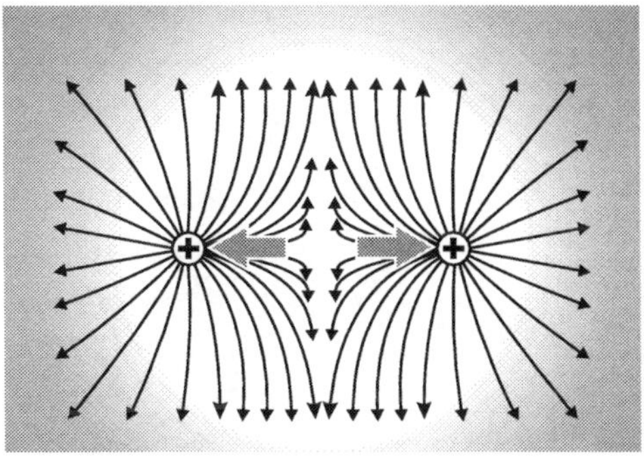

'ee-flow sinks' draw themselves toward the far-side happy gobbling ground and pull away from the nearside zone which they have depleted.

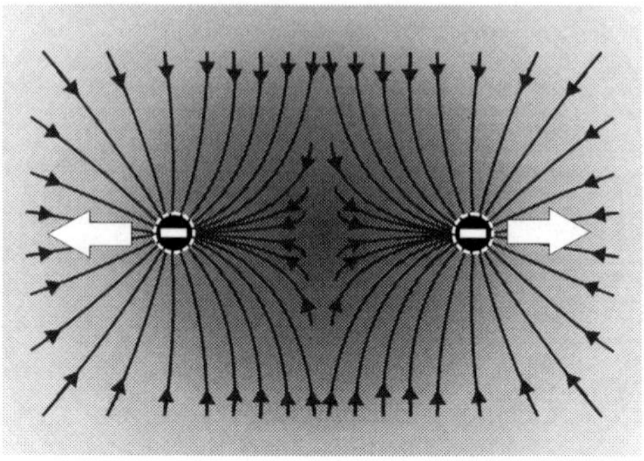

The action of sinks pulling themselves apart is not the same as sinks repelling one another. The latter falsely implies they are pushing themselves away from each other.

ee-flow geysers and sinks' attract each other because:

- An ee-flow geyser pushes against far-side ambient pressure and is propelled towards an ee-flow sink where the pressure between them is reduced.
- Conversely an ee-flow sink hoovers into an ee-flow geyser where the high pressure provides easier pickings than those available on the far side of the sink.

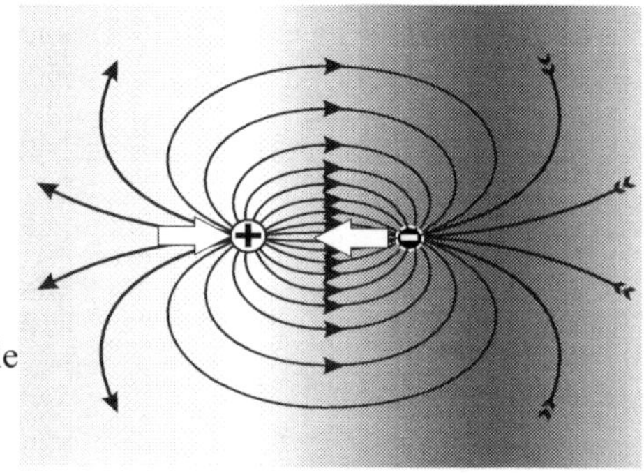

Geysers propel themselves towards sinks.
Sinks draw themselves towards the geysers.

Thus the rules for the forces of interaction between charged objects in the 'here-verse' become refined and therefore are redefined as:

1. Positive charged objects push one another apart.
2. Negative charged objects pull themselves away from one another.
3. Positive charged objects propel themselves toward negative charged ones.
4. Negative charged objects pull themselves towards positive charged ones.

In a so called 'electric field' between two oppositely charged plates:

- **A positive charge will push itself away from the positive plate.**
- **A negative charge will draw itself towards the positive plate.**

Coulomb Energy in the Antiverse

This work so far has described coulomb events in the here-side of the theoretical bi-verse.

In the antiverse option:

- A positive charge in the here-side (i.e. the 'here-verse' spouting out energy, is simultaneously drawing the same amount of energy through its portal from the antiverse.
- Similarly a negative charge in the 'here-verse' is simultaneously spouting out the same amount of energy into the antiverse.

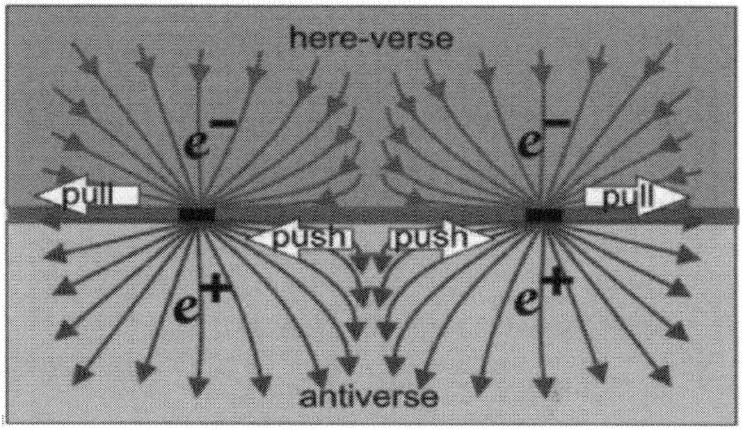

Positive charges in the 'here-verse'
become negative charges in the antiverse
and negative charges in the 'here-verse'
become positive charges, (albeit anti-negative-charges) in the antiverse

They are joined at the inter-dimensional equator/ hip.

This inter-dimensional function nicely resolves questions about how point sources vacuuming energy from each other is absolutely equivalent to sources spouting the same amount at each other.

Parallel Plates Illusion

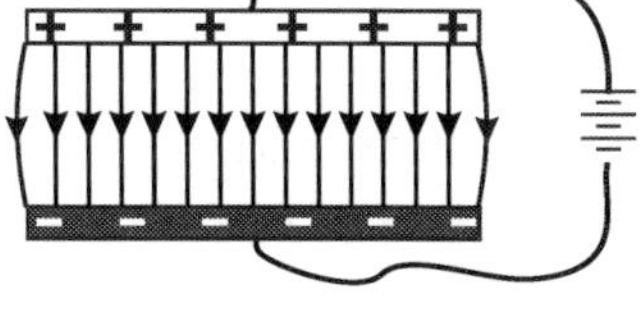

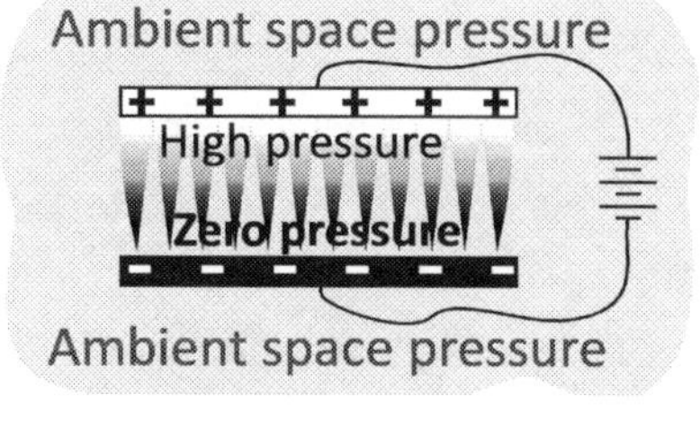

Graphics of streamlines are devious things. The sure straight streamlines in the conventional image of an 'electric field' flowing from a positive plate to negative one, carry no hint of any variations in pressure between the plates.

This is the 'parallel plate graphics illusion'.

The absence of pressure variation data in the lines in the left hand graphic, subliminally implies there isn't any *Space/ether* pressure.

The fluid dynamic model on the right shows that between the oppositely charged parallel plates, *electric energy* flows from a high pressure positive plate to zero or near zero pressure negative plate.

Inverse Square Law

The force of attraction or repulsion between charges is described by the inverse square law or rule.

$$F = \frac{1}{4\pi\varepsilon_0}\frac{q_1 q_2}{d^2} \quad \text{(where q = charge and } \frac{1}{4\pi\varepsilon_0} = \text{ a constant)}$$

This '*d squared*' type calculation has no spatial identity. It creates the illusion that a 3-dimensional event can happily be described by a 2-dimensional formula.

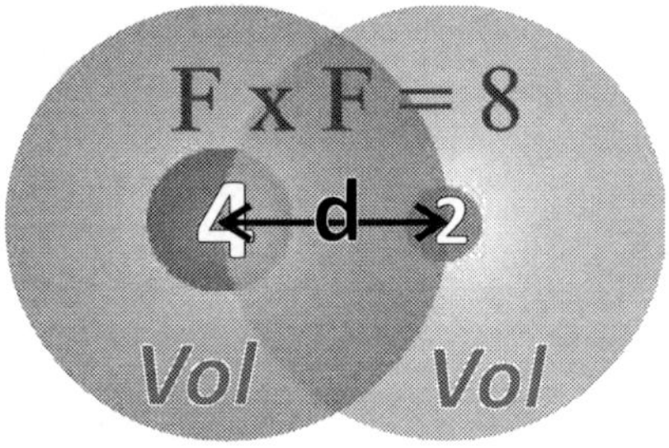

This type of mathematical illusion deludes the punter into believing there likely is a fundamental equation which describes the universe. *(X = 42 perhaps?)*

Sensibly, if the '*d squared*' is expressed as a volume '*d cubed*' divided by length '*d*' then we see the interaction of objects in three dimensional Space, separated by the distance '*d*'.

$$F \propto 1/\tfrac{V}{d} \propto 1/\tfrac{d^3}{d} \propto 1/d^2$$

By changing the constant in the inverse square formulae, the force between two charges can be expressed as:

$$F_q = C_x \frac{q_1 q_2}{\frac{V_x}{d}} \quad \text{(where } C_x \text{ is the new constant relating to some volume } V_x\text{)}$$

This 3D interpretation gives Space embodiment.

Me-flow verses Ee-flow

The essential difference between me-flow and ee-flow is pressure.

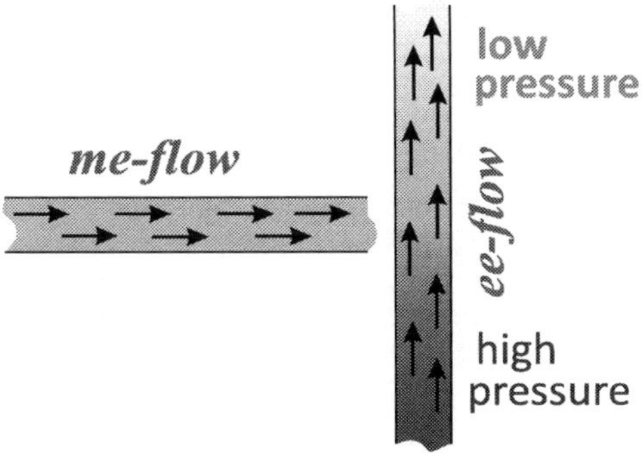

If a frictionless test object were placed in me-flow the pressure at its head would always be equal to the pressure at its tail.

- *Ignoring wall friction, this is analogous to pressures associated with a medium flowing at a constant speed through a pipe parallel to the Earth.*

If a frictionless test object were placed in ee-flow the pressure at its head would always be greater than the pressure at its tail.

- *Ignoring wall friction, this is analogous to pressures associated with a medium flowing at a constant speed up a pipe vertical to the Earth.*

Increasing me-flow from A to B is equivalent to ee-flow from A to B.
Decreasing me-flow from A to B is equivalent to ee-flow from B to A.

Charges in me-flow

- A positive charge in free Space spouts Space energy in all directions.
- It transports this energy from the antiverse into the 'here-verse'.
- When the positive charge moves through Space or Space moves past it, to stay put against the me-flow it reduce its head-on output and increases its tail-output or thrust.

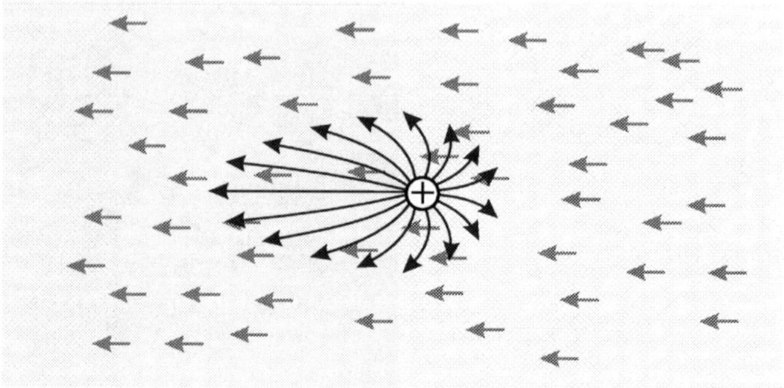

- Conversely negative charges intake Space energy from all directions.
- It transports this energy to the antiverse from the ‘here-verse’.
- When the negative charge moves through Space or Space moves past it, to stay put against the me-flow, it increases its head-on intake and reduces its tail-intake.

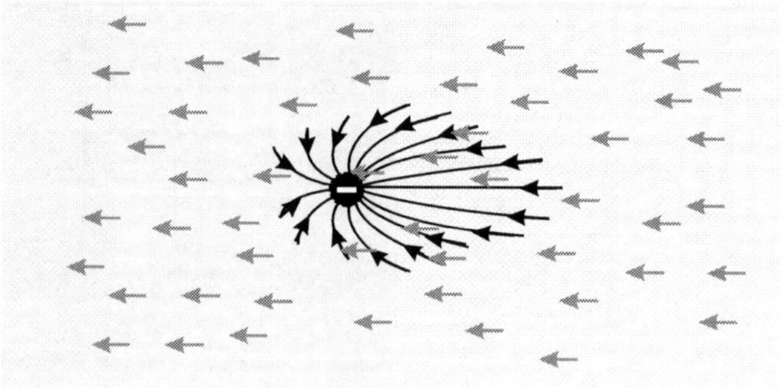

- The greater the energy of the me-flow the greater these adjustments are.
- Because ‘me-flow’ is a steady flow the pressure fore and aft of a charge is constant, so stationary adjusted charges are unmoved when submerged in it.

Charges in ee-flow

In ee-flow, the fore pressure is always greater than the aft pressure.

Positive Charges:

- A proton sandwiched anywhere between two oppositely charged plates always feels the ee-flow coming on at higher than ambient pressure but departing behind it at a lower pressure.
- It gets pushed downstream.

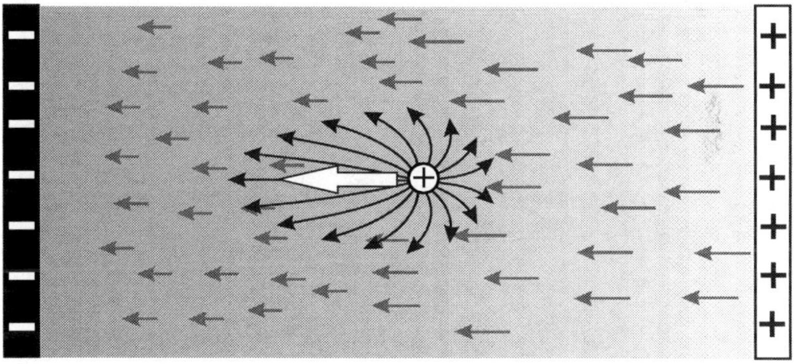

Negative Charges:

- An electron sandwiched anywhere between two oppositely charged plates always feels the ee-flow coming on at higher than ambient pressure but departing behind it at a lower pressure.
- It's suction draws it upstream.

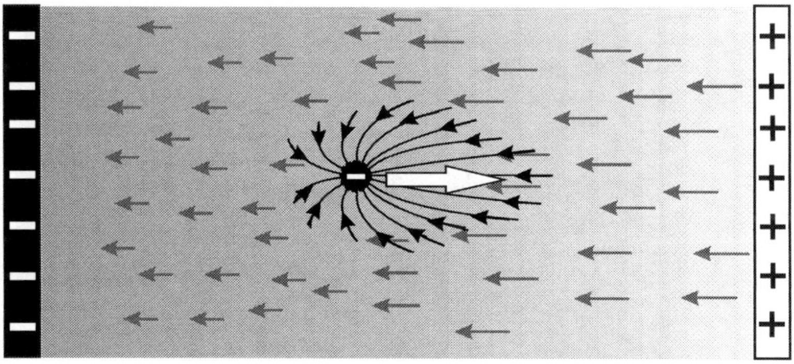

- Positive charges repel each other in the 'here-verse' while simultaneously their flip-side negative charges in the antiverse draw themselves apart, and vice versa.
- Mutually attracting positive and negative charges in the 'here-verse' are simultaneously mutually attracting their flip-side charges in the antiverse.

This combo of push and pull forces is what enables the absolutely equivalent fluid dynamic forces of positive and negative charges.

Charges are inter-dimentional dipole portals

Coulomb Winds in The Antiverse

- Coulomb winds between charges in the 'here-verse' flow the opposite way to their flip-side coulomb winds in the antiverse.

It was a choice to describe positive charges in the 'here-verse' as out-flowing energy and negative charges as in-flowing energy.

In the antiverse the reverse occurs. There the nucleus is negative sucking or vacuuming in orbiting positrons.

The total bi-verse Coulomb wind velocity is zero.

The total biverse magnetic wind velocity is zero.

The antiverse is the domain of the anti-charge-not antimatter.

This work continues mainly describing events in the 'here-verse', rather than always reiterating the flip-side parallel activity in the antiverse.

COULOMB FORCES IN ME-FLOW

CHAPTER 7

Moving charged particles can be deflected by so called 'electric fields' and 'magnetic fields'. One place to observe the latter is in a cloud chamber sandwiched between two identical strong electromagnets. These can have cylindrical iron cores of over 300mm diameter with thousands of turns or windings. There is in the main, a 'uniform magnetic field' throughout the gap.

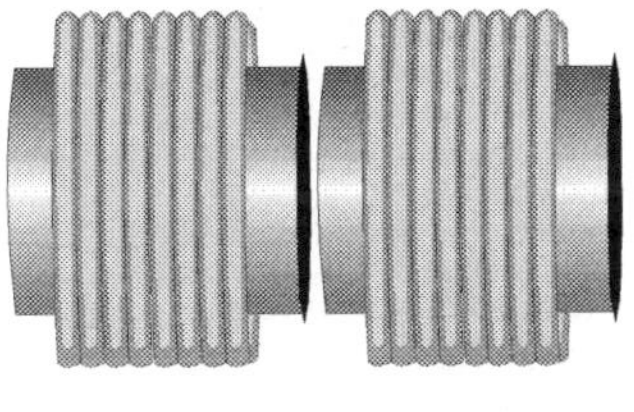

The cloud chamber was invented in 1911 by Scottish physicist and Nobel Prize winner Charles T. R. Wilson to study the properties of rain. Rain occurs when water vapour condenses on particles of dust forming droplets. A modern cloud chamber is an air tight container holding a gas which is cooled to super saturation. Here, instead of dust particles, the passage of charged particles ionises atoms or molecules of the gas which in turn cause other molecules to condense around them. The result is a trail of droplets (vapour trails), visible under lights.

All moving charged particles have unique signature trails. A 'uniform magnetic field' is essential for their identification.

The force on a particle of charge '*q*' and velocity '*v*' within a uniform magnetic field of strength β is given by:

$F = qv \times \beta$ (where '×' is the multiplier and the right hand screw vector operator which handles perpendicularitis.)

When $v = 0$ $F = 0$

So stationary charged particles are not affected by magnetism.

The cloud chamber today has largely been replaced by a superior detector called the bubble chamber, invented by American physicist Donald Glaser in 1953. This contains a liquefied gas such as liquid hydrogen which is brought under pressure above its boiling point. Here the high speed particles cause the adjacent molecules to boil leaving 'steam trails'

Conventional Model vs Fluid Dynamic Model

Conventional theory maintains that:

- 'Theoretical lines of force' flow through a cylindrical coil then radiate out and around it.
- Inside a well wound cylindrical coil and towards the middle of its length the 'magnetic field' is said to be uniform.
- Here the 'theoretical lines of force' run parallel to, and are drawn equidistant from each other.
- They theoretically are parallel, but since the magnetism is strongest closer the coil, they should not be shown as being equidistant from each other.
- The correct representation should show crowding near the coil.

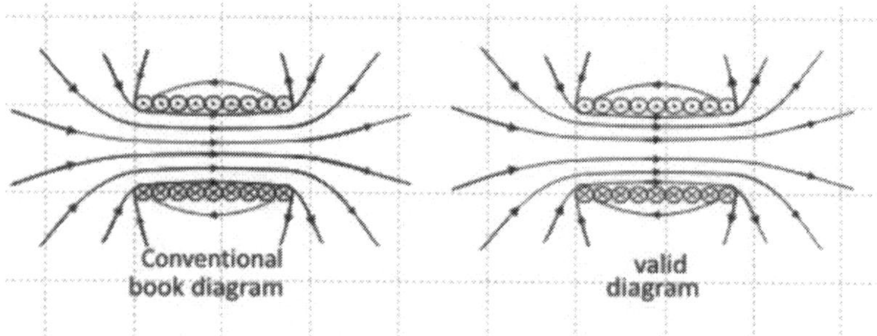

Stand a cylindrical coil upright and slip a page in midway.
Sprinkle filings down inside the coil onto page
Turn on the current and percuss.
The filings crowd towards the perimeter.

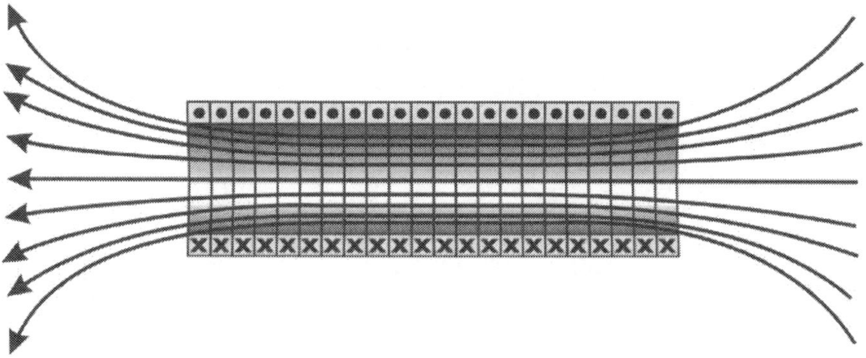

Fluid Dynamic Theory maintains that:

- Inside a well wound cylindrical coil there is a column of me-flow which rotates in the same direction as the current.

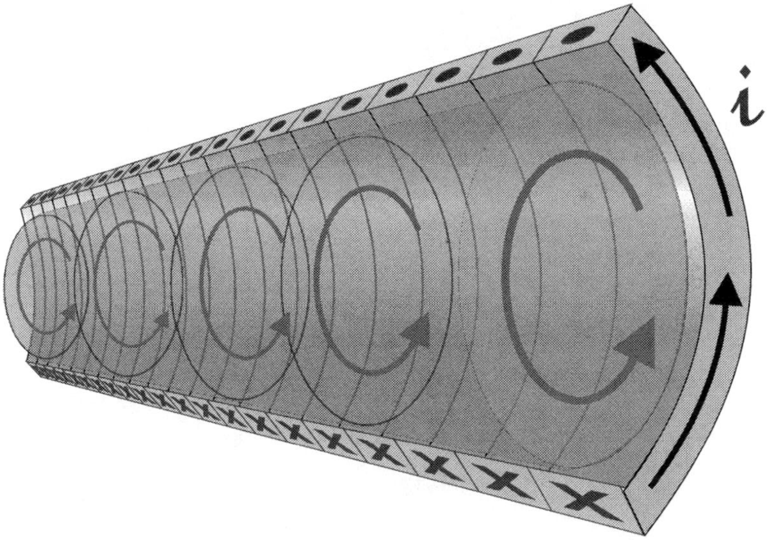

- Towards the middle of its length, every cross-section is the same. In this respect they are uniform.
- Magnetisable objects placed here are not drawn along the axis towards the poles.
- However because the strength of magnetism in coreless cylinders wanes with distance from the perimeter, objects are drawn radially outwards.
- The near impossible exception is if the objects centre of magnetism is balancing precisely on the column's axis.

Uniforms aint Uniform

Conventional physics applies the word 'uniform' to the magnetism inside a coreless cylindrical coil because the theoretical lines of force run parallel to each other.

It applies the same word 'uniform' to the magnetism in the gap between two matching cored cylindrical coils carrying the same current.

Both the above have parallel theoretical magnetic lines of force but each does very different things to charge particles moving through gaps between them. The diagrams below show the force on a charged particle, ***here a positron,*** passing through a gap between two current carrying coreless cylindrical coils.

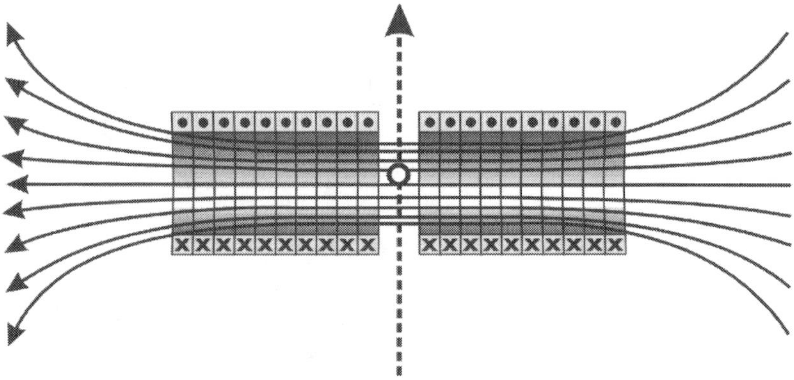

The perpendicular force on the charge becomes apparent when we look at Fluid Dynamics modelling where it crosses through the internal magnetic vortex.

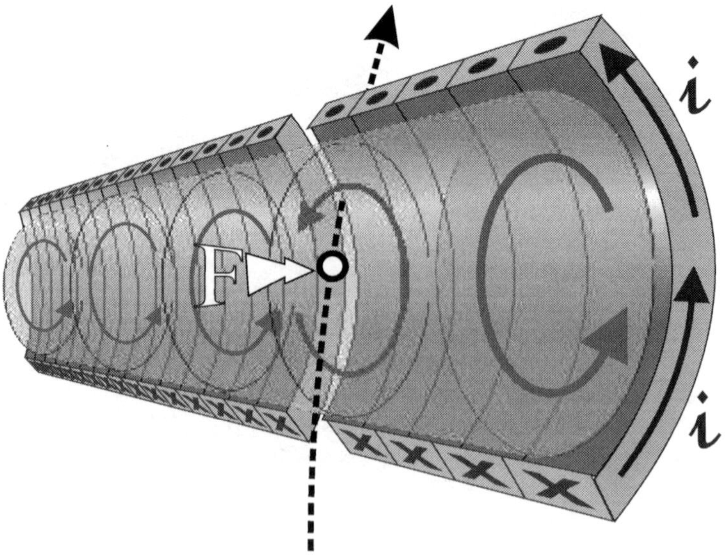

A moving charge has its own pressure zone wind of me-flow surrounding it. For simplicity this is represented as parallel wings.

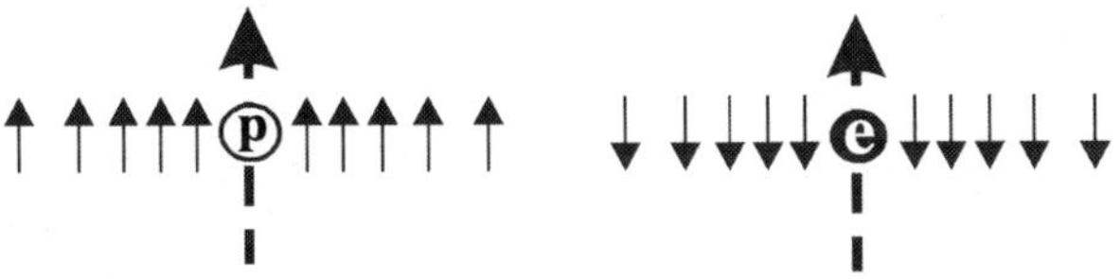

The force on the positron:

- Is initially zero as it enters the vortex, because its wings move perpendicular to the vortex wind.
- It then rises to maximum (perpendicular to motion) in the central region where its wings move more parallel to the magnetic wind of the circulating current.*(co-parallel attract, anti parallel repel)*
- Thereafter it falls back to near zero as it exits on the far side.
- The path is almost parabolic not helical.

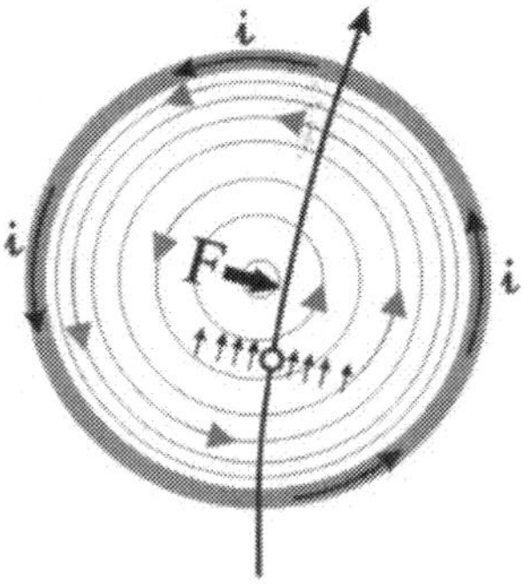

Coreless cylindrical coils <u>do not</u> impose a uniform force on a moving charge.

In the below diagrams
a conductor current flows though a magnetic vortex.

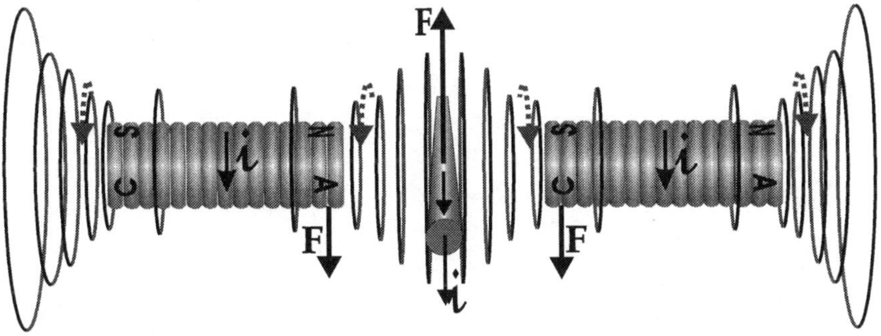

Only the parallel-to-conductor vectors 'Ve' of the coil's me-flow vortex operate.

- Co-parallel flows attract the straight conductor.
- Anti-parallel flows repel the straight conductor.
- When the current reverses the forces reverse.

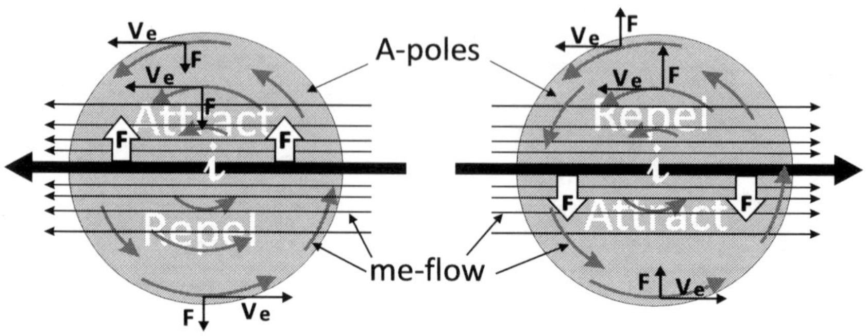

no illusion, no right hand rule, no fingers or thumbs, no perpendicularitis!

Aligned Multi Coreless Cylindrical Coils

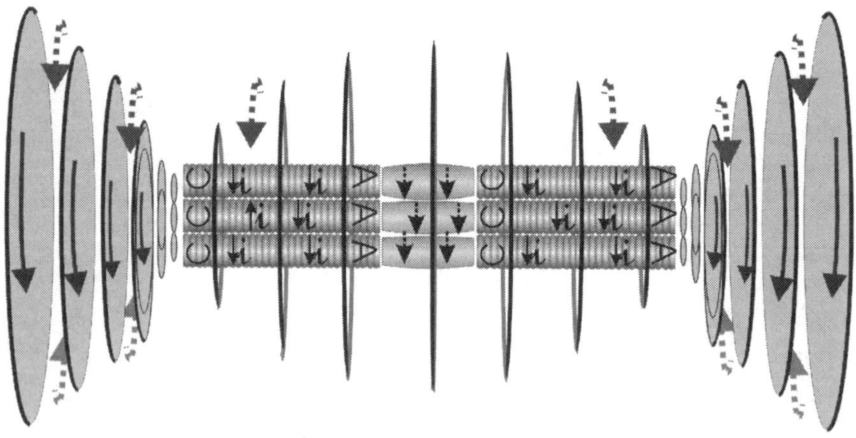

The far-side me-flow from the poles is shown here as being melded. Between the cylindrical coils:

- The vortexes unite across the gap with limited inflation.
- The shorter the gap is, the less they inflate.
- The more central the vortex is, the less it inflates.

Cored Cylindrical coils and Magnets

The presence of a core increases coreless magnetism by as much as 10,000 fold.

In order to noticeably bend the paths of charged particles, a coreless cylindrical coil would have to carry a current thousands of times greater than that required by a cored cylindrical coil.

The vortlets of two aligned magnets or electromagnets with the same spin interlink across a gap without merging.

When a charge traverses the gap:

- Its wind or wings of magnetism pass thru many vortlets, each of which simultaneously pushes and pulls at the passing wind.
- The sum of these vortlet forces, is what delivers the average uniform force, which acts perpendicular to both the vortlet axes and the motion of the charge.
- I believe there are very tiny fluctuations, hence the word 'average'.

It is the iron core which is responsible for the uniform force on a moving charge.

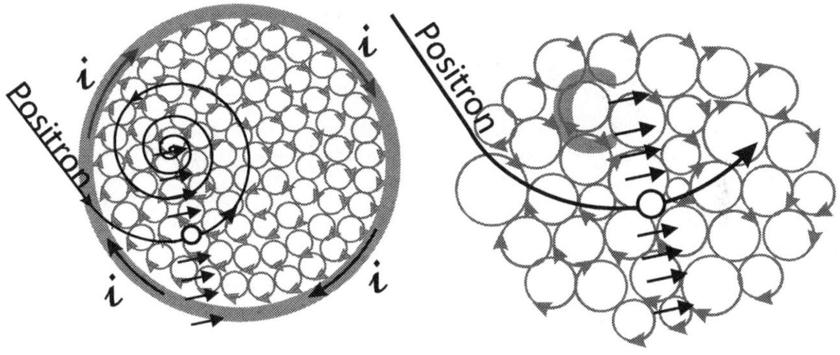

a positron moving thru a magnetic vortex

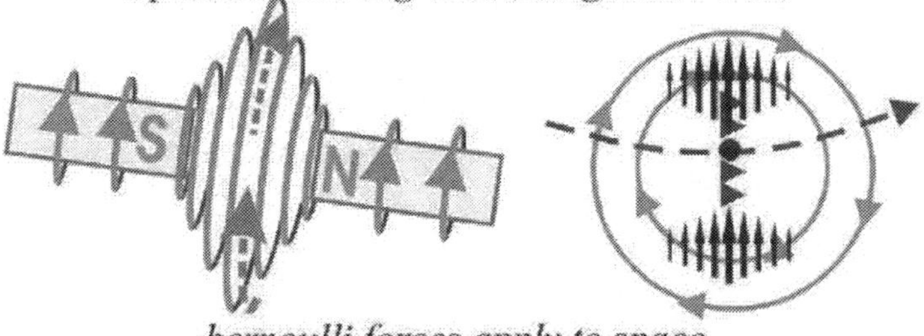

bernoulli forces apply to space
co-parallels attract; anti-parallels repel

Bremsstrahlung

Bremsstrahlung aka breaking radiation is emitted when a charged particle decelerates. Its frequency is determined by how much kinetic energy is lost.

The 1909 the Bremsstrahlung effect showed that depending on its proximity to a large nucleus like tungsten, an electron traveling at 25% light speed can loose a neck-breaking 90% or more of its speed over the distance of about 10^{-14}meters, by converting lost energy to X-rays.

Conventional physics sees that this is caused by getting T-boned as it passes thru transverse electric fields.

This is yet another consequence of irrational deductions from the apostles of perpendicularitis. where a transverse force absurdly causes perpendicular kinetic vector energy loss.

Bi-verse theory holds this loss is caused by the electron crashing into the dense Space surrounding the nucleus, (just like Space re-entry vehicles convert speed to heat.)

It views Space as medium of viscous energy which gets denser i.e. compresses as it dives into or out of a charges portal. The more portals there are in the nucleus the greater the density. (electron contributions are negligible)

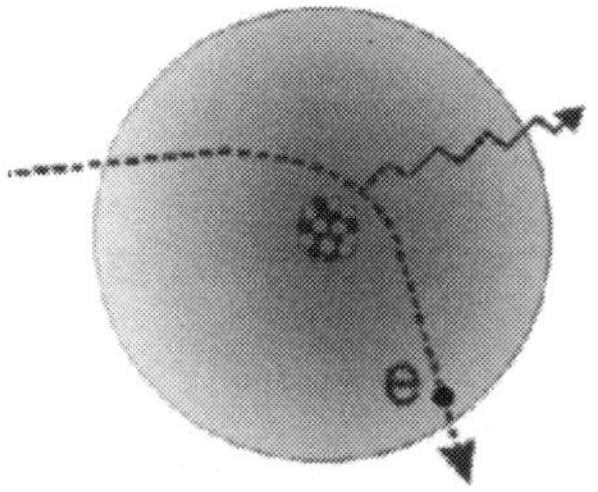

An electron portal voids coulomb energy to other dimensions/the antiverse at a specific fixed rate per nanosecond irrespective of the portals velocity. It does this from all directions.

When it moves in free Space, it experiences a Space energy headwind so it voids more from head-on (higher frequency) than from the tail (less doppler) while the transverse intake remains unchanged.

However when an electron portal passes close to a heavy nucleus it enters a increasingly denser Space, and it cannot adjust its forward intake fast enough. It chokes on the oversupply and cough out pulses of radiation.

It slows accordingly.

The nucleus's positive ee-field does interact with the electrons transverse intake curving its path.

The Shapiro Effect

The 1964 the Shapiro effect showed that radar pulses passing close to the Sun (short range) were slowing by 1/d, (hyperbola asymptote y=0), and far out (long range) by 1/ln(d), (asymptote y=1)

Ether theory can interpret these results simply by holding that ether forms a short range density lens sphere around Sun the when close to it, which at long range blends into the ambient energy density of free Space

Magnetobremsstrahlung aka Synchrotron radiation

In bi-verse theory magnetism flows parallel with its current so current loops spin their magnetism.

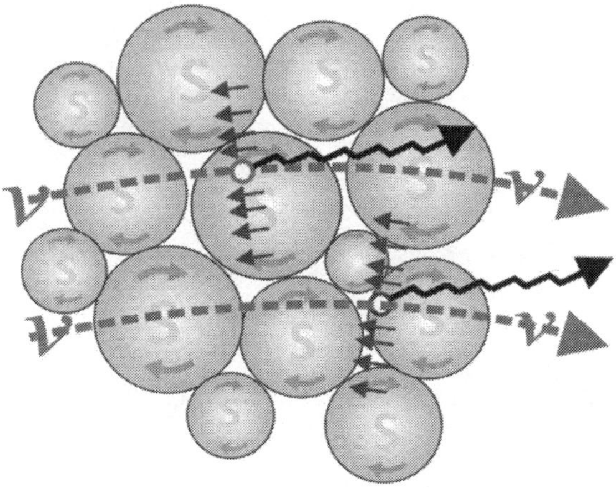

Magnets then are bundles of columns of molecular/atomic mini vortices (vortlets).

Traveling electrons spread their own wings of magnetism across many vortlets. Where parallel the co-flows attract and anti-flows repel.

The average vortlet force on the electron is half push+half pull (same direction).

This drives the electron via its magnetic wings into a greater head on zones of the vortlet's circulating anti-flow.

The electron can't adjust and rejects energy as X-rays.

Conventional Magnetism,

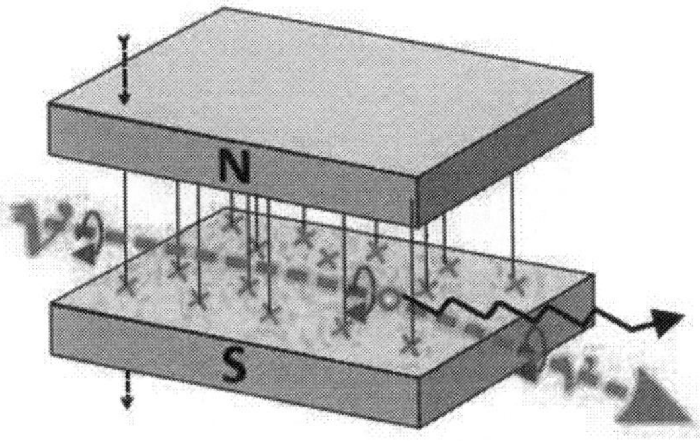

In the conventional irrational version of magnetism, an electron is driven (transverse to velocity) into another area of the theoretically perfectly uniform perpendicular magnetic field, where there is absolutely no physical or theoretical restriction to its speed, "and yet it slows", (to paraphrase Galileo).

Cyclotrons

Cyclotrons can employ 10 meter long liquid helium cooled racetrack super conducting coils or permanent magnets to accelerate and bend electrons.

The electron is forced into a greater racetrack headwind

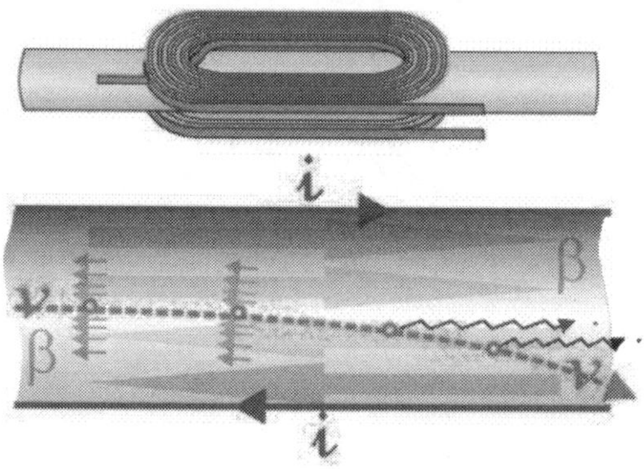

This causes synchrotron radiation.

Bremsstrahlung and the antiverse

In biverse theory the inside-out side of an electron is its positron in the antiverse and a proton is an anti-proton.

In the antiverse positrons orbit a negative nucleus where up-quarks are now down-quarks.

Here also Space gets denser as it dives toward the nuclei portals portal so the positron is simultaneously experiencing the same slowing force as the electron does in the here-verse, causing corresponding bremsstrahlung. '

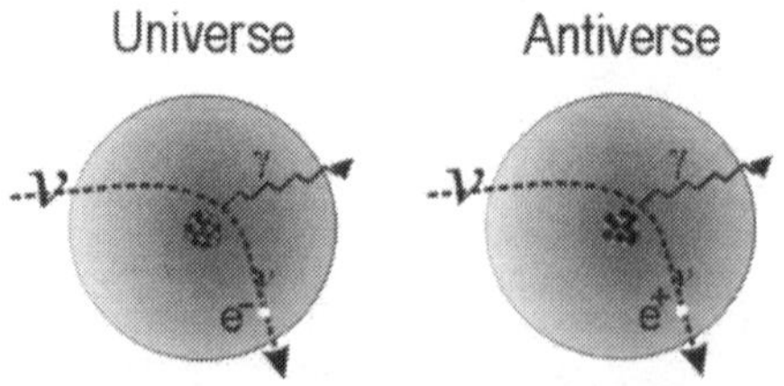

The above reasoning applies to radiation from adulators and wigglers.

SPACE EE-FLOW & ME-FLOW

CHAPTER 8

By now the correct direction of magnetism should be obvious to all barring flat-Earthers.

- The simple explanation of the perpendicular *(only to motion)* force on a conductor passing between magnets or across a magnetic pole is too logical to doubt the existence of, *and motion of*, magnetic vortexes.
- Magnets and coils spin Space so what is spinning?
- This compels us to rethink Space.
- This work is exploring the option that Michelson and Morley missed, namely that a medium of energy exists that when it is at constant energy, it doesn't affect the speed of light thru it.

Prerequisites of this Medium

Among other things it must:

1. Support attraction and repulsion of charges.
2. Support magnetism, letting magnetism affect the motion of moving charges but not stationary ones.
3. Not allow interaction of perpendicular magnetic flows.
4. Support electromagnetic radiation at the constant speed of light, and its transverse magnetic wave property.
5. Support the perpetual motion of masses through it until they are acted upon by an external force. (Galileo's sliding blocks)
6. Support the bending of light as it passes by mass.
7. Support gravity.

Speed Boat Analogy of Varying Current Induction

The speed boat analogue describes electrons as ***apparently frictionless***.

- A speedboat in a river current is programmed to stay stationary relative to the river bank.
- When the speed of the current increases, the boat must increase its thrust, however it is not quite quick enough.
- It gets swept downstream some distance before it increases thrust with ***higher frequency propeller revolutions***, which steadies the boat keeping it again stationary relative to its new down stream bank location.
- If the current slows, the boat speeds upstream until it slows its revs accordingly.

Continuing on;

A deaf observer, unaware of the speedboat's motor, sees it paused stationary in a visibly swift current and likely concludes that:

- The boat to water contact is frictionless.

When he/she sees that as current gets swifter, the boat is carried down stream, he/she likely concludes that:

- The contact has acquired friction.
- Since the boat has not changed physically the cause of the friction must be the speed of the water.

Thus he/she concludes:

- That the viscosity of water increases with its speed.

When he/she sees that as current gets slower, the boat moves upstream, he/she likely concludes that:

- The boat has magical acquired anti-friction.
  and accepts/lives with, the visual evidence.

Just as conventional physics accepts/lives with, no logical reason to explain induction.

Now when he/she puts his/her hearing aid in, he/she realises that
he/she got it very wrong.

Accordingly this work holds:

- The electrons self propulsion counters viscosity irrespective of the magnetic force.
- It only acquires apparent viscosity when the force varies its flow.
- The same criteria applies to positive charges.

Moving Charges In Space

The apparent resistance of stationary charges to increasing or decreasing me-flow pressure is equivalent to the apparent resistance from pressure on charges as velocity increases or decreases in stationary Space.

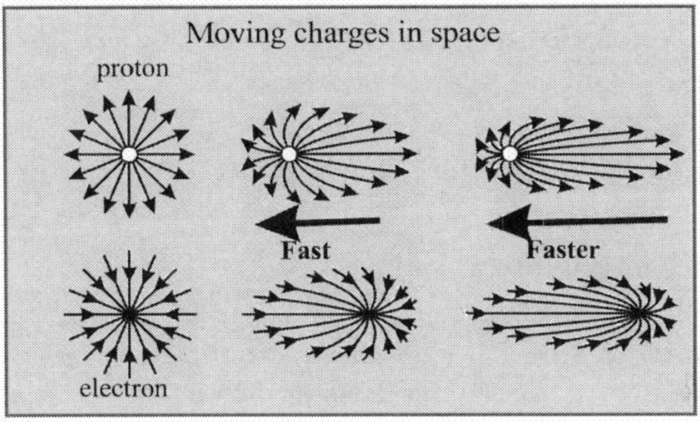

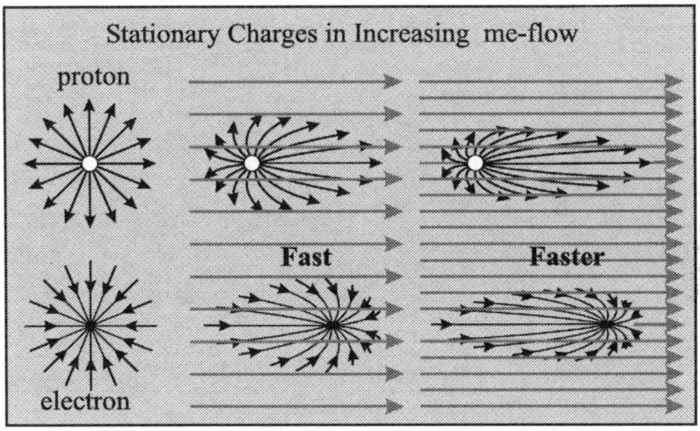

If say some gelatine were added to the river water increasing its density, it would cause a resistance to the boat similar to that of increasing the water flow-speed.

Positive charges in the here-verse pump out Space-energy into its ambient Space pressure environs. When they move a high pressure zones moves with them

Negative charges in the here-verse vacuum in Space-energy from its ambient Space pressure environs. When they move a low pressure zones moves with them.

When a negative pressure zone moves, the Space vacated behind it is immediately filled by the ambient Space, therefore it returns to ambient Space pressure.

The movement of positive pressure zones is me-flow

An electric current flowing in a conductor is the movement of positive Space pressure, which flows in the opposite direction to the electrons.

In the here-verse if electron zones move say left to right then positive Space pressure moves right to left, while its flip-side positron moves its positive Space in the antiverse left to right.

Parallel conductors in the here-verse have currents which attract each other via their me-flow. Similarly in the antiverse their flip-side reverse currents also attract each other.

So even though the bi-verse total me-flow is zero, the me-flow forces are combined.

Moving Charges In Conductors

Metals are good conductors of heat and electric current. Copper, silver, platinum and gold are some of the best. They have a valence of one, meaning each has one electron in their outer shell which is very loosely bound.

These flit from atom to atom, either briefly occupying outer orbits if vacated, or bouncing off densely negative shells.

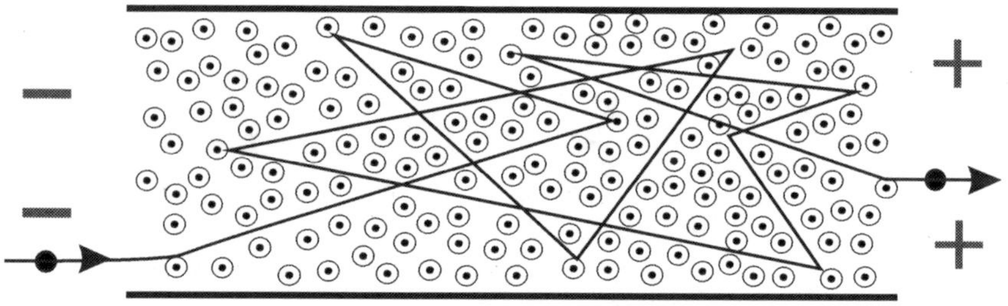

In copper for example, electrons spend 85% of their time in a termed *'electron gas'*, roving through the metal lattice between parent atoms which are left positively charged, *(cations)*. It is the attraction between cations and roving electrons which compacts the atoms.

Electrons are so small that many believe they are just points. In a copper wire the roving electrons tear along at a fantastic average speed of about six million kph, *(1/180th the speed of light),* temporarily adopting the outer orbit of, or hitting and rebounding from, about one in every two hundred atoms.

One millionth the speed of light in vacuo is about 1000 kph!

It doesn't matter whether a copper wire is long or short, roving electrons will still zip about 1600 kilometres back and forth every second. Each rover has its own companion negative zone.

Because the electrons are ultimately going nowhere, and/or as one is travelling left another is travelling right, the overall zone-flow is zero.

Things change dramatically when a current flows.

In a wire with a cross sectional area of one tenth of a square centimetre carrying a current of 10 amps, it takes a roving electron 13½ seconds to '*drift'* one centimetre. In that time it flitted to and fro covering a distance of about 325,000 kilometres, as did 8.5×10^{21} of its brothers, that is 8.5 million trillion trillion brothers.

Accordingly 8.5×10^{21} negative zones drifted one centimetre with them.

Thus an equivalent positive drift zone back fills.

Quantity amply compensates for the snail pace drift velocity.

Electron drift is responsible for me-flow in and around a conductor.

Infusing Viscosity into Magnetism

A conductor is a string of molecules between/thru which free electrons drift upstream into /against downstream electric energy flow *ee-flow*.

The back-fill positive pressure zone simultaneously moves downstream, extending beyond the conductor's molecules.

Its viscosity extends its influence into surrounding free Space.

This influence fades proportionally with distance $\pi d^2/d$ from the conductor.

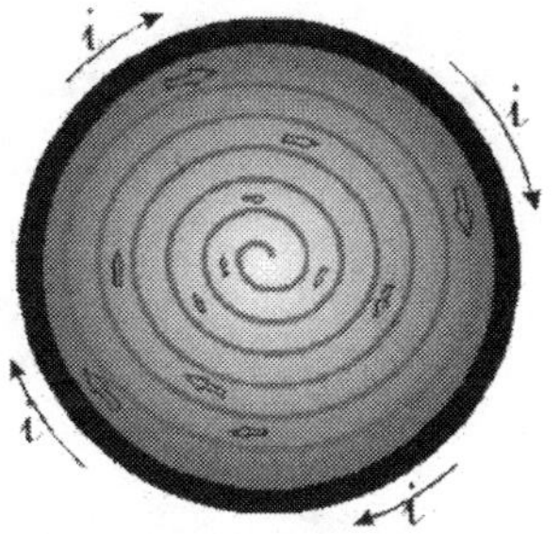

When the current flows in a circular loop the inner perimeter positive back-fill zone is intensified by the concavity, it drags via viscosity the central Space around with it. This drag fades toward the radius centre.

- Diagonally opposite zones augment the central pressures
- The conductor wrapped zone becomes a sticky vortex.
- The center becomes slow spinning dot/ point.
- Solenoids have columns of this sticky vortex
- Viscosity extends the column energy/spin beyond the coil.

- Free of constraints its pressure zone expands in all directions.
- Most goes spreads forward
- Some feed back down the outside of the coil.

More details in appendix 1.

A finite current coil is submerged in its vortex of magnetism.

Cylindrical Coil Simulation

A short tube with spinning impellers set back from each end provides an extremely slow simulation of how a cylindrical coil or magnet rotates its surrounding magnetic energy.

- An electric motor in the middle of the tube drives the impellers.
- All the fluid in and around the tube rotates in the direction of the impeller drive.
- This is detected by small test impellers.
- Those in front rotate with the impeller where the outer flow is greater than the inner fl
- Those around the waist rotate in the opposite direction because here the inner flow is greater than the outer.
- There isn't any movement of the fluid in the direction of the axis. All motion is rotation in the planes perpendicular to the axis.

- ***If a test impeller has a mini-motor it would generate a mini-vortex or vortlet which streams back to the pole***

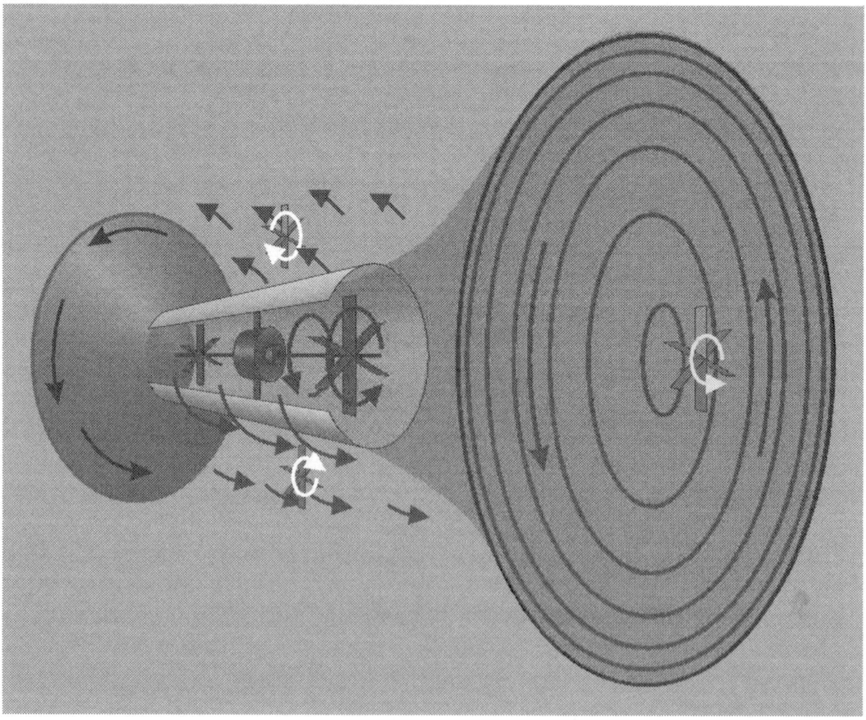

Me-flow Traps

In the early 1800s, der wunderkind Carl Friedrich Gauss showed mathematically, that if an infinitely long cylindrical coil were wound with perfectly fitting square wire, the internal magnetism could not escape to the outside.

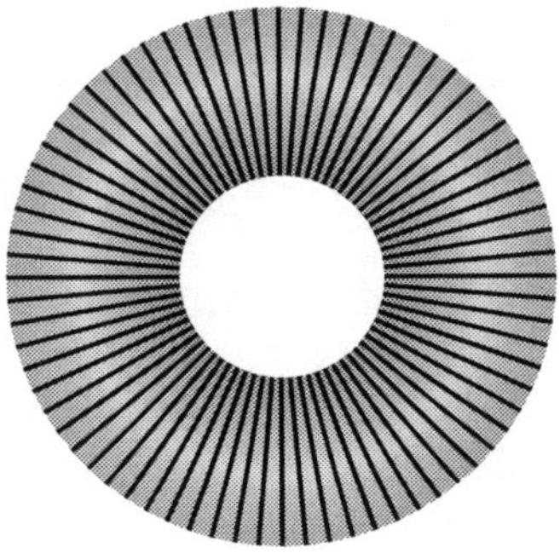

Similarly, all the magnetism is contained inside a perfectly wound toroid.

INDUCTION

CHAPTER 9

Generator and Electric Motor Type Induction

Fluid dynamic modelling has given a sound reason for why there is force on a conductor crossing a me-flow vortex.

The converse is when a conductor moves at velocity '*v*' through a me-flow vortex causing a current to flow.

The fluid dynamic interpretation of why this happens is similarly sound.

The fluid dynamic interpretation of <u>why</u> there is resistance to this motion is piercing.

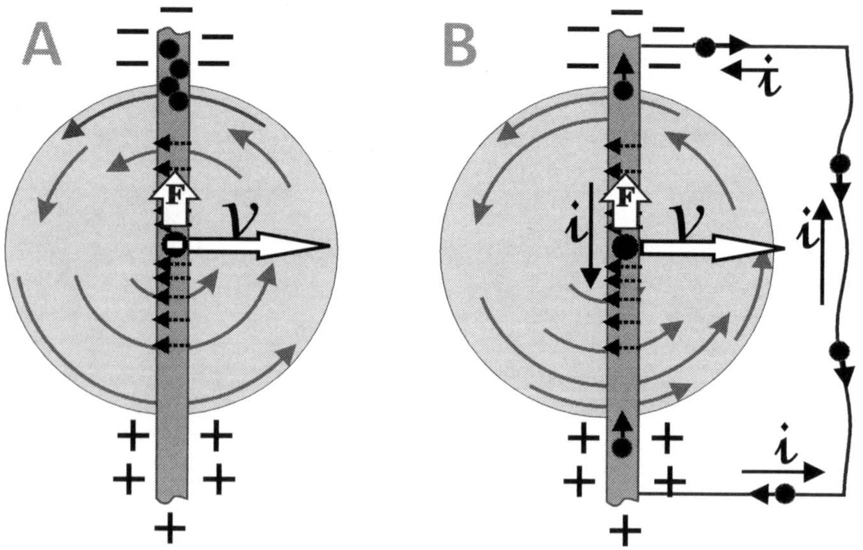

- In the conductor 'diagram A', free electrons are attracted up by the co-parallel sections of the coreless generated me-flow vortex above, and pushed up by the anti-parallel sections of the vortex below, establishing a potential difference between the ends.
- The force on each electron is so minute that it hardly moves the electron, however there are millions of them per centimetre drifting up which collectively causes a readable potential difference.
- They acquire horizontal energy from moving conductor and vertical drift energy from the magnetic vortex.
- When a circuit is connected 'diagram B', the a current flows against the electron drift.

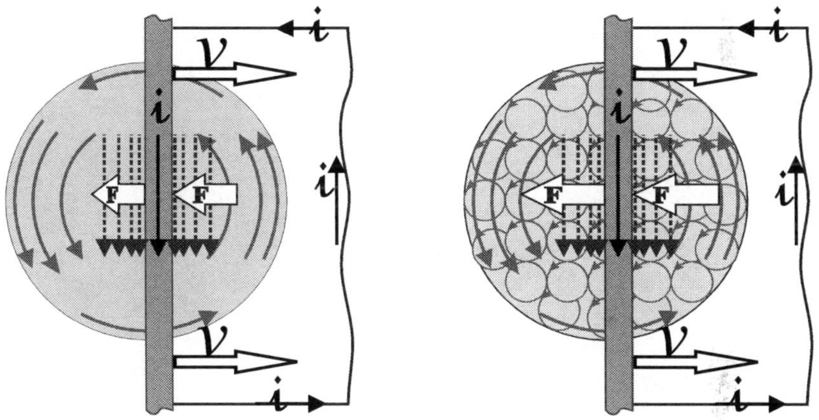

- The current travels with its own magnetic wind which now interacts with the north-south wind vectors of the vortex.
- This results in forces which simultaneously pull and push the conductor back.
- If the vortex is core generated, the current and resistance operate via each vortlet.
- For the velocity to remain constant a forward force must be exerted on the conductor.

This unifies the motion and its resistance and justifies the existence of magnetic vortexes.

Note that perpendicularitis physics can offer no explanation for induction apart from "it happens". Its reason for the resistance is the old-one-size-fits-all excuse "To comply with the laws of conservation of energy a force resists".....

In diagram 'C' and 'D' the vortex is moving and the conductor is fixed.

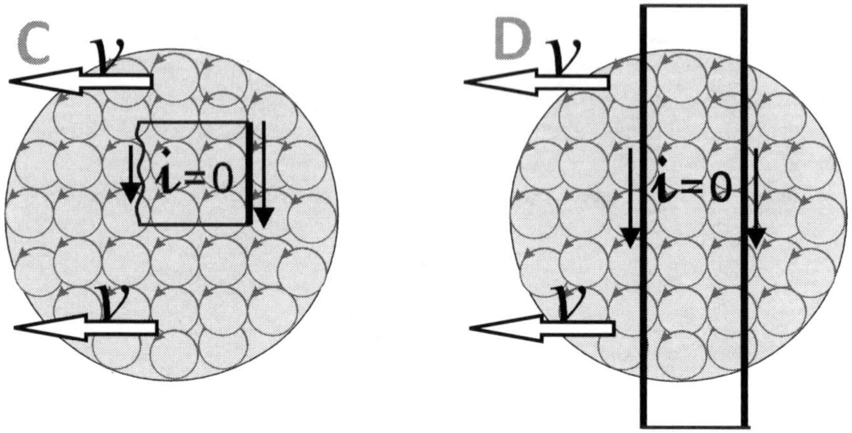

- When parallel sections of a circuit are submerged in uniform magnetism their currents cancel.

Loops and metal plates moving out of magnetic vortlets are subjected to a resistance force.

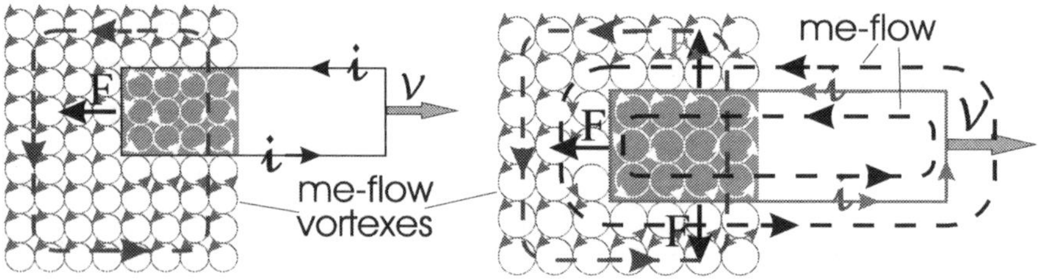

- The strength of the current circulating in the loop is proportional to: length x velocity *(lv)*
- The strength of the current circulating in the metal plate is proportional to the rate of change of the submerged magnetised area of the plate:

$$i \propto \frac{dA}{dt}$$

Transformer type Induction

Induction occurs when electrons, = self propelled, coulomb jet- fuelled, particles/portals, (*apparently frictionless),* are not quick enough to instantly cope with a change in me-flow..

As the strength of the magnetic vortex increases or decreases, it induces a current to flow in an axially aligned circular conductor.

The fluid dynamic interpretation of <u>why</u> this happens is unique.

Its basis is the fluid dynamic model of an electron which must intake a specific amount of energy per nanosecond irrespective of its velocity.

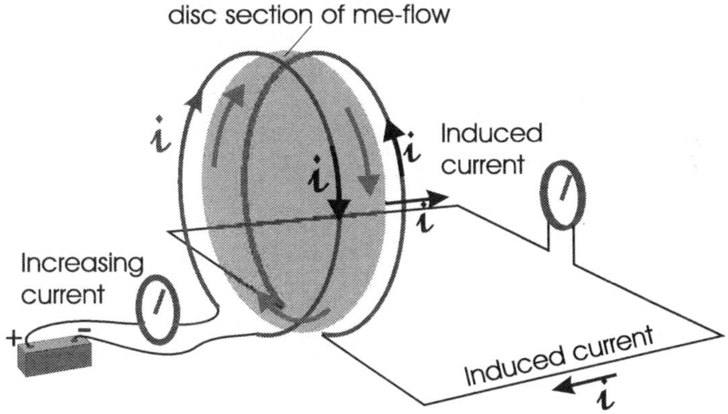

- The current circulating in the primary loop has a co-axial co-rotating companion me-flow vortex which surrounds the primary loop therefore engulfs the secondary loop.
- When current in the primary loop is steady, co-rotating me-flow in the plane of the secondary loop is steady.
- Here electrons in the secondary loop have adjusted to the me-flow by increasing their head-on intake and reducing their rear intake so that their drift velocity is zero.
- Consequently there is no induced current.

- When the current in the primary loop increases, the energy of the me-flow vortex instantly increases.
- During the me-flow increase the electrons in the secondary loop can't increase their fore intake and reduce their aft intake fast enough.
- Lacking 'traction' they are drift back as the electrons strive to adjust.
- This results in an induced electric current flowing the opposite way through the secondary loop.As the current in the primary stabilises, the fore and aft me-flow intake of the secondary loop electrons balances to suit the new conditions and they stop drifting.

- Reducing the current in the primary reverses the process.
- The electrons now overpowered, draw themselves upstream into the waning me-flow.

Similarly for positive charges

- A positive charge in steady me-flow has adjusted its forward and rear thrust to remain stationary.
- When the me-flow increases it restricts the charges forward ee-flow pressure
- It cannot increase its rear-thrust quickly enough to overcome the rising me-flow and drifts downstream until it readjusts thrusts and becomes stationary again.
- This results in a positive charge moving <u>with</u> the increasing me-flow.

These processes are mimicked by a bar magnet approaching or withdrawing from a loop or coil

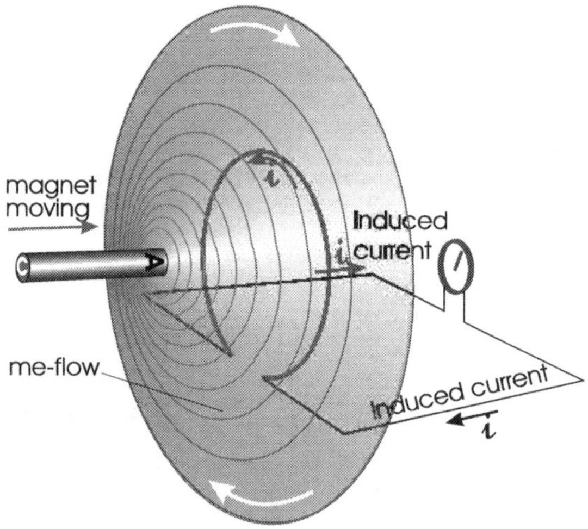

Cured of perpendicularitis we see magnets and current loops have vortexes of magnetism within and without them.

- When stationary magnet is in front an axially aligned conducting loop the electrons in the loop circuit have adjusted their for and aft intake to accommodate the me-flow.
- When the magnet approaches the loop the magnetic vortex energy or wind increases around the loop.
- The electrons in the loop can't quite cope with the increase fast enough and get swept down-flow.
- This now generates its own current opposite to the increasing loop current.

- This is called the induced current which has its own magnetism which now opposes the magnet's magnetism and repels the magnet.
- When the magnet withdraws the reverse happens.
- When the magnet is again stationary its magnetic wind is constant and the electrons hover.

Force is required to overcome the induced resistance and either push the magnet in or pull the magnet out.

Induction is governed by a quantum restriction that only a fixed/limited quantity of coulomb energy can pass thru a portal per nanosecond irrespective of a charges velocity.

Coulomb forces are a combo of push and pull acting together on either side of the bi-verse.

This fits the jet boat analogy of induction where the boat represents the portal bridging across the inter-dimension equator.

If the speed boat has a propeller mounted on the bow to draw the boat upstream the analogy represents an electron.

If the propeller is mounted on the stern thrusting the analogy represents a positron or a positive charge.

viz. turboprop engines.

Inductance or Self Induction

Energy is required to start a body moving in free Space or to alter its momentum. The resistance to the applied force is called the inertia which is proportional to the mass.

Similarly energy is required to start a current flowing in a conductor or to change its rate of flow.

- The transformer type induction which induced currents in coils also operates on any conductor increasing or decreasing its current.
- The electrons cannot adjust the directions of their in and out flows fast enough.
- This is results in what is called a back-current or back-emf, since it opposes the direction of increase or decrease respectively of the current.
- Therefore the back-emf resist the change in current.
- This resistance to change is called self-induction or inductance.

With every change in current there is resistance, therefore in alternating currents the frequency determines the amount of resistance per second.

A conductor has an inductance of 1 henry when the current is changing at the rate of 1 ampere per second inducing a back voltage of 1 volt.

1 henry = 1 volt per ampere per second = 1volt-sec/amp

That part of the resistance of the circuit due to inductance 'L' is called the inductive reactance, and is proportional to the frequency 'f' of the alternating current. The back voltage or feedback current occurs every time the current changes direction.

inductive reactance $X_L = 2\pi fl$ (volt-sec/amp)/sec = volt/amp

The total resistance in a circuit carrying an alternating current is called the impedance 'Z ', and is the square root of the sum of the squares of ordinary resistance 'R' and the inductive reactance 'X_L'.

Impedence $Z = \sqrt{R^2 + X_L^2}$ volt/amp = ohm

Whereas the inertia is proportional to the mass, the inductance is proportional to the shape or the geometry of the conductor. This is because certain shapes require the current to do more work than others to build or diminish their me-flow.

The graphic symbol for an inductor is a cylindrical coil, since cylindrical coils require more work than most other shapes to alter the current through them.

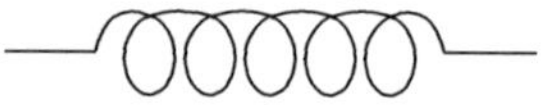

They must **wind up** the me-flow vortex inside them and all around them.

In the middle of a well wound cylindrical coil, in the total absence of iron in the core or nearby, the inductance is given by:

$$L_0 = \mu_0 n^2 l A \quad \text{henrys}$$

where μ_0 = permeability constant *(i.e. magnetization constant)*
n = number of turns, l = length, A = area, *(i.e. lA = volume)*

It takes time for the current to over come the inductance before it can settle down to a constant flow.

- When a current starts flowing through a cylindrical coil, as it builds so does the co-rotating me-flow vortex.
- When the current becomes constant the vortex is constant, and is maintained so until the current changes.
- If the current is reduced or switched off the vortex unwinds, this time in the opposite direction to the reduction.
- This then creates a back-current which tries to prevent the current turning off.

Both winding and unwinding take time so inductors are used to smooth out sharp changes in currents. *(e.g. spikes)*

Go to Appendix 3: for resistors, capacitors & circuits.

Induction in the antiverse follows the same processes for flip-side charges.
Magnetism flows against electrons in the here-verse and
with its flip-side positron (anti-electrons) in the antiverse.
they concur

CAPACITORS

CHAPTER 10

- The corrected direction of magnetism justified the embodiment of Space.
- Thereafter the pressures of Space winds presented a reason for electrostatic or coulomb forces.
- This in turn began the argument for the existence of other dimensions.

These hypotheses now permeate all electromagnetic phenomena, modifying conventional hypotheses.

Capacitors and *'displacement current?'*

An electric field *'ee-flow'* is always present between the positive battery terminal to the negative terminal .

Conductors can carry some of this field to a capacitor

- Electrons are the centre of their negative pressure vacuum zones.
- The greater the group of electrons the greater the vacuum zone is.
- The positive plate in a capacitor is caused by a deficiency in electrons which have crept/leant/migrated towards the positive battery terminal.
- Positive charges are not free to move.
- The presence of a dielectric insulator allows the capacitor plates to get closer to each other without a spark/electrons jumping the gap.

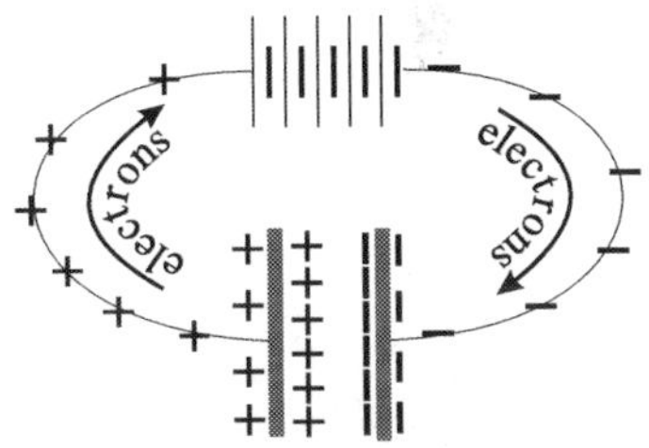

- However the electrons have/are large negative zones which do stretch over and bridge the gap.
- When they move towards the gap they push the opposite plate electron away.
- As a current flows downstream electrons and their negative pressure zones drift upstream
- Consecutive moving electrons interlink these zones.
- Positive pressure from the battery/ *ee-flow* backfills the vacated Space at light speed.
- ee-flow then bridges the gap between capacitor plates.

- ee-flow is viscous and draws ambient Space along with it.
- This is me-flow.

ee-flow across a capacitor,

- Progressively increases electrons in the negative plate until they have pushed *mini-microns* away the maximum amount of free electrons possible in the positive plate.
- Trillions of them get the shove.
- They stay there stored until discharged/short circuited when they rush home,
- The capacitor has become a little cell of the battery.

There is no ee-flow or me-flow across a fully charged capacitor.

A current is ee-flow with electrons moving in it.

- It moves either side of a capacitor but only ee-flow 'physically' bridges the gap.
- Maxwell called this 'sail without a ship' a displacement current.
- I prefer calling it a jump current.

Rectifying Capacitor Graphics

Graphics of streamlines are devious things. The straight streamlines in the conventional image of an 'electric field' flowing from a positive plate to negative one, carry no hint of any variations in pressure between the plates.

These graphics demonstrates these pressures.

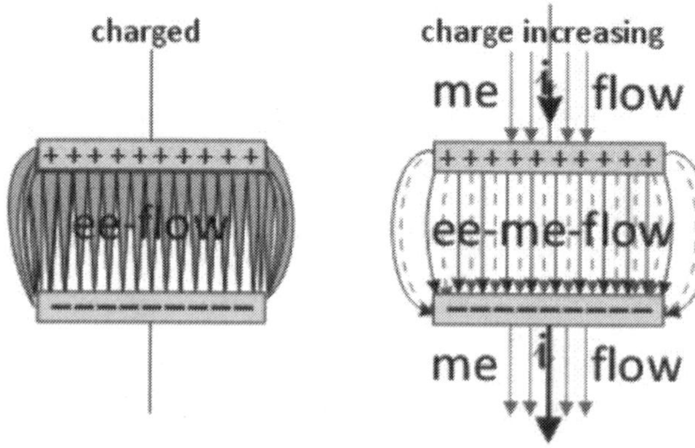

The left diagram fully describes the ee-flow pressure variation.

Capacitors Charged/Charging

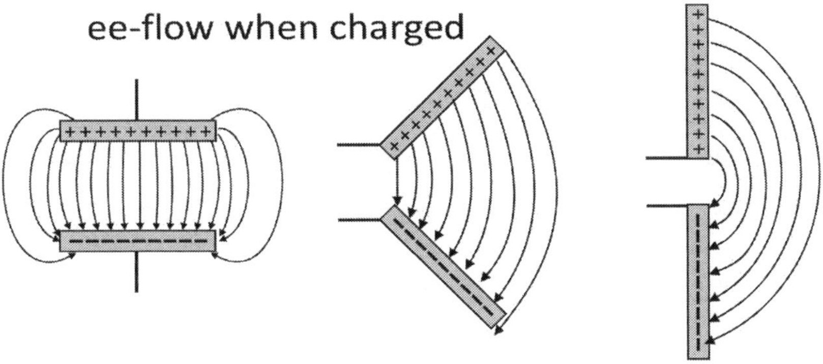

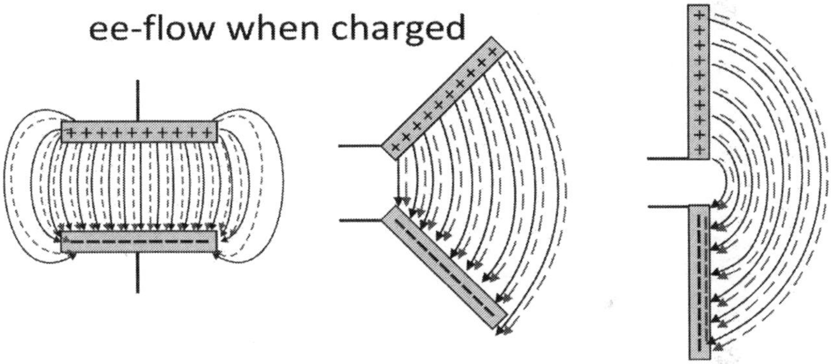

Pinching

As noted a current in a conductor is self compressed by its co-parallel me-flow.

- The lattice of the conductor resists this which prevents it from collapsing.

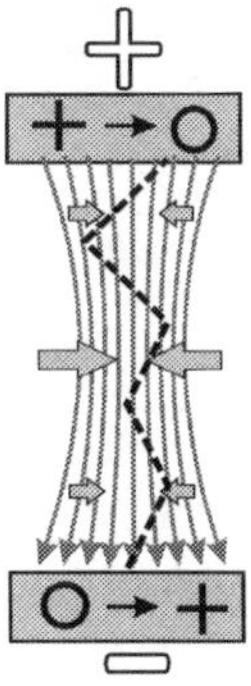

This compression also applies to jump currents.

- However without lattice support some pinching occurs.

When electrons spark/jump across a gap they have co-parallel me-flow and no lattice to resist compression.

- With large sparks, powerful pinching occurs which can squash conducting material between the plates e.g. aluminum cans.
- This has industrial applications.
- Lightning can cause catastrophic pinching.

In 1899 Tesla produces a 41 metre long spark. from his coils.

Spark-Gap Transmitters

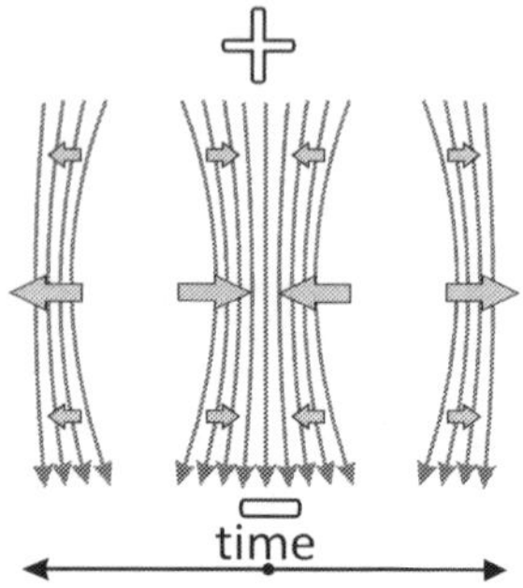

From 1887 to 1918 was the wireless age of spark-gap transmitters which produced pulses of em-waves called dampened waves.

The sparks caused magnetic pinching which splits off at light speed between sparks

A telegraph key sent messages in Morse code.

Oscillators

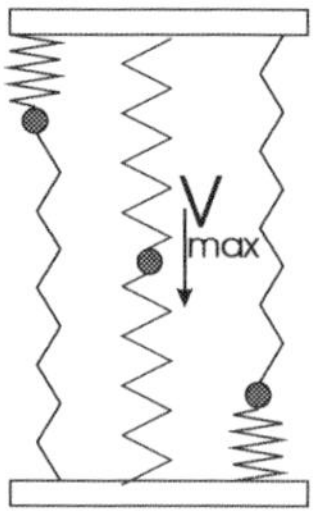

Out in Space away from gravity a mass oscillates between two fixed springs.

- It leaves one node and rapidly increases speed as it tears through the halfway point, before slowing down towards the opposite node.
- Its velocity is zero at each turn around and maximum at midway.
- Whereas the displacement of the spring corresponds to Hooke's law x = -F/k

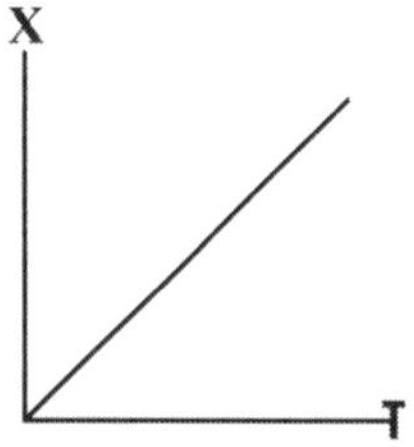

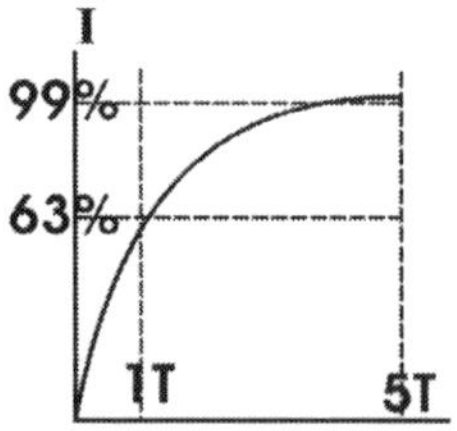

- The charge in a capacitor initially increases rapidly then slows as the as the capacitor fills.
- *Initially in one fifth of its charging time it is already 63% of the supply voltage*

LC (Inductor and Capacitor) Oscillation Cycle

The LC oscillator which I call 'the tick-tock circuit is the heartbeat of a radio receiver. A variable capacitor adjusts the frequency of the tick-tock to coincide with the frequency of the desired incoming radio wave. This is amplified, modified and sterilized then sent to the speaker.

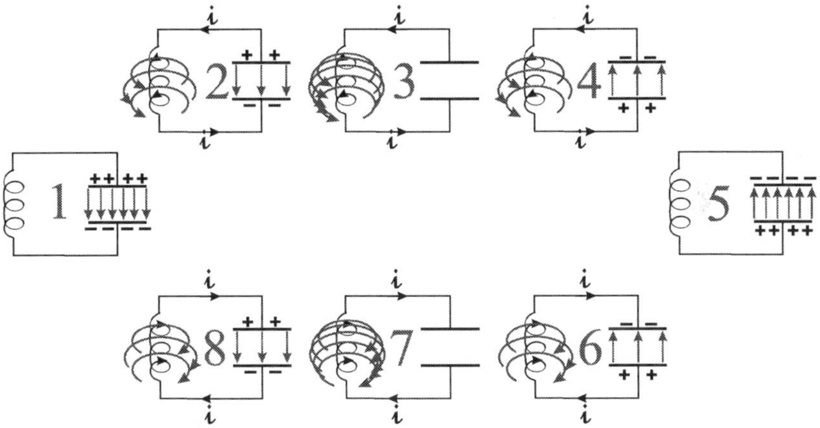

1. The capacitor is fully charged.
   The capacitor discharges through the inductor winding up the me-flow in the direction of the current.
2. Capacitor is half discharged and the inductor is half wound.
   Capacitor continues discharging and winding up the magnetic energy.
3. Capacitor totally discharged and the inductor totally wound up.
   Now the inductor starts to unwind but is still rotating in the same direction so the current still flows in the direction of the rotating me-flow .
4. The inductor is half unwound while the capacitor is half charged in the opposite direction to (1).
   The inductor continues to unwind as the capacitor charges.
5. The inductor is completely unwound and the capacitor is totally charged in the reverse direction to (1).

6, 7,8,1. The reverse cycle.

Note that in '3' and '7' even though there is no charge on the capacitors the current, which the rate of change of charge, is still flowing.

Fluid Dynamic Resonance Cavity

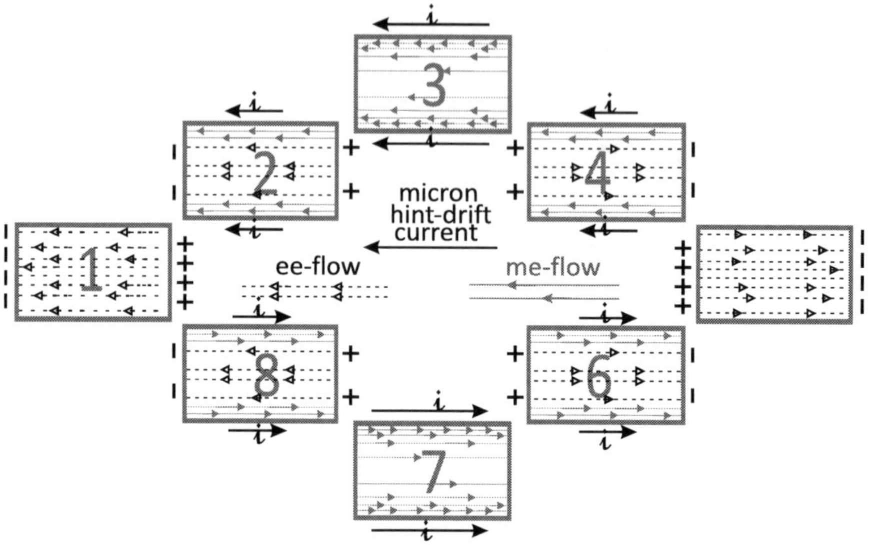

1. At full capacitance and zero *me-flow.*
2. Discharge left to right building up co-parallel *me-flow.*
3. Capacitance zero; *me-flow* and current at maximum.
4. The *me-flow* and current falls as *ee-flow* increases.
5. At full capacitance and zero *me-flow*.
6. 6.7.8.1 The process is reversed.

- Millions of electrons only need to overall hint at budging microns to charge or discharge the end plates.
- They behave as incompressible fluid vibrating hints of motion in the side and end walls of the cylinder.
- Consequently the frequency of oscillation <u>does not</u> depend on the length of the cylinder.

The frequency of this sort of resonating cylinder is given by:

$$f = \frac{1.22c}{r} \qquad c = \text{speed of light},\ r = \text{radius}$$

Electromagnetic Resonance Systems

Systems with energy stored in different parts like the spring system or the LC oscillating system are called 'lumped systems'.

Systems where all the various energies occupy the same Space such as acoustic resonators, (*e.g. organ pipes)*, and electromagnetic resonance cavities are called 'distributed systems'.

Cavity Resonator Frequency

The frequency of this sort of resonating cylinder is given by:

$$f = \frac{1.22c}{r} \qquad c = \text{speed of light}, \; r = \text{radius}$$

Go to Appendix 3: for resistors, capacitors & circuits

In 1899 Marconi sent the first radio transmission across the English Channel.

ELECTROMAGNETIC WAVES

CHAPTER 11

There can be nothing more spooky than the direction of magnetism encircling a current; there is no logical reason for this at all. Even if we consider that it might circle its conductor somehow, then why will it only circle according to the right handed rule. Why not left handed???

We have accepted this spooky absurdity and improperly included its classification in 'classical physics'.

Despite this, we have built great machines while not needing to how they really/absolutely work.

Chapter five showed a true classical way to explain magnetism according to fluid dynamics of Bernoulli.

- This corrected direction of a currents magnetism justified the embodiment of Space.
- Thereafter the pressures of Space winds presented a reason for electrostatic or coulomb forces.
- This in turn began the argument for the existence of other dimensions.

These hypotheses now permeate all electromagnetic phenomena, modifying conventional hypotheses.

Maxwell Waves

It must be in our genes to build towers of dazzling hypotheses on the sands of illusion or delusion, which generations are doomed to polish.

- Aristotle built his physics tower in successive density spheres of water, air and fire around Earth.
- Ptolemy built his on the Sun circling the Earth.
- Faraday built his on it the iron filings patterns.
- Maxwell built his on top of the Faraday and Oersted foundations .

Faraday had discovered that changing the current in a conductor correspondingly changes its companion magnetism *(his flux)* which induces a reverse current in a nearby conductor which caused a mutual repulsive force.

Moving a bar magnet moving towards a conductor loop showed that the increasing magnetism through the loop as it got closer, induced a current in the opposite direction, which in turn generated a force *emf* against the magnet's motion.

Similarly the force resists the bar moving away.

Maxwell asked how does varying the charge across a capacitor also produce a change in magnetism since no current crosses the gap?

Moreover*; if A causes B can B cause A?*

This was 25 years before Thompson discovered the electron.

Not only did he have no electrons to work with but he was severely handicapped by Faraday's perpendicularitis which claimed that magnetism circled/**'curled'** around a current and/or 'a time changing electric field'.

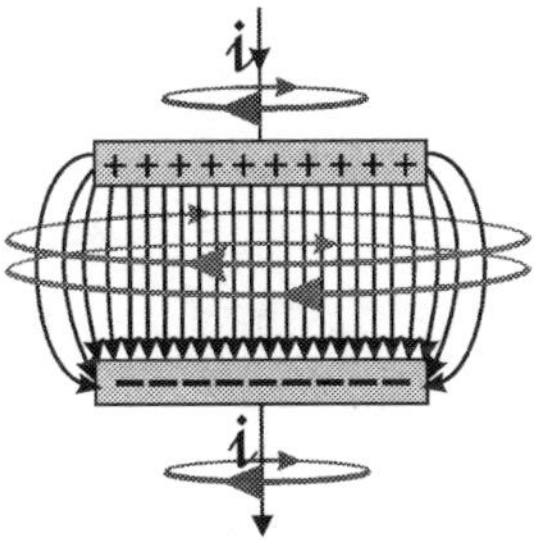

He proposed that the change in magnetic flux was fundamentally produced by a perpendicular electric field not only in the conductor but also in the Space surrounding the electric charges.

He dubbed this charge continuity 'a displacement current'.

Furthermore he proposed that a changing magnetic flux in Space produces in turn another perpendicular electric field in Space which in turn produces yet another perpendicular changing magnetic flux in Space ad infinitum.

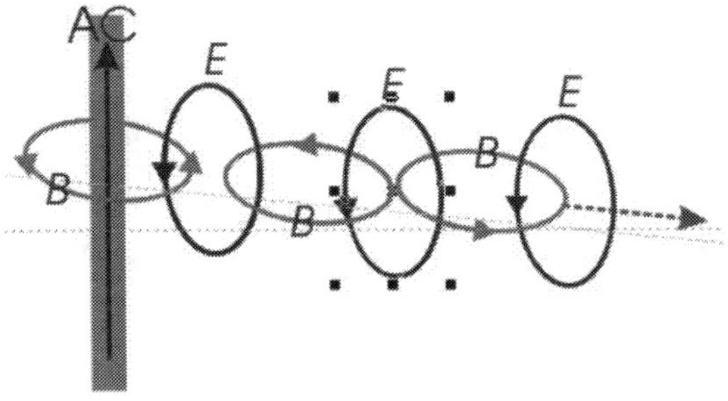

Me-flow Magnetism = m-waves

In bi-verse theory magnetism's Primary magnetic energy-flow called 'Pme-flow', is co-parallel to its current.

One of the most familiar devices for launching em waves is a dipole.

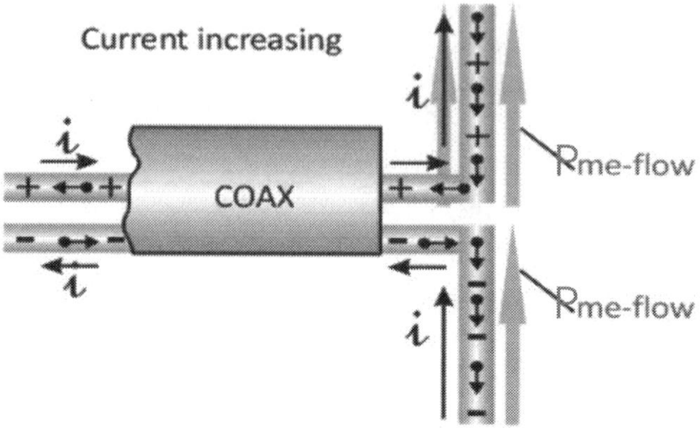

- The current flows into one post as it flows out of the other.
- It draws ambient Space co-parallel with it, *Pme-flow.*
- The dipole capacitor fills quickly at first then it gradually slows and stops.
- The current now reverses,(*it flows out of the capacitor 'say back to the battery'.)*
- The Pme-flow also reverses.
- The process is repeated

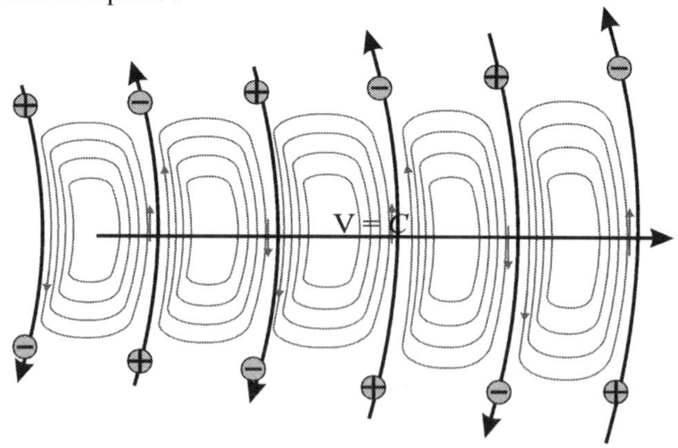

- The dipole's alternating current is responsible for consecutive opposite rotating Pme-flow loops forming eddy current which try to expand centrifugally.
- Expansion in the direction of wave propagation is constrained by consecutive loop-eddy groups each increasing contact pressure at frequency intervals.

- Transverse to wave plain expansion is not constrained.

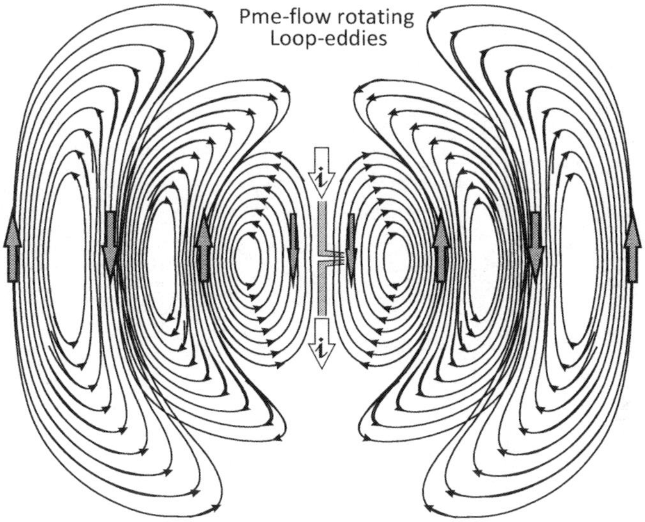

- In the plane of wave propagation the co-parallel Pme-flows between consecutive loops mutually attract each other.
- Simultaneously they augment their mutual rotations.
- They become consecutive counter rotating loop-eddies.
- The wave expansion is constrained along the axis of propagation by group cohesion.
- Where the groups meet,the pressure waxes and wanes across the wave front.
- ee-flow is generated when a wavefront's associated Pme-flow passes thru a point.
- Electrons in the pathway are increasingly jolted one way then decreasingly the other. *(viz chapter 9 Induction.)*
- ***They are submerged in an oscillating electric field.***

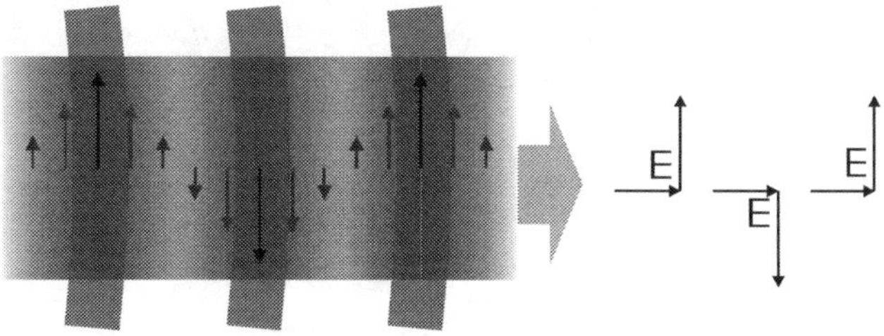

- To say that induced aerial currents generate more em waves is like saying a wind causes a fan to spin which generates more wind.
- Maxwell's virtual aerials in Space perform this delusion.

Transverse Magnetism in em Waves = Tme-flow

When two opposing magnets or coils are locked beside each other, since Pme-flow rotates with its coil current, their poles are linked by spiral Tme-flow vortexes like slinky toys.

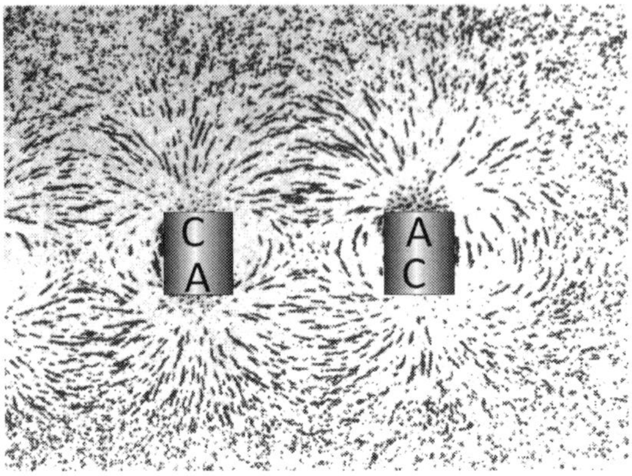

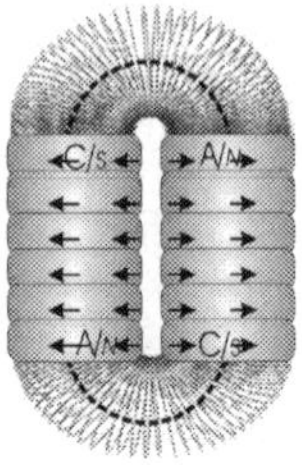

- Aligned inverted single coils have limited external magnetism beyond their slinkies.
- They are unaffected by *(neutral to)* external magnetic fields and vice versa.
- *Note;* Perfectly wound toroids have no external magnetism. (Gauss)

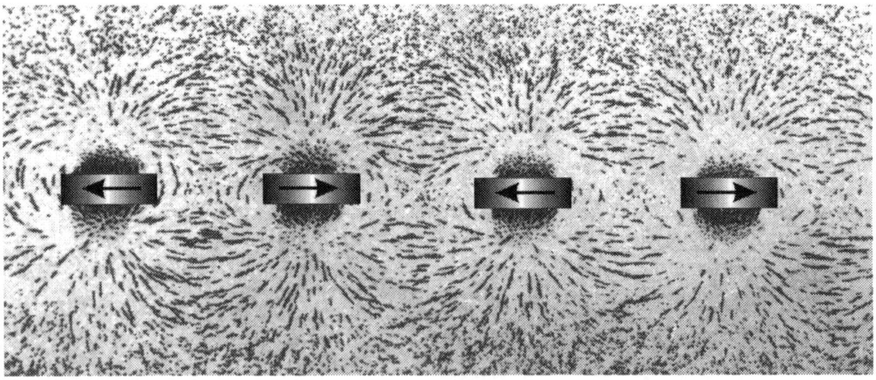

When they are aligned their close-by slinky Tme-flows interlink.

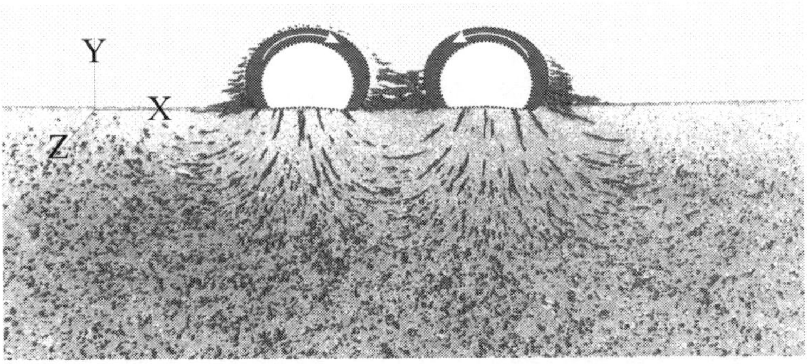

The loops are the faces of consecutive opposite magnetic poles

- The filing pattern show Tme-flow bridging near both sides of the loops but fading to neutral beyond this.

Accordingly the so called 'electromagnetic-wave' is purely a magnetic wave, i.e. a 'PTm-wave', having two components of me-flow:

1. Primary me-flow *(Pme-flow)* are the loops between wavefronts.
2. Transverse me-flow *(Tme-flow)* links slinkies ***thru and across*** consecutive pairs of loops.

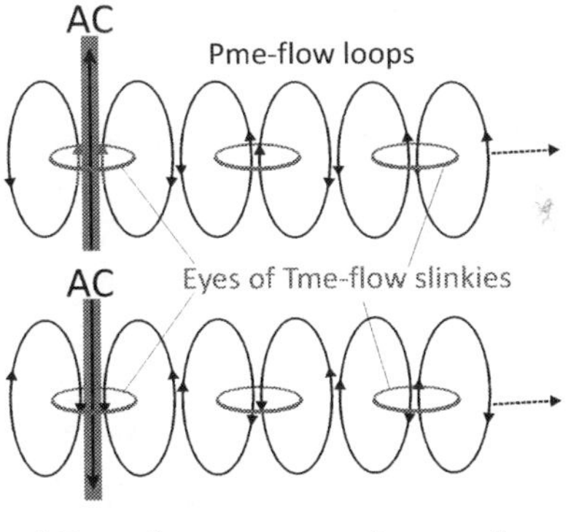

- Slinkiest wrap around Pme-flow compressions at the wave fronts.
- The slinky pairing begins with the first current direction.

The transverse Tme-flow is on both sides of Pme-flow loops, not one-side as in Maxwell's em-waves

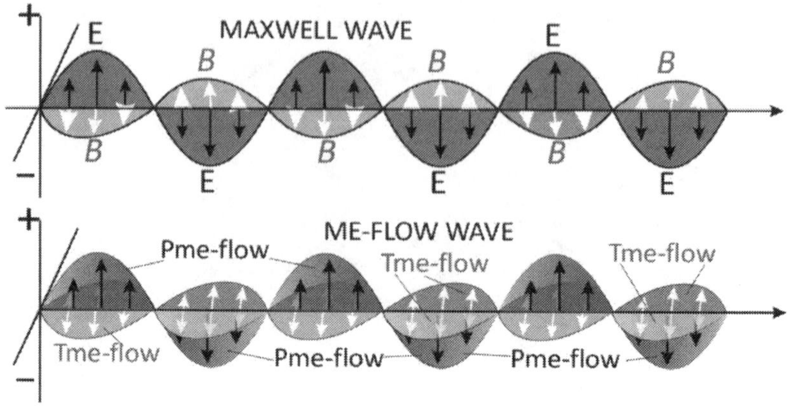

Handedness *chiral*

- The handedness of Tme-flow is meaningless since the Pme-flow vortex wraps around its wavefront.
- Science concentrates only on right-handed side for consistency. *(Like cars navigating maps with right hand drive rules)*
- Had science chosen a left handed spin then - Cos θ would have applied to calculations instead of +Cos θ.

If an algorithm was tuned to right-handedness then it would report the <u>invalid</u> mono-polar option below.

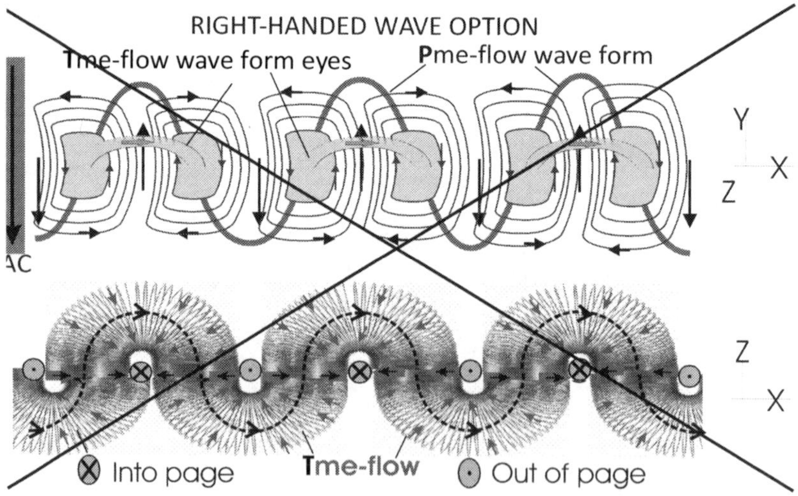

Loops below are shown circular for graphic convenience.

Right handed

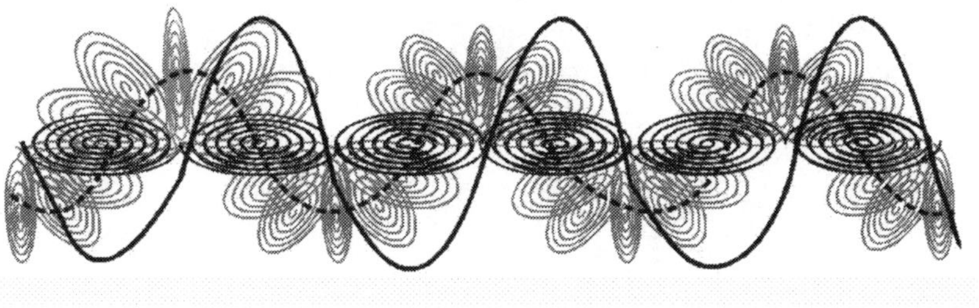

Bi-verse theory is a bi-polar model.

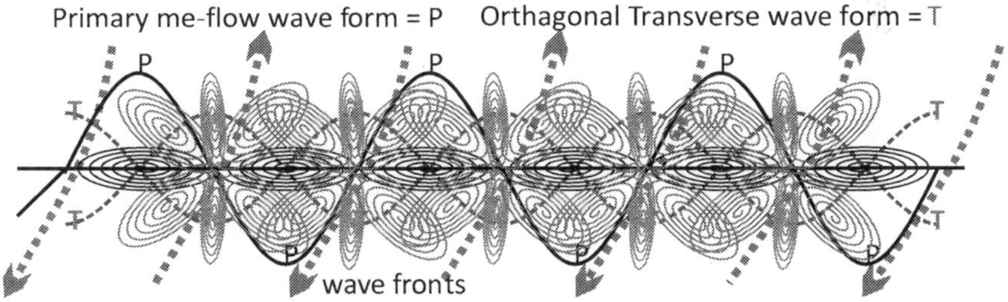

The counter-rotating Pme-flow loops enhance mutual rotation.
The Tme-flow slinkies enhance wave cohesion.

Antenna

Open ended musical wind instruments have flared ends to prevent *(or minimise as far as possible)* the feed back from the sound wave as it is set free.

Similarly wave guides and coaxial cables engineer their open ends to launch electromagnetic waves.

One of the simplest launchpads for em-waves is a trumpet antenna.

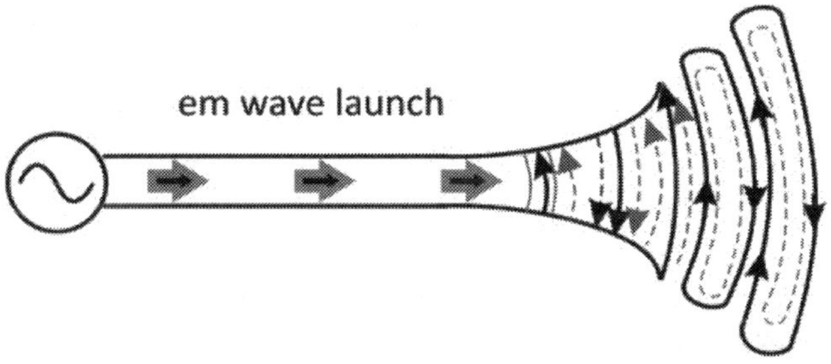

Parabolic Dish Waves

The parabolic antenna was invented by German physicist Heinrich Hertz during his discovery of radio waves in 1887.

Pme-flow zones from wave-guides or dipoles can be projected from the focus of a parabola onto its dish where they are reflected.

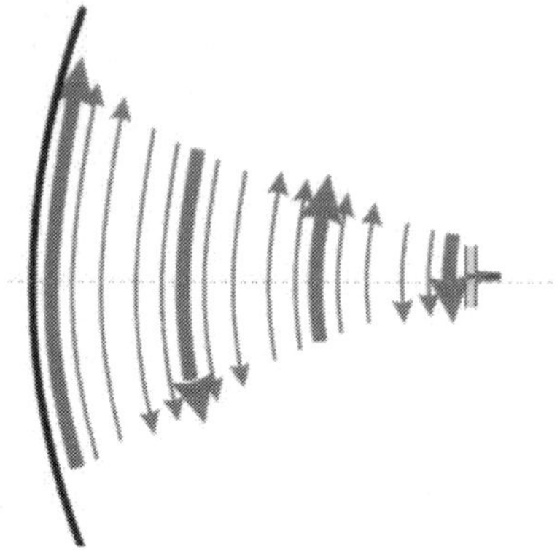

The reflected me-flow wave fronts are perpendicular to axis of propagation.

$|F\ A_x B_x|$ is constant

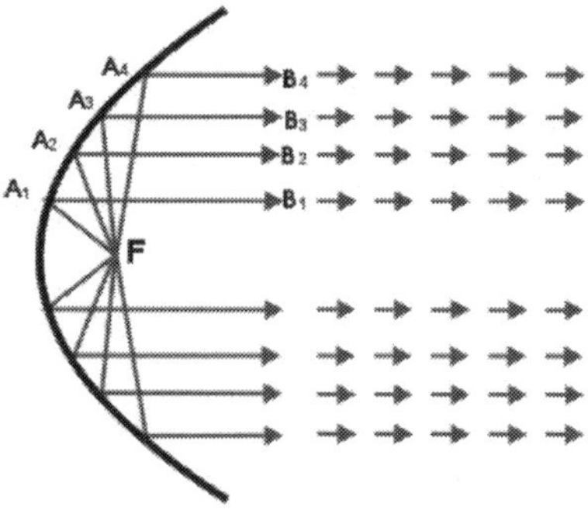

The co-parallel flow elements across the wave zone are cohesive.*(Bernoulli)*

This cohesiveness concentrates/pinches the wave, enhancing long range transmission.

Aerials

Since the days of early broadcasting, because the radio waves were transmitted vertically polarized, receiving aerials had to be also vertical.

Today microwaves may begin their journey polarized, but as their wavelength is so small they get easily twisted by the atmosphere and blocks of data arrive randomly polarized at mobile telephones which filter polarize and amplify them.

The above dipole wave diagrams have described consecutive m-waves as having compressed/dense counter rotating Pme-flow contact zones.

- The Pme-flow waxes on then wanes off these wave-fronts.
- When the compressed/dense front of group wave loop enters an antenna it submerges electrons which can't immediately cope with any increase in flow and they are induced to drift/creep mini-microns along with it.
- This causes/induces a current to flow the other way.
- When the back of the compressed/dense group loop arrives it induces the electrons to move the other way causing a reverse current.
- Trillions of them move back and forth creating tiny oscillating currents which can be amplified.

Transmission Lines

Transmission lines carry information in the form of impulses generated by turning a current on and off or forwards and backwards repeatedly.

The first carriers were single wires which carried messages in Morse code. *(Samuel Morse was a successful portrait painter and sculptor who branched out.)*

The messages needed boosting about every 32 kilometres which required repeater stations. These were originally all manned, however Morse developed an automatic battery powered repeater which took over much of the work.

As faster oscillations were developed, providing faster transmission of information, the wires became aerials which generously radiated, therefore leaked energy from their entire length. Improvements were needed.

Paired wires had their own problems principally with impedance and cross-talking. Twisted pairs were better.

Then along came coaxial cables which were faster, cheaper, less noisy, less cross-talky,... etc. Installation of long distance coaxial cables began in 1946.

Coaxial Cables

Like the wire, their transmission signals also needed boosting which the coax only required every 10 kilometres.

The transatlantic TAT-6 coaxial cable has built in solid state boosters every 9.5 kilometres.

These are powered from a companion high voltage cable delivering energy from both coasts. This coax can carry 4,200 two-way voice circuits simultaneously.

These wonders will be, and are already being, outstripped by fibre optic cables whose transmission rate is faster still, and totally immune to electric or magnetic noise. They can transmit data a few thousand kilometres without boosting.

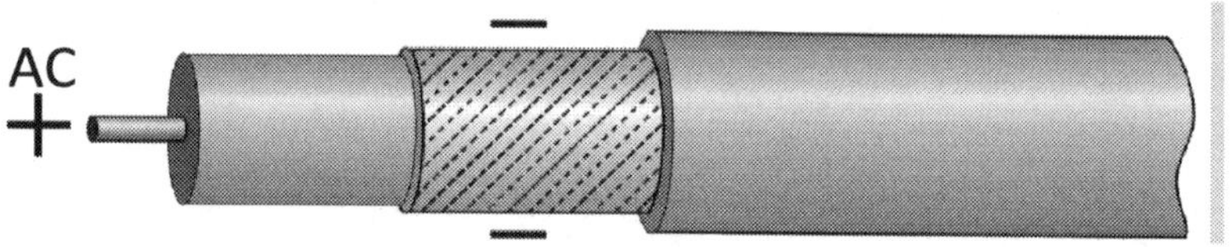

A coaxial cable has a central core positive charged conductor surrounded by insulation plastic which is itself surrounded by a negative conducting sheath which it in turn is encased in insulation plastic.

It is a long thin capacitor where the inner insulation plastic acts as a dielectric between opposite charged surfaces.

When the *AC* current is switched on:

- Free electrons in the central copper core oscillate (drift mini-microns) back and forth along it.
- Simultaneously free electrons in the sheath oscillate the other way.
- Because the current is alternating, tiny Pme-flow eddy currents form in the center of the core wire and force free electrons to the surface as carriers of a skin current.
- Because of attraction to the positive core, the sheath electrons travel along the inside surface of the sheath.
- The cable has become a long series of capacitors with ee-flow alternating from core to sheath.

The cable has very little resistance along its core or its sheath and it has infinite resistance across the gap, however coaxial cables are resistance rated.

This is because when an alternating current is applied across it, a multitude of currents move back and forth generating, cancelling and reversing capacitance while their companion Pme-flow generate fluctuating inductance. All this gets bottled into impedance therefore resistance to current.

A coaxial cable might have a resistance rating of say 50 ohm. This is based on the concept that the cable is infinitely long.

If it were infinitely long then if a current started flowing for a moment then was stopped, a column of capacitance would continue down the copper core ad infinitum.

If the cable is of finite length the column could only go to the end where it gets reflected, thereby smothering the section of the column which is still oncoming.

Consequently cables of finite length need to have some sort of outlet at the far end to allow the current energy to flow on and not be reflected.

This is achieved either with an antenna, which releases the energy into the atmosphere and/or Space, or a resistor across the ends which acts as a load and dissipates it.

Fluid Dynamic Coax

- The alternating current from a Rf oscillator generates Pme-flow eddies.
- Pme-flow eddies create transverse Tmeflow eddies.
- Eddies from opposite sides of the coax have the same spin.
- They bend toward each other and meld, encircling the core like torus slinkies.

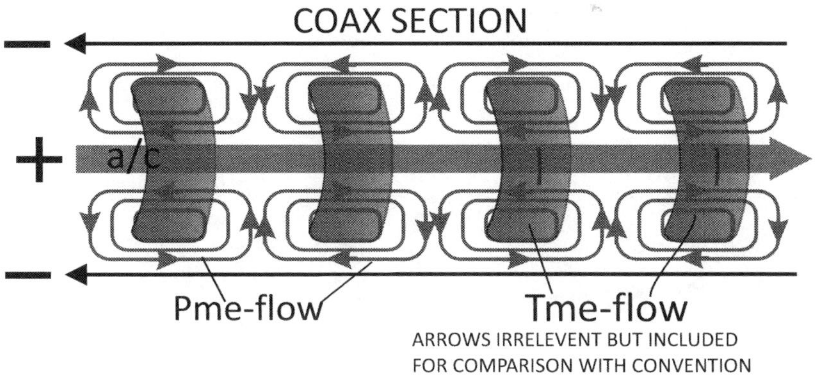

Doughnuts/toruses of me-flow contact along the coax @ light-speed.

- They carry electric information along the cable until liberated by a dipole antenna and sent into ambient Space as waves.

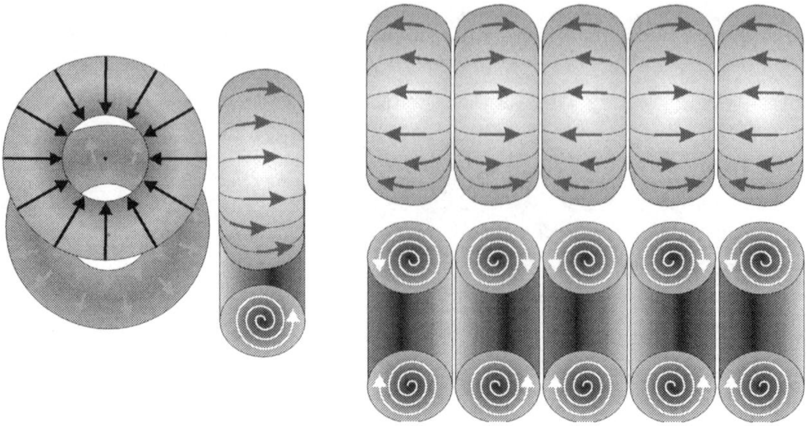

Fluid Dynamic Wave-guide

Wave-guides are practical only for signals of extremely high frequency, where the wavelength approaches the cross-sectional dimensions of the wave guide

- Wave-guides confine wave propagation to one direction
- The basic shapes are circular or rectangular
- Circular guides confine wave bandwidth *(range of frequencies)* to max 1.3601:1
- Rectangular guides have a much higher bandwidth
- The width to height ratio is usually 2 : 1..
- Its wavelengths are usually about 3.7 times longer than the height.
- The shape for the guide confines the m-wave between the larger surfaces of roof and floor.

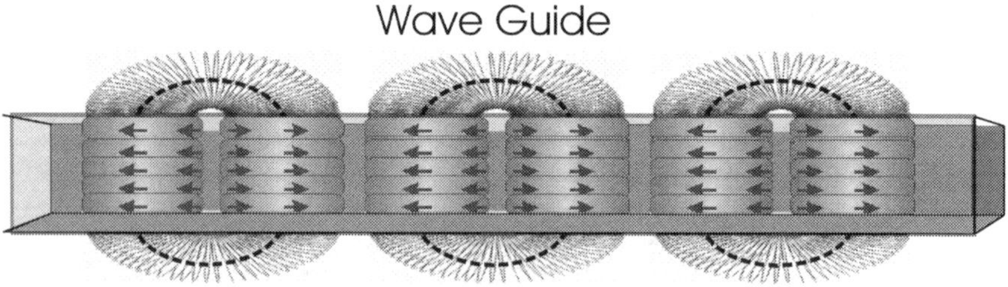

A sinusoidal electromagnetic oscillator generates successive columns of consecutive counter rolling me-flow down a metal wave guide.

- These induce alternating currents in the copper roof and floor which help contain the wave.
- The interface of consecutive columns are co-parallel therefore they attract.
- They have viscosity therefore friction
- The consecutive counter-spins assist each other.
-

Circular Polarized Light

Pme-flow waves and Tme-flow waves are orthogonal.

- At a given point two orthogonal waves are in phase.
- If the waves are superimposed the resultant wave lies in a plane.
- It is linearly polarized.
- The amplitude may vary but the waves must be in phase.

If the phase is shifted by a quarter of a cycle,

- The superimposed wave becomes a circular helix wave.

If the phase difference were anything other than a quarter cycle
or the amplitudes were different,

- The resulting wave would be elliptical polarized.

A polarizer plate slows the light along one forward axis.

- Left circular polarized light results from light passing thru a a quarter wave polarizer which when viewed head-on rotates
- The wave is counter clockwise

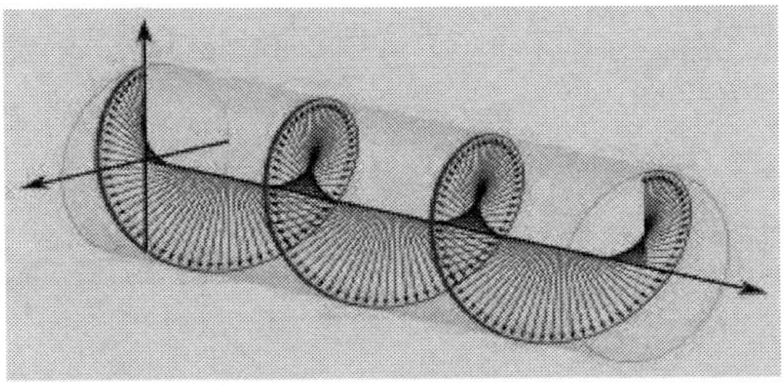

If circular polarized light passes thru a half wavelength plate,

- The helix becomes right circular, rotating clockwise when viewed head-on
- . Tme-flow flips 180° when reflected, so left circular flips to right circular.

- Scalar Waves

A scalar has quantity but no direction vector like for example 'pressure' or 'heat'. Science says they have scalar fields.

They can have no inherent vectors or have counter direction/cancelling vectors

Electric Scalar Waves

Electric scalar waves confusingly have no Tme-flow vectors.

The correct direction of Pme-flow helps to unravel their behaviour.

Abstract:

Two perpendicular conductors have perpendicular magnetisms which seem to ignore each other, however the energy must be still there. This is more so with x,y,z,conductors.

Local Space here has omni-vector-magnetic potential energy which alternating currents can oscillate.

Two anti-parallel conductors repel each other via their central anti-parallel Pme-flow which generates eddy currents which try to expand centrifugally..

When two axially aligned conducting loops have mutual alternating currents their Pme-flow inside each loop, oppose each other.

The counter revolving magnetism gets compressed as the current increases, and decompressed as the current decreases.

This produces longitudinal non-Tme-flow waves whose peculiarities science and the alternative-science are exploring.

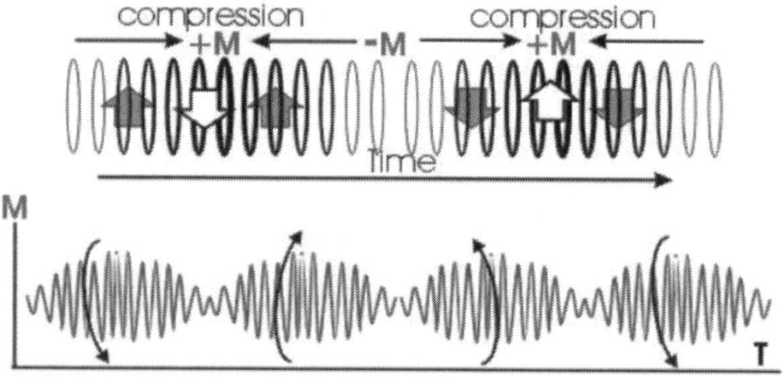

- These are trains of counter long axle rotating spirals of Pme-flow.
- When they are seen graphically, they are trains of counter long axle rotating spiraling ellipsoids of magnetism.

Caduceus Scalar Waves

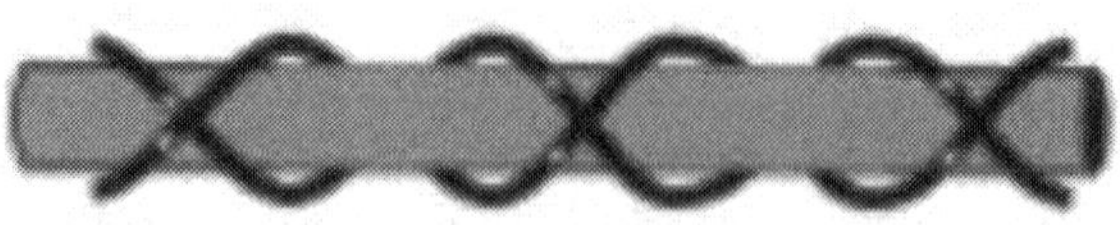

The Caduceus staff aka 'the Herald wand' with two entwined serpents was carried in Greek mythology by Hermes while delivering messages from the Gods.

The serpents cross over each other at right angles.

The electric caduceus staff has multi-strand insulated conductors wound around a ferrite rod which also cross over at right angles.

This wand employs both of the above abstract properties.

Just as a dipole antenna launches PTme-flow waves, the caduceus coil antenna launches circular counter rotating/spiralling Pme-flow waves which repel each other.

They vibrate like longitudinal waves thru space.

Some Physics Properties:

- Pencil beams of longitudinal waves can send Morse code messages to critically aligned other caduceus coils.

- The waves can penetrate Faraday cages, possibly because they induce rotating currents which in turn produce induced circular Pme-flow-waves on the other side.
- The wave then piggybacks its way thru the cage.

Faraday cages absorb ordinary m-waves like sound proofing. They excite electrons like aerials do, but here in all directions, which causes interference.

Once a Pme-flow loop is broken it self destructs.

Tesla Waves

The electric scalar wave was 'discovered ?' by Nikola Tesla around 1890

Tesla wanted to create a global, wireless communication system broadcasting sport, music, news, etc *ie the ethernet.*

He believed Marconi waves could only travel 'line of sight' so would be limited globally. He did not predict they would soon be bouncing back to Earth from heavy-side atmospheric layers.

Tesla thought he could develop wireless scalar waves to travel thru the Earth and overtake Marconi's growing financial successes.

Wardenclyffe Tower

He built a 57 meter tall wooden Wardenclyffe tower on Long Island New York State.

- The tower had 5 ton cupola which held a waist of AC loop currents of various invention which squeezed out pulses of compressions *(longitudinal waves)*

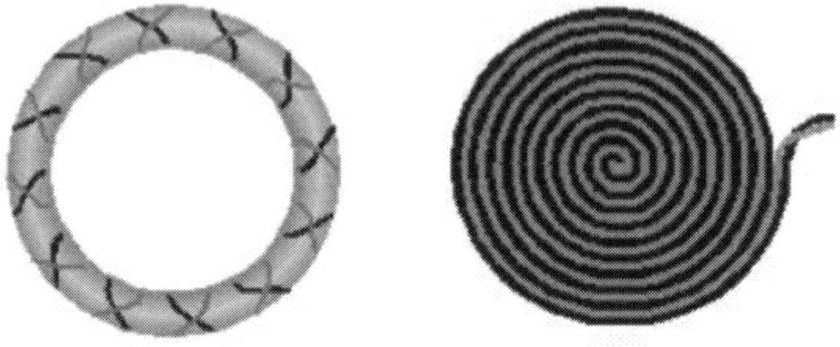

- These waves were tuned to travel out and power/voltage-supply to distant appliances, then return to the tower generators via the earth.
- Long underground tunnels beneath the tower would home them in.

The project/folly gobbled up fortunes as well as his own patents. He gave Westinghouse his royalties from alternating current motors *worth billions*, always believing his 'next one' would be the mega-jackpot.

Tesla died a pauper in 1943.

Biological scalar waves

Biological scalar waves are a natural form of energy shaped like an hourglass. All your cells and your DNA operate on this hourglass scalar shape.

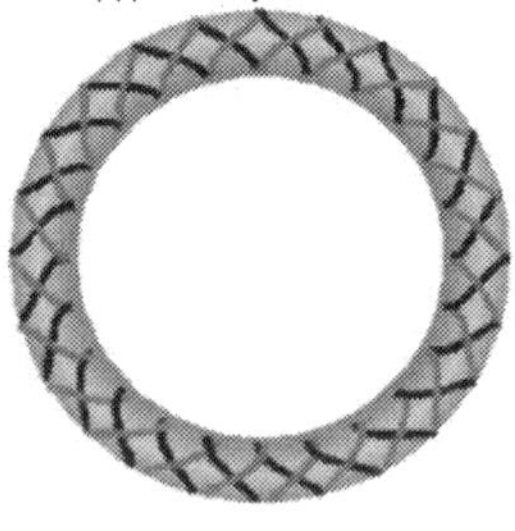

There are many alternative devices/modes of these designed to benefit health
viz Tesla Metamorphosis.

PHOTONS & QUANTA

CHAPTER 12

Black Body Radiation

In 1900 understanding cavity radiation was the outstanding scientific puzzle of the time. The white hot glow in the depths of a coal fire is one example of this radiation.

Various blocks/ovens of different metals with internal cavities of any shape and size had a small hole drilled into each of them.

When they were heated to the same temperature, say 2000° C, the various shaped surfaces radiated heat at different rates, however, the cavities of each block radiated heat through the drill hole at precisely the same rate.

The Ultraviolet (UV) Catastrophy

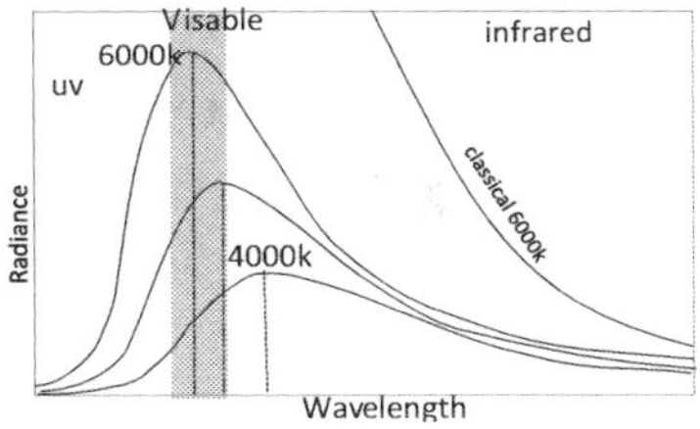

The ultraviolet (UV) catastrophe, also called the Rayleigh–Jeans catastrophe, is the prediction of classical electromagnetism that the intensity of the radiation emitted by an ideal black body at thermal equilibrium goes to infinity as the wavelength decreases.

The German physicist Wilhelm Wien (1864-1928) got very close to a solution for wavelengths with the formula:

$$R_\lambda = \frac{c_1}{\lambda^5} \frac{1}{e^{c_2/\lambda T}}$$

Where c_1 and c_2 are constants determined by fitting theoretical formula to experimental data, and *'e'* is the *emissivity* which depends on the material and the temperature.

This worked well for short wavelengths but not so well for long wavelengths

Planck and Quanta

German physicist Max Planck (1858-1947) tweaked Wien's formula realising that the cavity heat was a form of electromagnetic radiation caused by excited electrons jumping up and down.

To explain the common radiation of the cavities Planck postulated that it had to be quantized. To this end, and to fit experimental results, he daringly speculated his famous assumptions:

1. An oscillator can only have energy given by:
   $E = nh\nu$ (where E = energy, n = an integral number, ν and ν = frequency.)
2. The oscillators do not radiate energy continuously, but only in jumps or quanta. Therefore if the quantum number 'n' changes by 1 then:
   $$\Delta E = \Delta nh\nu = h\nu$$
3. An oscillator in one of its quantum states or so called 'stationary states' neither emits nor absorbs energy.

Planck announced his findings in 1900,

Major support for his theory came in 1905 when Albert Einstein demonstrated that electromagnetic radiation, including light, has the characteristics of both a wave and a particle which was consistent with Planck's theory.

Planck was awarded the 1918 Nobel Prize in Physics.
Einstein got his gong in 1921.

In 1916 Robert Millikan made the first good measurement of Planck's constant. Today it sits at:

$$h = 6.625 \times 10^{-34} \text{ joule-sec.}$$

This very small number, though significant at the atomic level, pales to insignificance in the macro world. For example, a 1 kilogram weight bobbing up and down when suspended on a spring, can only have single jumps of quanta energy of the order about 10^{-30} joule, which is so minuscule it is undetectable.

Also in 1900, Planck resolved one of the difficulties associated with radiant energy posed by a 'black body'. This theoretical body absorbs all the radiation that falls onto it, then after reaching some equilibrium temperature, proceeds to re-emit energy as fast as it absorbs it.

He concluded that these radiations were emitted in discrete amounts called quanta or theoretical photons each with an energy equal to Planck's constant *'h'* multiplied by the frequency of radiation.

$$E_2 - E_2 = h\nu$$

Quantum mechanics and Relativity

Quantum field theory merged quantum mechanics with 'special relativity'.

Every fundamental particle has its own field which exists everywhere in the universe but as a zero value except where a particle would probably be.

That is where the field has a high value = excitation.

It is so far incompatible with general relativity re gravity-dark matter-dark energy-matter antimatter asymmetry- neutrino oscillations
matter antimatter asymmetry.

Wave-Particle Duality

Wave-particle duality evolved from the observed reaction of charges (more specifically) electrons when struck by light.

How could an electron absorb so much energy from a wave to make it jump to higher orbits, if it was itself only a point charge???

This work holds that electrons have no matter. They are sink-portals which evacuate Space energy from the 'here-verse' to the antiverse at light speed.

Their effective range encompasses several atoms which means they can absorb energy broadly, from incoming me-flow of wave fronts .

They are not just points! They absorb quanta of energy concentrating it towards points.

All particles are portals. Accordingly if photons were particles/masses, they would be portals and therefore have anti-wave photons in the antiverse.

Quantum theory speculates that a photon is it own anti-photon.

The fact that no experiment can measure both the wave and the particle behaviour simultaneously is called the principle of complementarity.

Quantum theory accepts this particle duality and calls it superposition.

"My dyslexia can't help confusing superposition with superstition."

Spooky

All theories on electromagnetism, which includes light, are grounded on the perpendicularitis illusion, where magnetism not only illogically encircles its current but does so according to right handed rules.

Right handed magnetism is one 'spooky' which this work ghost-busted. It preprepared generations of physicists to accept present and future scientific spookiness.

Quantum Spooky has its origins in extensions of the 1801 double split experiment. One hundred years later, modern scientists playing with the apparatus (as they do), confirmed that light could in fact show both wave and particle characteristics.

Moreover around 25 years later still it was discovered that electrons, then atoms then molecules could behave similarly.

Entanglement And Hidden Variables

- The Einstein Podolsky Rosen thought experiment had proposed that local hidden variable could explain how entangled particles communicate.
- In 1964 John Stewart Bell proposed a criteria test to examine the possibility of hidden variables (which Richard Feyman considered a pointless exercise since quantum absolutism declared it already knew there were no variables).

- However Physicists Alain Aspect, John Clauser and Anton Zeilinger persevered which resulted in the development of experimental tools that laid the foundation for a new era of quantum technology.
  They shared the Nobel Prize in Physics 2022.

Some Bell loop-hole tests showed that communication between pairs of entangled particles or photons had happened at 10,000 times faster than the speed of light. An impossibility beyond expectations..

Bell test' experiments showed that hidden variables could not explain the entanglement of particles.

__The answer is elsewhere__
__Perhaps in the bi-verse.__

Double Slit Experiments

In 1801 with the Thomas Young double slit experiment showed that light formed interference patterns on screens demonstrating wave behaviour.

Particles passing randomly thru two slits form two groups on the screen, whereas waves form interference patterns.

This favours the Huygens model of light over the Newtonian particle model however:

A photon gun can be tuned to shoot photons one at a time thru the slits.

At first the screen pattern shows a random particle pattern. but as the photons numbers increase the randomness merges into an interference,diffraction pattern

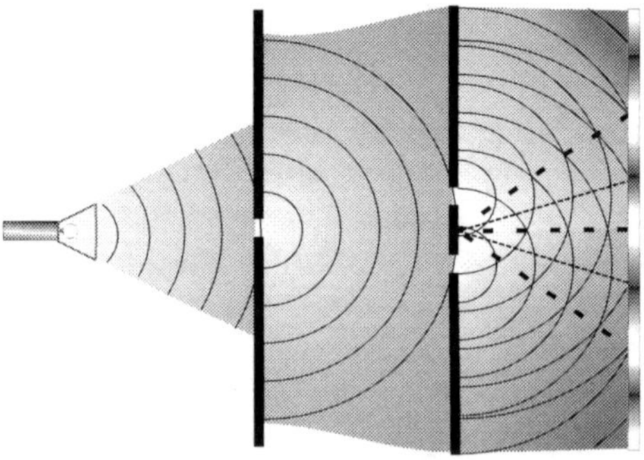

This shows even a single photon has a wave function.

If a detector is placed in front or behind a slit to record *'which-way'* a photon passed thru, the screen now shows only two groups, no interference pattern.

The watched boiling pot stops boiling, but boils again when not watched!!

Theoretically then:

- Either one of the photons or both have become unable to interfere.
- So Either one of the photons or both has lost or canceled its wave function.

•

The Double Slit Eraser Experiment.

A laser fires photons thru double slits then thru a beta-barium borate (BBO). crystal which splits them into ½ sized *(longer wavelength*) coherent entangled photon pairs, then onwards via detectors to a screen....

- If undetected a wave pattern is recorded
- If the *'which-way'* is detected at either A or B only two groups are recorded and *there is no interference pattern.*

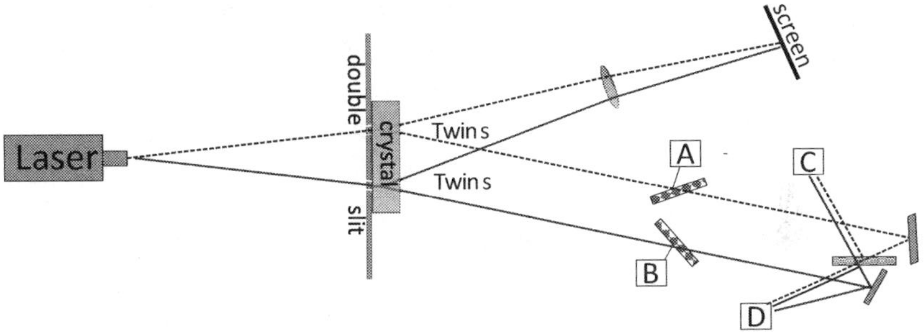

- If we tweak more and delay detecting *(delayed choice)* by distance until after the brother photon has hit the screen, there is still *no interference pattern.*
- When the photons can proceed thru an arrangement of split screens and mirrors *(the quantum eraser)* to recombine at detectors, the diffraction appears again on the screen. *There appears to be interference. (There are some disputed opinions here concerning the screen pattern additions).*

•

Bi-verse Quanta

There is an old joke between an English man, Scotsman and an Irishman about the greatest inventions ever. The Englishman says it's the steam engine, the Scotsman says it's the ship, and the Irishman says "no no no you've got it all wrong! It's the Termos flask!

???

You put something hot in and it stays hot, and you put something cold in something cold in and it stays cold, but how does the Thermos flask know what to do with it?

Paddy has no concept of a vacuum and/or what it can do; he has no real understanding of Space.

Bi-verse theory bridges the inter-dimensional equator between 'here-verse' Space and antiverse Space.

Bi-verse Energy

The first law of thermodynamics is "Energy can neither be created nor destroyed; it can only change from one form to another."

- *Science doesn't say what form Coulomb or electrostatic energy was before it became the so called 'electric fields'.*
- *Whatever it is/was, it must travel thru the charge, which defines charges as being energy portals.*

Bi-verse theory holds that the same energy that flows out from positive charges and in to negative charges returns to the positive charge, either thru some amorphous subspace, like Brane theories' predicted fourth dimension, or as I predict thru a structured antiverse.

Accordingly the first law of thermodynamics becomes:

- ***"Energy can neither be created nor destroyed; it circulates."***

- A particle or event in the 'here-verse' is its anti-particle or anti-event in the antiverse.
- They are joined the inter-dimensional equator. Time is mutual.
- An electron 'here' is a positron 'there' and vice versa. *A posilectron.*
- Coulomb energy spirals into, thru and out of a portal with the same angular momentum.
- ***Coulomb energy is inter-dimensionally entangled with it self.***

In the Bi-verse

- "Energy will only circulate in the most efficient way."
- Uninterrupted 'here-verse' energy goes to infinity.
- At every instant it is returning from infinity via the antiverse along branches which fork back to the trunk.
- Some speculate back to the "singularity".
- Interrupted energy returns directly from the point of interruption.
- ***It returns home in the most efficient way,"***
- ***Preferably thru the 'here-verse' dimensions when available.***
- ***Total kinetic energy is zero.***
-

Circulating m-wave Theory

An m-wave ie. a PTm-wave going east has a sister PTm-wave going west in the antiverse.

- In the bi-verse they combine to form an inter-dimensional standing wave.
- Orbiting electrons are bi-verse standing waves.

When bi-verse standing m-waves pass thru Cooke's double slits:

They interfere with each other unless their *which-way* is detected or *marked)*

'Marked' suggests adding something like a circulating filter, whereas the detection has really extracted some almost imperceptible delta information, from the wave.)

Whether marked or detected, (± Δ) the return antiverse wave losses its ability to interfere with the 'here-verse' wave.

This modifying cancels the standing wave and cancels the waves ability to interfere with other standing waves.

All things are bi-verse events which can behave as either particles or as waves, but cannot be simultaneously observed in the 'here-verse' to be both a particle and a wave.

Conventional Light

Light is a construct of multi varying wavelengths.

- Light of each wave length can have multi orientations.
- Each wave front is a construct of multi Huygens thin wavelets having the same wavelength as the wave.
- They are comprised of individual packets of energy called photons.
- Photons have energy and momentum but their rest mass is zero.
- Photons can be observed as being waves or wave-less particles.
- Or as some hold,particles whose waves interfere or don't interfere with other waves.

Bi-verse Theory of Light

In Bi-verse theory em-waves are purely me-flow waves which have orthogonal properties:

- Primary Pme-flow loops
- Transverse Tme–flow. slinkies.

As the wavelength of a PTm-wave gets shorter it becomes compacted into a slinky toroidal ball science calls a photon.

Perfect toroids have no external magnetic field (Gauss)

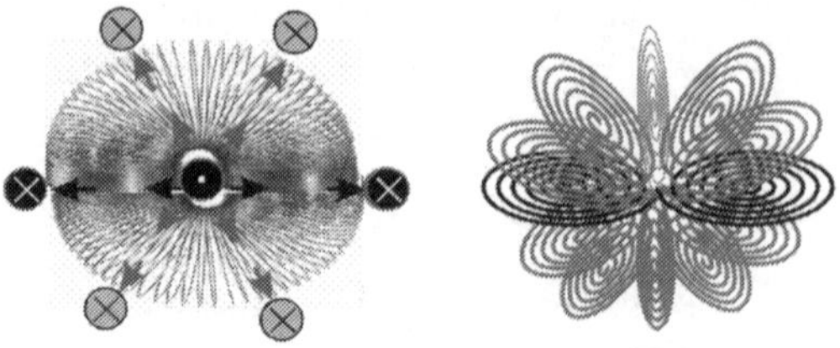

- A photon wave has 2 counter-revolving Pme-loops with transverse interlinking Tme-flow slinkies on each side.
- ***It is a Gaussian torroid therefore has totally confined magnetism and therefore imunity from external fields.***

Pme-flow wave loops oscillate in all orientations.

- Its slinky Tme-flow synchronizes with each Pme-flow
- ***All Pme-flow orientations of the same wavelength are in the Tme-flow slinkies of other Pme-flow orientations.***
- ***They synchronise/harmonise***
- Photons can split into smaller photons of longer wavelengths.

Polarized Light

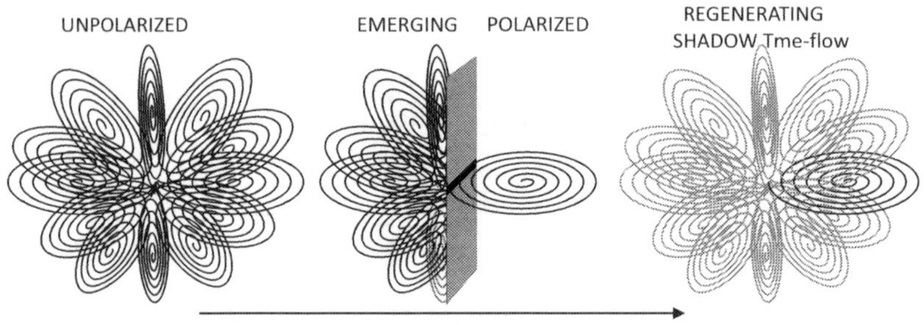

When light passes thru a horizontal polarizer,

- It looses all oblique Pme-flow

If it also looses its Tme-flow orientations then as the Pure Pme-flow emerges,

- It immediately regenerates them.
- The Tme-flow always persists.

The Aharonov–Bohm effect

Though all the me-flow in infinitely long coils is trapped in its internal vortex, this doesn't stop me-flow from infringing into the solenoid core.

Electrons zipping past the outside dip their wave wing tips into the massive internal vortex.

They get a nudge-spin passing on one side and a reverse nudge-spin on the other side.

This suggests that the light speed on one side increases while it slows on the other side, which staggers their phases; viz. interference patterns on the screen.

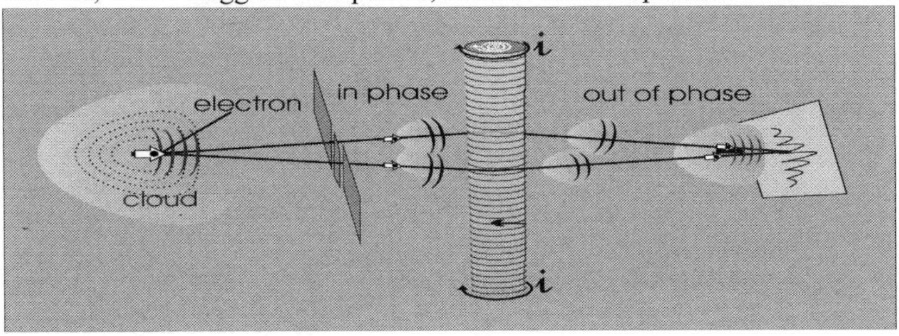

Because they are entangled the over all wave speed is light speed.

Foote; this is easier to apprehend the effect when we hold that magnetism flows co-parallel with its current, not irrationally perpendicular to it.

This challenges quantum solutions which embrace vector potentials.

The Mirror Box

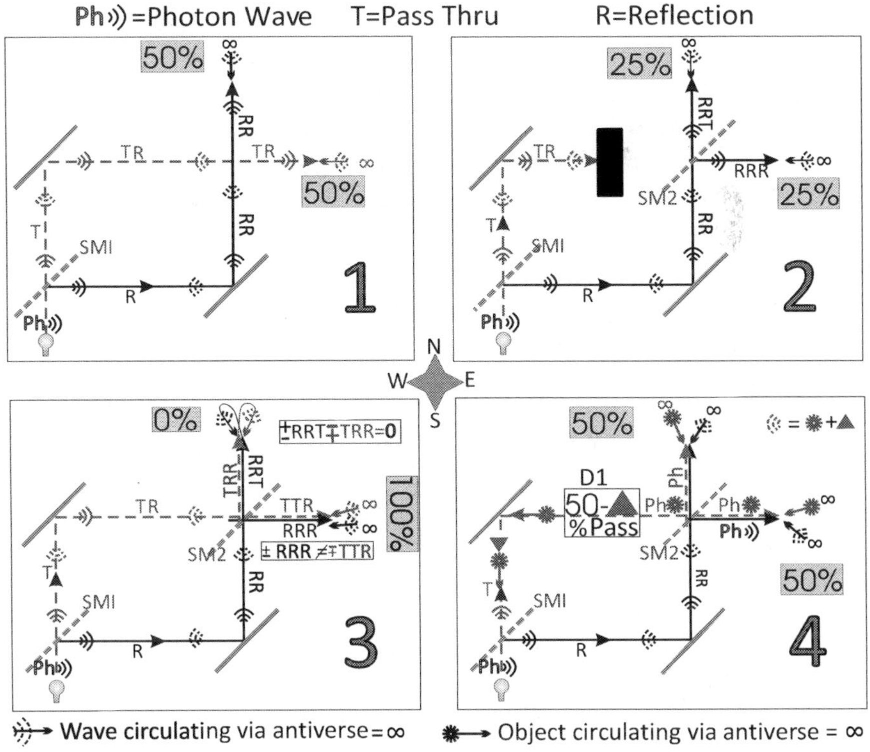

Every obstacle affects the photon

In 1637 Fermat proposed what is known as his last theorem where

$a^n + b^n = c^n$ only works when n = 1 or 2

After 358 in 1994 Andrew Wiles proved it and got a Nobel Gong.

In the addenda of my edition of 'A brief history of time' 1988, Hawkins leaves us the 'mirror box puzzle' petitioning us from time, to ghost-bust this spook.

In the Biverse Mirror Box:

All thru passages have an equal affect. **T**

All reflections have another type of equal affect. **R**

In box 2

)))→SM1 → 50% photon waves → Block→antiphoton waves →SM1
↘ 50% →SM2 → 25%→E→∞→Av→SM1
↘ 25%→N→∞→Av→SM1

In Box 3

- North-bound photons from SM2 prefer to circle back and head south via the neighboring photon's pathway. They return at an opposite phase and cancel each other's journey all the way back to SM1.
- The pathways have been affected the same obstacle sets---RRT = TRR.

From SM2 →100% photons →E

They have different obstacle sets ---RRR≠TRR at SM2

Therefore → 50%→N→∞→Av=*antiverse*→SM1
↘ 50%→E→∞→Av→SM1

Light will only interfere with light of the same polarization.

In Box 4

)))→ SM1→50%→N→E→D1.

Photon waves are detected at D1 which takes some data 'Δ' from their wave function which ceases to be complete. Partially depolarised it now cannot interfere with other photon wave functions at SM2.

From D1→Δ→ SM1

(Ph-Δ)→SM2 → 50%→N→∞→Av→SM1
↘ 50%→E→∞→Av→SM1

OR

)))→ SM1→50%→E→SM2 → 50%)))→N→∞←Av→SM1
↘ 50%)))→E→∞←Av→SM1

NB. *If photons passing thru the detector maintain their unadulterated wave function then at SM2 they would interfere like those in box 3 - viz 100% & 0%, which they don't.*

Bi-verse circulation presents this solution to spooky duality????

THE ATOM

CHAPTER 13

The statement "most computer models predict" is often a sure sign that the speaker really hasn't any definite ideas about what's happening.

The statement "such-and-such is not fully understood", means that there are considerable anomalies which confound all versions of interpretation.

This fluid dynamic theory of atomic structure is without doubt "not fully understood." It is after all based on this work's radically correction of magnetism which hopefully makes the atom more transparent.

It holds that all the atomic forces are caused by the fluid movement or fluid transfer of Space energy.

This energy emerges from electrically positive sources and disappears or de-materialises via electrically negative ones.

This necessitates the real or virtual existence of other 'not even partially understood' dimensions.

Quantum theory brilliantly provides the mathematics of the atom.

This chapter follows some fluid dynamic logical threads to suggest or guess at the mechanics of of the atom.

Atomic Scale

- If an atom were the size of a football stadium the nucleus would be the size of a marble in the center.
- Since electrons have a mass 1836 smaller than protons, they would be pinheads rocketing around the stadium perimeter.
- They travel at around 2000 kilometers per sec.
- ie 5.000 laps per second.
-
- These pinheads, though minute have huge centrifugal forces.
- It is the nucleus's task hold them in orbit.

- The diameter of the Hydrogen atom is around 1/1000 the wavelength of visible light.
- It has in its nucleus one proton with one electron in orbit.
- The radius of the electron orbit is around 5000 times the diameter of the nucleus.

Heisenberg Uncertainty

The uncertainty principle is very German. Es schmeckt nach Sigmund Freud's analysis of self-harm and suicide.

- The less a patient is restrained, the greater their behavior to self-destruct.
- The more a patient is restrained, the less their behavior to self-destruct.

Gobbling Goblins

- Fluid dynamics holds that the electron is a sink gobbling up a specific quantity of Space energy/ether per nanosecond irrespective of its velocity.
- It must consume this energy at the same rate as what gushes out from the proton, and it does this in the most efficient way possible.
- It is always moving into the densest zones of energy it can find, therefore away from a less dense ones.
- However once there its very presence reduces that density, therefore it then heads towards the next dense zone, which it instantly gobbles and therefore also de-pressurises.
- ***It can never be anywhere specific, because at any instant it self destructs its reason for being/going there.(Heisenberg uncertainty)***
- Obviously no identical (self respecting) goblin will want to go where it can't get fed.
- Nor will it want to occupy any orbital which limits its intake.
- However there is sufficient food in an orbital for a sister electron if it spins in the opposite way to the incumbent brother electron.

Together they ingest the most energy with the least effort.

Electron Waves

1801 Thomas Young's double-slit experiment demonstrated wave properties of light

In 1927 Davisson and Germer used the experiment to show the wave properties of electrons.

When the separation of the vertical splits is less than or near equal to the approaching electron's wave length, the wave reconstitutes on the far side of the splits with constructive and destructive interference, just like the light does.

When the slit separation is a little larger, the wave continue through each as a circular wave with a strong centre.

They are separated too far to interfere with each other which helps define the effective peripheral range of the electron.

Experiment shows that the wavelength of an electron is inversely proportional to its kinetic energy. The faster it goes the shorter the wavelength and the higher the frequency.

A theoretically stationary electron has an infinitely long wavelength.

The Orbiting Electron

Orbiting electrons are controlled by the Coulomb force. This work has described this force as the result of pressure differences between the outputs of the nucleus and the sinks of the electrons midst a mid-pressure atmosphere of Space.

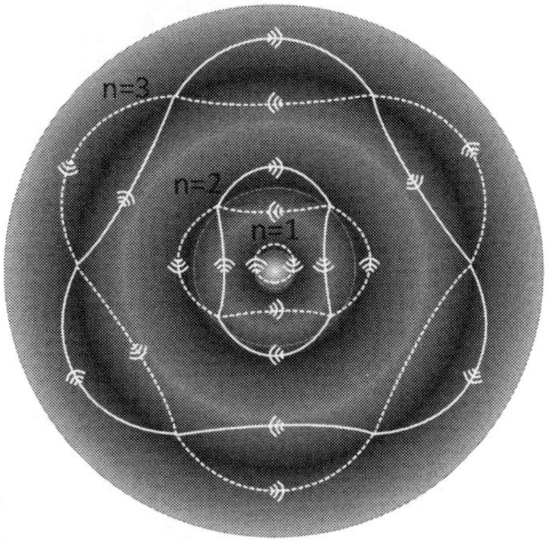

Some observable properties of electrons are:

- Each electron orbits the nucleus in orbitals which have known probability shapes.
- They spend 95% of their time in these electron clouds.
- Integer multiples of the wave length can only occupy specific orbits.
- They can't be where interference is non constructive.

Convention theories that the electrons are going so fast (≈ 1% c) that each two revolutions pair and overlap and constructively interfere with each other thereby become standing waves.

Bi-verse Quantum Orbits

abstract

Since electron orbits are probable and therefore approximate, it's a big ask for two consecutive revolutions to perfectly align to become standing waves.

Bi-verse theory holds that electric fields are quantized a***nd speculates that the electron forms a standing wave with its antiverse wave.***

Nuclei generate spherical waves with density spheres where gobbling electrons prefer to feed.

- They roll in and out of these zones as they feast.
- If the atom heats up they can roll up and over into an inner wave-front and gobble their way at higher energy.
- When the atom cools they slide back over a low pressure hump into an inner orbit simultaneously releasing energy in the form of a photon.

Spin

When convention refers to a particle as having 'spin' fluid dynamics interprets as **'what enters or exits the portal spins'.**

As energy voids into the electron portal it swirls, (say clockwise) just as water swirls into a bath drain and correspondingly is spurts out in the antiverse with the same angular momentum.

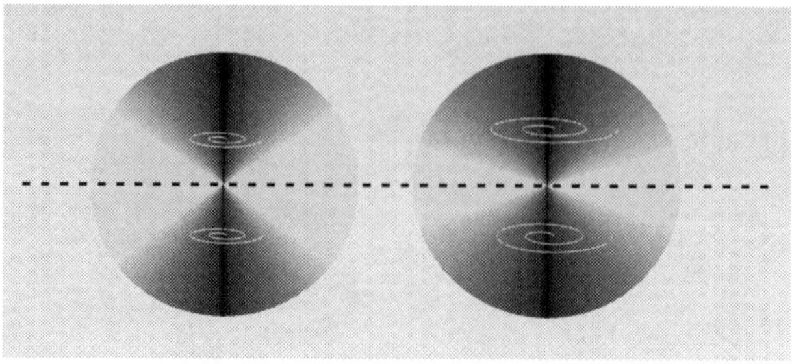

What enters a 'here-portal' exits in the antiverse with the same angular 'Space-momentum'.

Does the portal itself spin with the flow, against the flow, or not at all is ???

It is known that both protons and neutrons have a so called half spin like that of the electron. The proton spins like Hero's aeolipile.

As the proton goes faster it spins more rapidly. There is some evidence to suggest that at high speed the core of the proton spins faster than its cloudy crust.

Intriguingly the Earth's core and magma spins slightly faster than the Earth's crust rotates.

Orbitals

In neutral atoms, for every proton in the nucleus there is an electron in an orbital.

Whatever happens to an atom's electron is a reflection of something happening in the nucleus and vice versa.

Schrödinger's equations describe the probable shapes of the orbitals.

Electrons in atoms first fill the spherical '1s' orbital. *(all the 's' orbitals are spherical bands)*

Next the '2s' orbital fills.

After the '2s', they progressively fill some of the dumbbell shaped lobed 'p' orbitals.

Here the ee-flow from additional protons generates too much local pressure to be dispersed spherically.

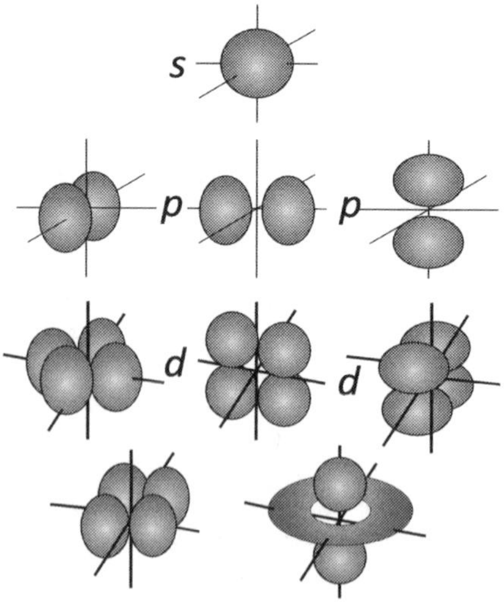

The proton now resorts to spurting out geysers of energy.

The equations tell us these geysers come in pairs in the shape of lobes, and are distributed on opposite sides of the nucleus.

- Each double geyser generated by a singular proton can support one electron.
- Each double geyser generated by two protons can support two electrons if their spins are opposing.
- After filling some of the 'p' lobes, one then two electrons fill another 's' orbital band, i.e. the '3s' orbital.

These electrons more or less re-establish the spherical shape to the atom.

The nucleus now generates more geysers, turned lobe shaped by hungry electrons patching Coulomb leaks.

These poke through the '3s' barrier band.

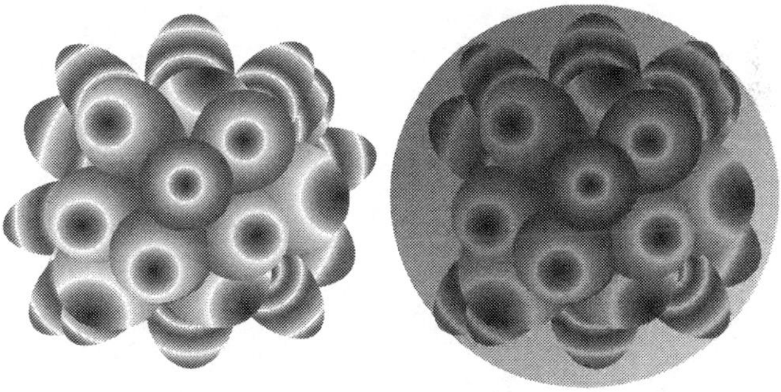

This process of lobes getting covered by spheres, continues with variations right up to the '7s' orbital. *(See the Aufbau Principle and Hund's rule.)*

Quantum Bar-Codes

Swarming around the atom's nucleus at close to 1% the speed of light is a race of electrons in strange yet predictable *(according to probabilities)* orbitals, each one being unique with its unique bar-code of quantum numbers.

While there are righteous mathematical formulae for these probabilities, the picture mathematics presents is abstract.

This work sees quantum theory as a digitalized version of the cosmos.

The atom is a fluid dynamic entity where diverse tides ebb and flow in complex interference and harmony.

Over 2½ millennia ago Pythagoras conceived the 'harmony of the spheres', where all the planets orbit a central fire. The planets then were separated from each other by distances which corresponded to the harmonic lengths of strings. He conceived a musical universe.

Overview

The nucleus radiates energy spherically at a constant rate with constant flowing geysers of energy bursting forth. Spinning electrons rove around the nucleus gulping this high pressure energy where it leaks out.

This results in a lumpy electron cloud surrounding the nucleus.

To exist in this volcanic environment everything spins, which blends the whole into a symphony of economic ee-flow energy.

Electron Capture

It has been previously said that a moving electron gets more of its intake from head-on.

When it approaches a proton it blends its intake with the proton's output which results in forces which pull the particles together with electric force, which when the electron is captured, becomes the centripetal force on it.

Positively charged atoms *(positive ions)* capture electrons with either their lobe or spherical 's' orbitals.

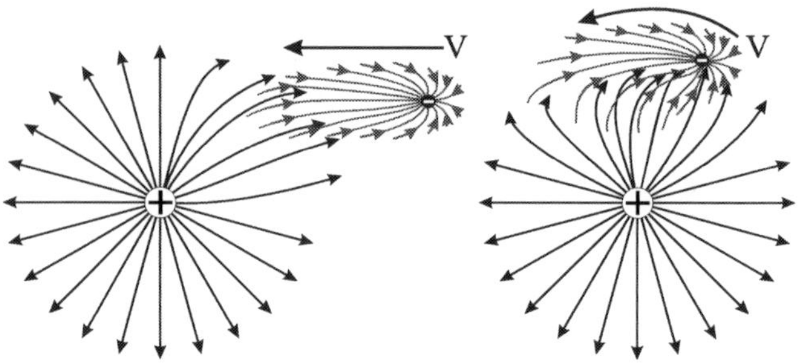

Annihilation

When enough positrons *(i.e. millions)* are collected in the here-verse and are revolving in a cyclotron's storage ring, they can be accelerated to near light speed and directed to collide head on with a near light speed stream of opposite moving electrons from the main cyclotron ring.

They are mutually attractive, however according to fluid dynamic theory, collisions only occur when their opposite moving wavelengths conjoin at the instant of impact.

Here also attraction is augmented by the spiral output from the positron matching and melding with the like rotating spiral intake of the electron.

Most of the collisions are imperfect leave an assortment of flotsam and radiation in the here-verse, and presumably also in the antiverse.

Perfect collisions annihilate each other and become pure radiation.

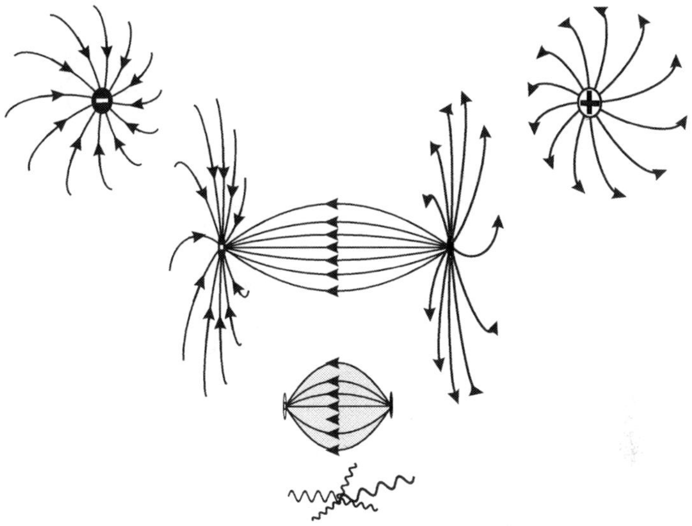

Good positron emitting isotopes are created in cyclotrons then used in Positron Emission Tomography (PET) scans.

Electron Microscopes and Tunnelling

The wavelength of light from infrared through to ultraviolet is far too large to see a tiny atom.

X-rays are small enough but too energetic. Instead of diffracting around a nucleus they crash into it and knock bits off'

High energy electrons are able to see large molecules like cells but can't go beyond objects smaller than 10 nanometres i.e. 10^{-7}metres.

Tunnelling microscopes allow electrons from a tiny hovering probe to tunnel or sneak into the surface of atoms. The rate at which they tunnel is proportional to the distance between the object and the probe. Depth resolution is around 0.01 nm and lateral resolution is around 0.1 nm.

Since electrons are crossing over the gap a measurable current is flowing. By adjusting the hovering distance to maintain this voltage, as the probe moves over the surface, details of the atomic structure are revealed.

This is another clear example of electrons being waves of ee-flow, not matter.

Photo-electric Effect

In 1887, the very year that Michaelson and Morley claimed they executed the *ether,* scientists playing with light beams, *(as they do),* discovered the photoelectric effect. Light shining onto a negative charged metal disc caused a current to flow across a gap into a surrounding positively charged ring. Aspects of this created turmoil in the wave-works.

A decade later it was found that light rays excited the newly discovered electrons with enough energy to jump across the divide to the positive ring. The electrons carried the current.
Wave theory could not explain why:

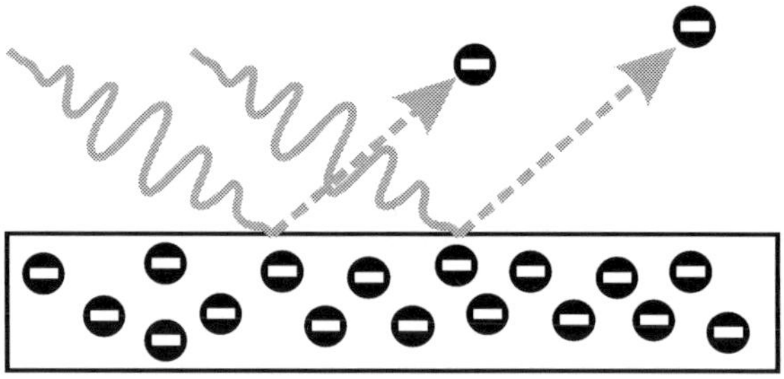

- Increased intensity of light increases the amount of electron emission but does not increase their energy.
- Each surface has a characteristic cut off frequency below which the emission ceases, irrespective of the intensity of the beam.
- Theoretically electromagnetic light, feeble or strong, would need to be soaked up by an area of the surface equivalent to more than a few atomic areas, thereby collecting enough energy to excite the electron to escape. This should take some finite amount of time; however there is no detectable time lag. This is exemplified when the photoelectric material is a gas. Here the energy for the electron must be sucked out of the beam by its single mother atom but this is insufficient.

It was theorised that if light were comprised of bundles of energy called photons, then these little bouncing snooker balls would be capable of breaking electrons out of the pack.

Fluid Dynamics and Surface Electron Ejection

Electrons are not just dots. They are perpetually motivated light speed portals whose motion is driven by hunger. In the here-verse they always drawn towards the biggest positive carrot near it, and/or away from a negative empty larder.

In metals free electrons zip about at fantastic average speeds of about six million kph. These little Tarzans with electrostatic arms swing from atom vine to atom vine taking quick spins around one in 200 nuclear trees.

All the time they are ingesting in gulps a fixed quantity of Space-energy from multiple directions. They are totally integrated with their immediate and close by neighbourhood and interact at light speed with it.

The arrival of light rays brings rain into the jungle.

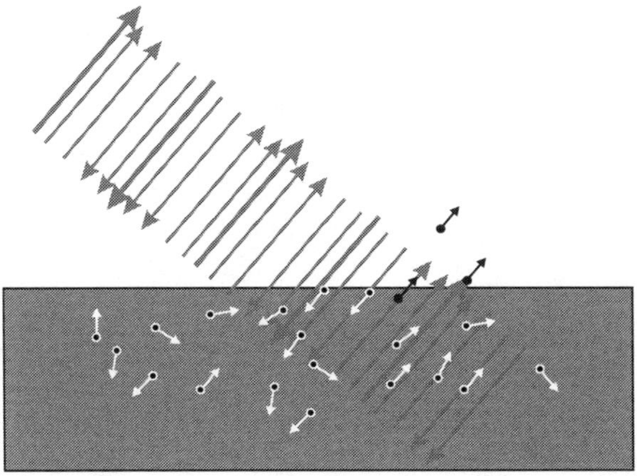

Non-polarized light is a mixture or jumble of orientations of polarized light.

This work has described an electromagnetic wave as a longitudinal wave travelling through Space which is at ambient Space pressure.

It is comprised of primary Pme-flow loops and transverse to it is Tme-flow

The Pme-flow is parallel with and part of the wave front and is transverse to propagation.

Light is a high frequency m-wave whose wavelength is around 1000 times the diameter of atoms.

- Depending on the phase of the incident wave, the free electrons either pull themselves up or down in the passing transverse Pme-flow zones.
- Most of the time the host atomic lattice contains them, but sometimes, when they are traveling in a path which coincided with me-flow up-phases of the frequency of the light wave, they get a boost which enables them to reach escape velocity.

The direction of distribution of emitted electrons peaks in the direction of polarization plane.

The time lag between the incidence of radiation and the emission of a photo electron is very small, less than 10^{-9} second.

Atomic Light Absorption and Emission

- The radius of an atom is around 10,000 times the radius of its nucleus.
- It is around 1/1000 the wavelength of light.
- At any instant, polarized light covers a strip one thousand atoms wide.

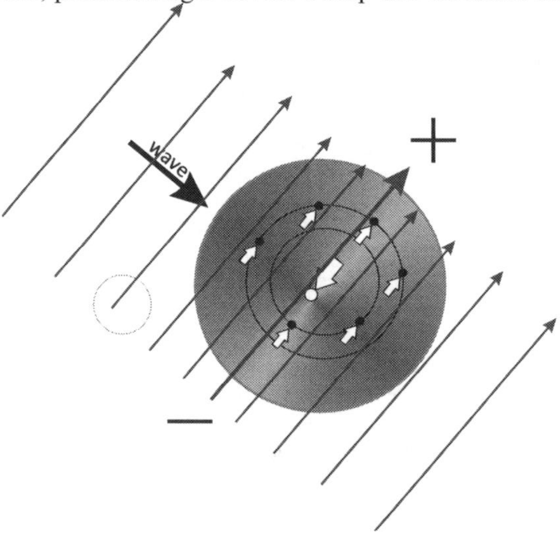

- When a Pme-flow zone of polarized light penetrates an atom, electrons are forced one way and the positive nucleus the opposite way.
- Depending on the phase, electrons on the positive side of the wave get forced away from the nucleus which is simultaneously force the opposite direction.
- Simultaneously electrons on the negative side get pushed towards the nucleus which is pushed towards them.
- The atom is stretched as electron orbits become more elliptical.

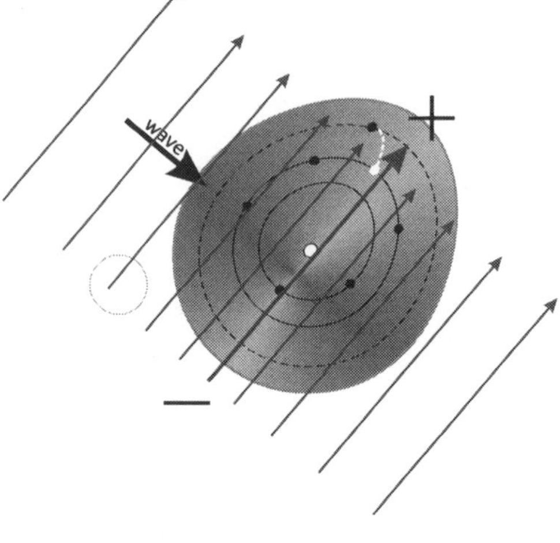

- When the frequency of the polarized light is in sync with the forces in an atom, the atom will gain enough energy to allow an electron to jump outwards to a higher energy orbit, typically in a few nanoseconds or less.

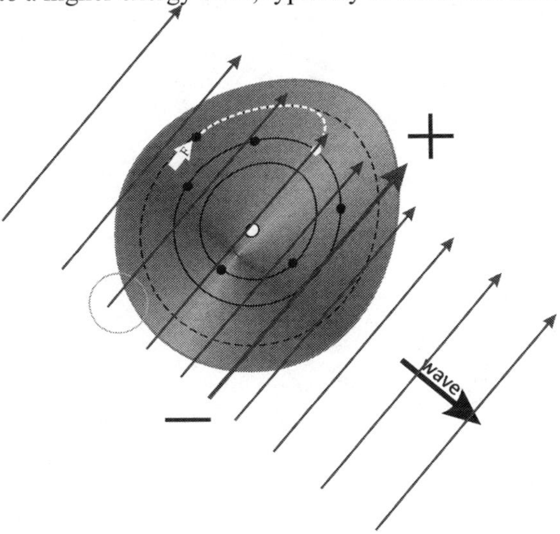

- The Pme-flow wave-front zone acts weaker on the atom as it passes through.
- The electron loose energy and flips back into its old less elliptical but stable lower energy orbit.
- To do this it must shed the energy it had gained and it does this in the form of a tiny light wave photon.

The incident light wave carries on albeit having lost a fraction of energy to the atom.

The direction of distribution of emitted electrons peaks in the direction of polarization plane.

Possibly:

- Phases might combine their effects;
- One phase could push down and the electron could then rebound in sync with the incoming up phase *.perhaps?*

Faraday Effect

In 1845 Michael Faraday discovered that polarized light in a transparent medium like air or water, rotates as in passes through a cylindrical coil in the direction of the coil current.

The vortex affects a rotation bias in the medium's quadrillions of electrons which is affects the light beam path.

Even though Faraday or even Maxwell might have thought differently, post Michaelson and Morley it was presumed that since there was now no medium, the Faraday effect could not effect light in a vacuum. Knowing this no theorist would bother testing it.

This work holds the photon is a torroid and imune to transverse external me-flow forces, however here it is totally submerged in rotating space and possibly has a detectable nano-rotation.

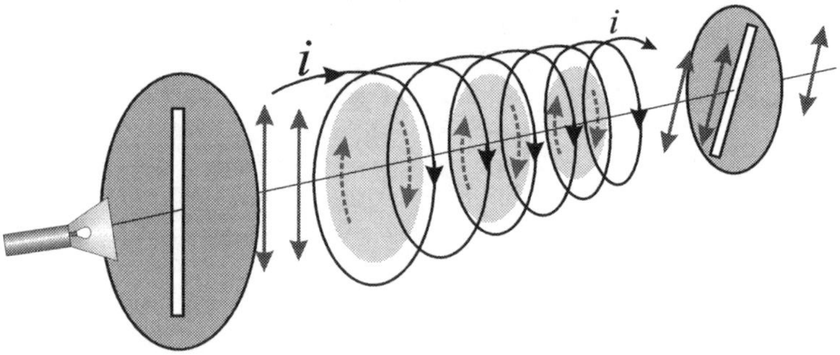

In 1933 Max Delbrück attributed the scattering of gamma rays to the polarization of the vacuum by a coulomb 'field'.

This lends weight to the idea that the Faraday effect could influence a 'pure?" vacuum.

This chapter on 'the atom' has been confined to analysis in the 'here-verse' and represents only half the work being done.

The equivalent work is being done on the underbelly of the charges in the antiverse

NUCLEONS AND NUCLEAR FORCES

CHAPTER 14

The Nucleons

So far we have only looked at the source of positive charge being a proton which delivers high pressure ee-flow to the here-verse.

The previous chapter showed that a high speed electron has a wavelength too large to <u>see</u> the atom, however it is deflected and/or scattered as it hurtles past or into a nucleus.

The nucleons, protons and neutrons, have properties of mass, charge (ee-flow) and magnetic moment (me-flow). The latter two are as we have seen are spread out in Space around the nucleon therefore electrons with their own ee-flow and me-flow deviate as they wizz through it.

When they smash into a nucleon it either recoils or scatters into various bits of energy.

These events provides data which has helped to define the inner workings of the protons and neutrons.

Electron scattering reveals that:

- The positive charge within the proton is distributed in a smooth cloud.

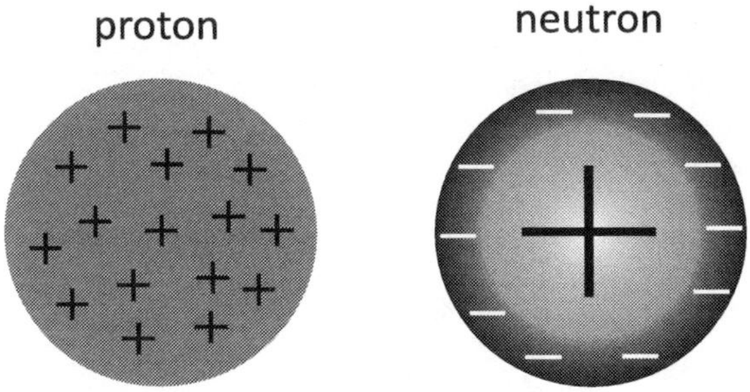

- The neutron has a positive core surrounded by a shell of negative charge. It might have a light positive fringe.

There is some discrepancy between these results and those of Robert Hofstadter, which describe these nucleons as each having a positively charged dense point, surrounded by two intermingling layers of meson cloud.

The neutron appears electrically neutral to all other charged objects until it gets very close to them.

- As it closes on positive charges it becomes polarised which enables it to bond with them. *(via quarks)*
- Evidence of this polarisation, (though indirect), is found in fusion reactions.
- Fast moving neutrons will bounce off heavy nuclei because the neutron has not had sufficient time to marry its polarised face with a reciprocal face on the nucleus. Slow neutrons have enough time to engage and meld.

The only place one finds a neutron in nature is when it is embedded in a nucleus. Artificially it can be bumped free of some nuclei where it is anything but stable. In free Space it will fragment in the mean time of about 900 seconds.

Nor is it stable in certain radioactive nuclei, where it fragments in a process known as beta-decay.

Even when embedded in nuclei its elements are in constant flux, since it is constantly influenced by its nucleon neighbours and the motions of orbiting electrons.

Electrons and the Strong Force

- High energy electron beams are used to probe the inner sanctums of nuclei and revealed much of what is known about them.
- According to conventional physics they achieve this because they are unaffected by or immune to the strong force, but this is 'not fully understood'.
- Since only slow neutrons can meld with some nuclei and fast ones can't, it suggests that the strong force needs time to hook its fish.
- Similarly it seems plausible the much higher energy electron beam also won't take the nucleon's bait
- Recall that induction was the consequence of the electrons inability to adjust quickly enough to change.

The Strong Force Facts

(1 fermi (fm) = 10^{-15} meter)

- Protons and neutrons have diameters of about 1 fermi.
- The strong force holds protons and neutrons together in the nucleus while not letting them get too close.
- At closer distances than 1 fermi the force becomes repulsive.
- At further distance than 2 fermis the force dies off rapidly.
- Protons and neutrons have diameters of about 1 fermi.
- The strong force only extends 2 fermis beyond a nucleus.
- Both the proton and neutron have a spin quantum number of ½.
- The proton's magnetic moment is 2.8 times larger than the electron's.
- The magnetic moment of the neutron is slightly smaller than the proton's.

- If there is an even number of nucleons in the nucleus it doesn't spin.
- If there are an odd number of nucleons in the nucleus it does spin.
- The radii of nuclei found in nature range from 2 to 8 fermis.

The accepted conventional analogy equates the strong force as powerful spring which prevents nucleons from getting too close or to far apart.

The Strong Force Hydraulic Analogy

In the following graphics the T-sections represent protons and the union represents the neutron.

The graphics show how general plumbing could generate forces similar to the strong forces operating between nucleons.

The T-sections here are submerged in a fluid,
be it liquid or gas

- Fluid is pumped up the vertical pipes into a T-section whence it splits and gushes out equally in opposite directions.
- When two T-sections are constrained to always align, the increase in near-side pressure causes them to repel each other with a force which multiplies exponentially as the distance between them closes.

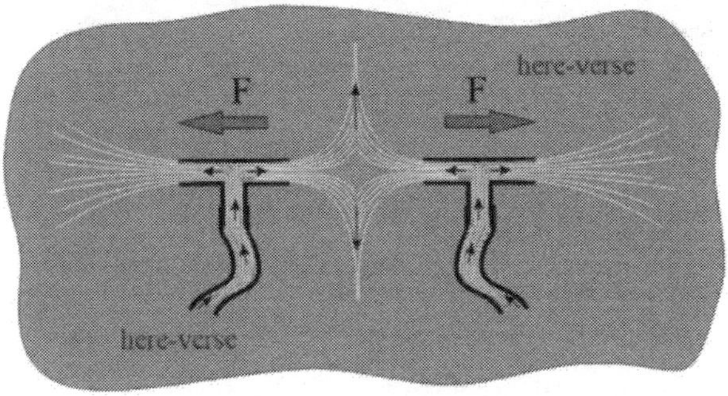

- When a union is inserted between T-sections which has exits into a lower pressure zone *(like in the antiverse),* the near-side pressure is now lower than the far-side pressure
- The gushing far-side fluid now thrust the T-sections together.

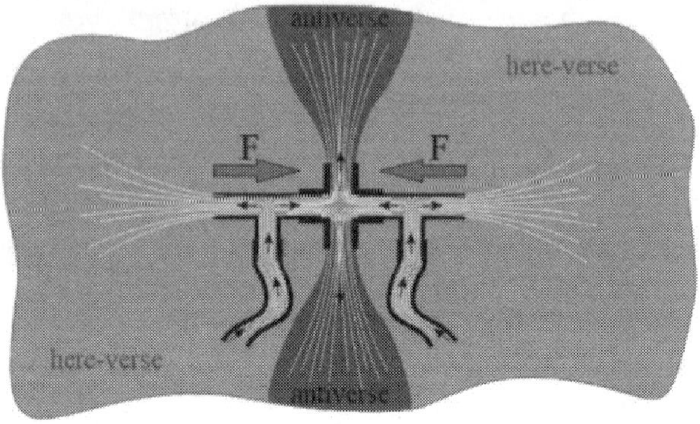

- If the T-sections get too close together, pressure in the union backs up and the T-sections repel each other again.
- In between these events there is a point of equilibrium about which T-sections oscillate.

This is a strong force type hydraulic spring.

In the here-verse:

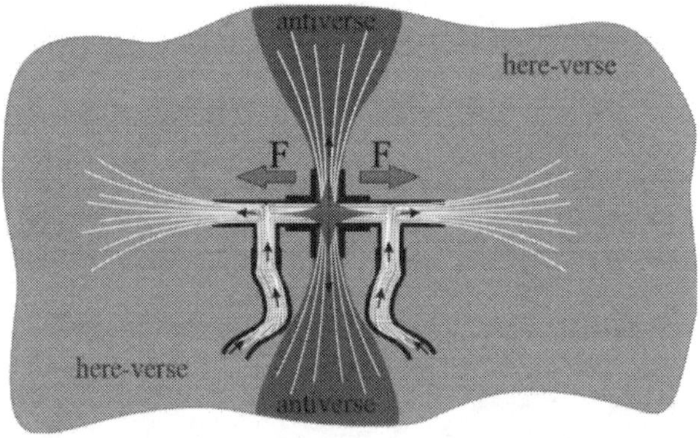

Up-quarks which dominate the proton draw Space energy/*ether* in from the antiverse (viz. up the hoses) and pump it out into the here-verse.

- Two stand alone protons have too much out-pressure to get close to each other.
- A neutron sandwiched between them aligns its side down-quarks to inhale some proton pressure sending it into the antiverse.
- This enables the protons to get very close to each other.
- If they get too close the neutrons down-quark chokes limiting the protons proximity

They now find their equilibrium attract/repel point which they minutely oscillate around.

Antiverse

In the antiverse the hoses vacuum the Space energy/*ether* out of the anti-proton pair

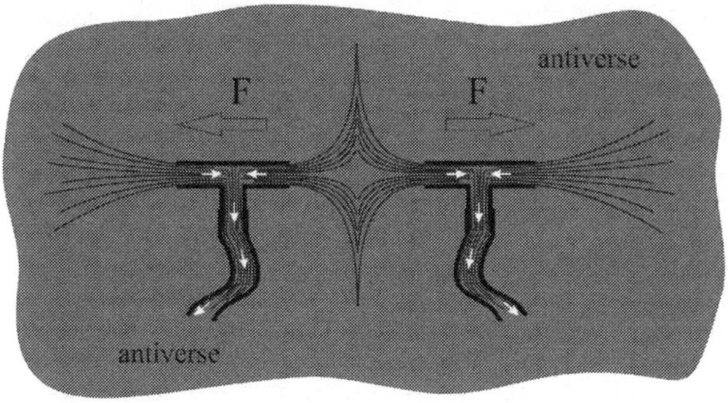

When an anti- neutron is sandwiched between them it directs Space/energy/*ether* from its anit-down-quark into the anti-protons.

The anti-protons draw themselves into the higher pressure feast.

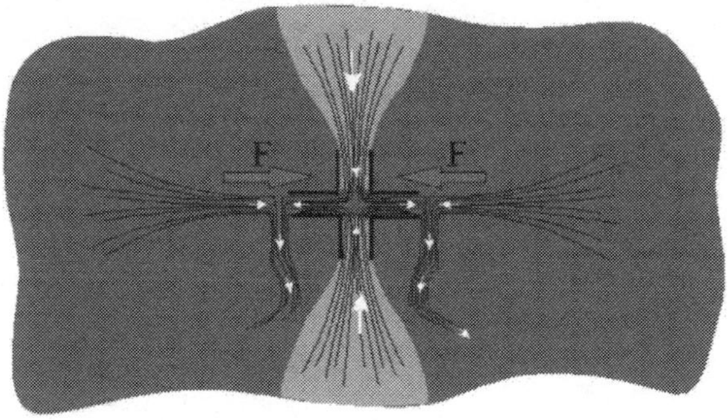

When they overindulge the passage chokes

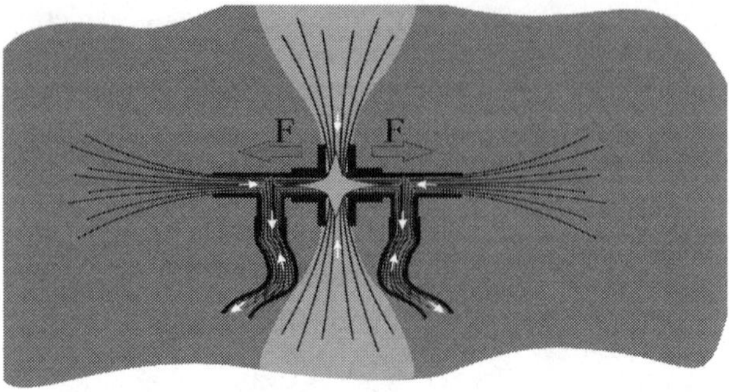

The strong force is a combo of here-verse and antiverse forces

The Volcanic Nucleus

Conventional physics has connected to electromagnetism the weak nuclear force, whose range extends for only 1/10th the radius of a neutron.

This work, with its unique interpretation of electromagnetism theorises here-side nuclear decay so:

- Nuclei are an assembly of parts either gushing and vacuuming energy.
- In total they are positive charged therefore gushing, which they try to do in the most efficient and harmonious way possible, but nothing is perfect.
- The bigger the nucleus gets the more jet-flows and geysers their parts have to organise.
- Sometimes heavy nuclei can't quite cope, and the jet-flows experience both constructive and destructive interference.
- This can result in violent explosive eruptions on or under the surface of the nucleus.
- One type of eruption ejects an electron which is recognised as a beta wave moving close to the speed of light.
- Another type of eruption is so violent it can launch a helium-4 nucleus which is recognised as an alpha wave moving at 1/20th the speed of light.
- Convention attributes these eruptions to the 'weak force'.

Ejecting the flotsam quells the magma.

QUARKS &ANTI-QUARKS

CHAPTER 15

Under certain conditions the underbelly/inside-out of 'here-verse' portals can be materialised, or can even randomly pop up in our observable 'here-verse' as anti-particles or so called antimatter.

Apart from positrons these have extremely short life spans, *(usually less than a billionth of a second).*

Fluid dynamic theorises that these anti-portals are actual inversions of 'here-verse' portals themselves.

They are joined at the hip.

They are inter-dimensional biverse dipoles

The Electron-Positron-Portal

The following theoretical biverse' graphics show one half of particles in our 'here (1 2 3 D) verse' and their other half, separated by a plane, in the 'there' 4 5 6 D) antiverse.

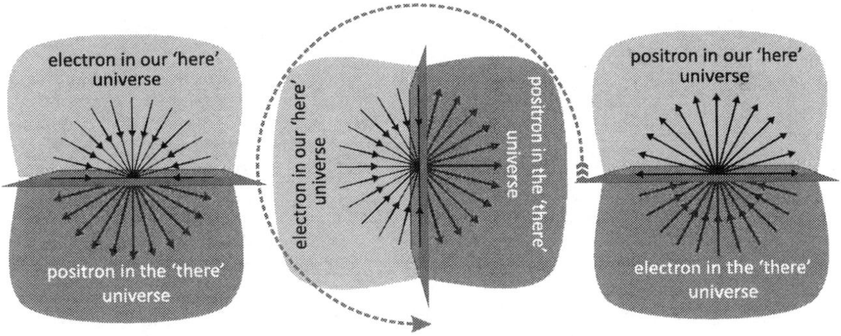

The electron-positron-portal voids energy from the electron-side in the 'here-verse' and delivers it through its positron-side into the antiverse.

- Certain conditions and forces can cause it to invert and appear in the positron-side up in the 'here-verse'. Simultaneously an electron has flipped into the antiverse where is seen by an observer there as an anti-positron.

The Subatomic Particle Zoo

Physics in the 21st century presents us with a population of scores of subatomic particles sporadically created from ultra high energy collisions which can only exist for a flicker of time e.g. milliseconds and much less.

Many particles are detected in the upper atmosphere caused by cosmic ray collisions with air. Many are detected in indiscriminating cyclotron collisions.

This work holds that particles are portals through which jet streams of energy spiral. When particles collide they either ricochet or fragment. These fragments are mini-jet-streams. Some of these remain joined at the hip with their anti-mini-jet-streams therefore maintaining opposite charges.

Part of this group are the leptons with a charge of -1. These are the electron, the muon which is 207 times heavier, and the tau, 3500 times heavier.

Leptons are 'here-verse' particles of lepton portals with their anti-electrons, anti-protons, anti-muons, and anti-taus in the antiverse.

Some collision fragments are disturbed enough to disconnect from their anti-siblings therefore possibly becoming conical or ellipsoid whirlwinds or eddies in either the 'here-verse' or the antiverse. This might describe lepton-neutrinos and anti-lepton- neutrinos which have no charge and little if any mass, but lend their identity spin to particles they bond to.

The most prodigious of the fragments are the quarks.

The most 'normal' of these are up-quarks and down-quarks.

First Generation Quarks-Standard Theory

- All the bonding of nucleons is governed by the strong force between the so called first generation quarks.
- Protons are comprised of 2 up-quarks *(each charged 2/3 positive)* and 1 down-quark *(charged 1/3 negative)* giving a total a charge of +1.
- Neutrons are comprised of one 2/3 up-quark and two 1/3 down-quarks giving a total charge of zero.
- Up-quarks are geysers which expel energy.
- Down-quarks are sinks which void energy.
- Up-quarks and down quarks are strongly mutually attracted.
- They are absolutely not fully understood!

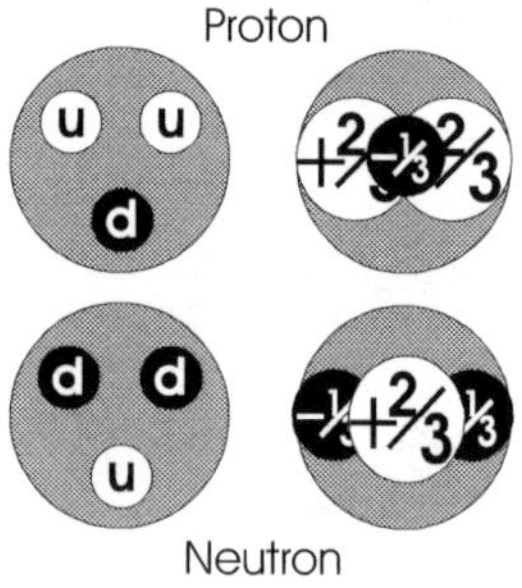

The strong force between quarks is so great that they can never exist by themselves, so our knowledge of them is based on the statistics of their performance in 2-quark *mesons,* or 3-quark *baryons.*

The following diagrams attempt to explain experimental results with logical fluid dynamic theory.

Bi-verse Graphic

In the bi-verse all particles in the 'here-verse' (our one-verse or universe) are their antiparticles in the antiverse. Under certain high energy conditions particles can flip inside out and expose their antiparticle underbelly in the 'here'

Nucleons and its Anti-nucleons

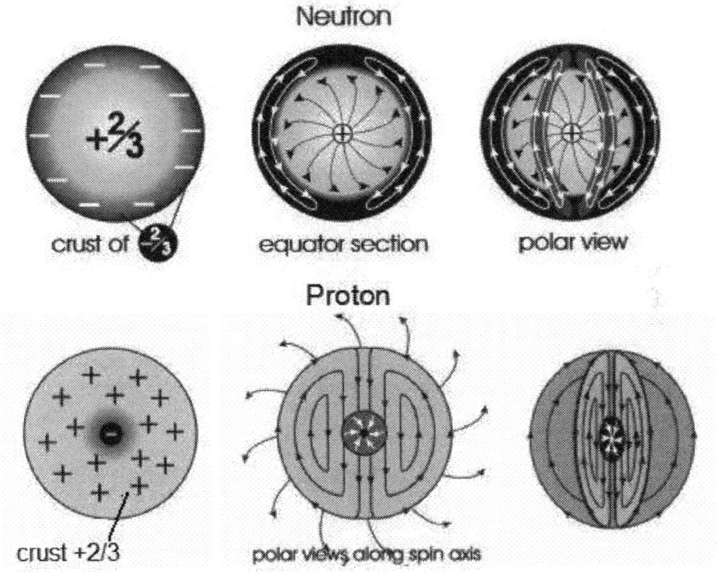

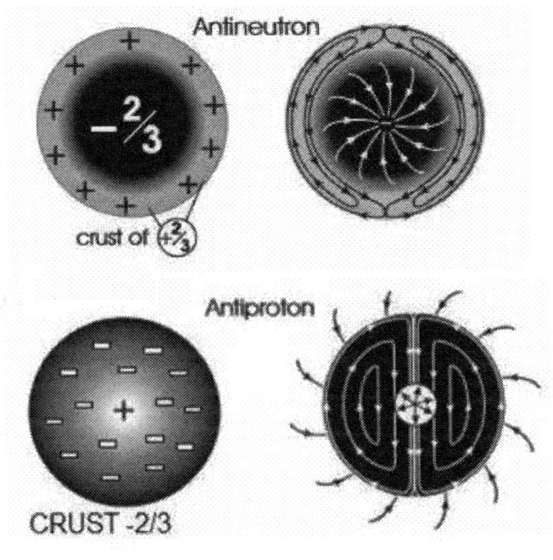

Bi-verse Cross Section Simulations

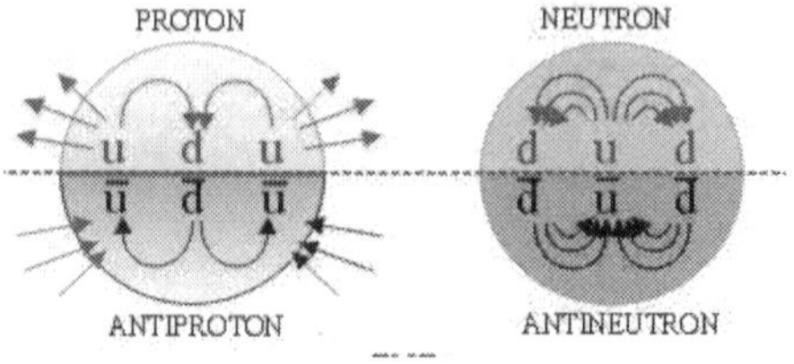

The dotted centre line in the bi-verse graphics represents the inter-dimensional equator between the 'here-verse' and the antiverse.

. These graphics provide some hint as to the possibility of weak polarization of nucleons, as for example, the neutron.

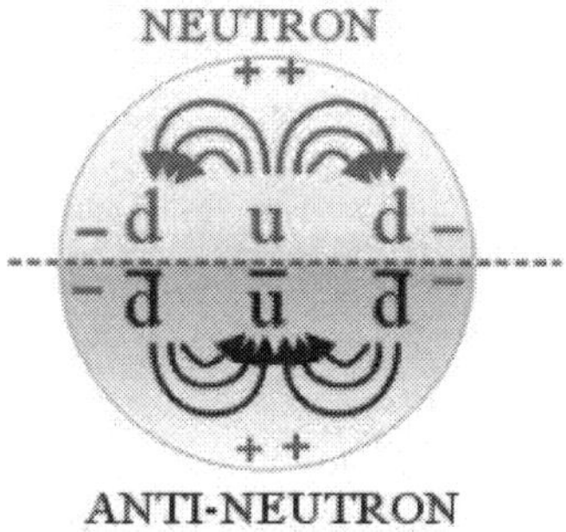

The opposite charged quarks in nucleons are electrically attracted to the central quark causing compression. If the one or both of the outer quarks drift apart the electrostatic attraction draws it/them back in.

If one or both come in too close, the exit to the other dimension gets choked and the back pressure pushes it/them back.

The quarks push-pull in a state of constant oscillation/vibration.

The polar opposite process to the 'here-verse' push-pull is simultaneously happening pull-push in the underbelly antiverse.

Breaking this Proton-Neutron-Proton Helium 3 nucleus down to its quarks.

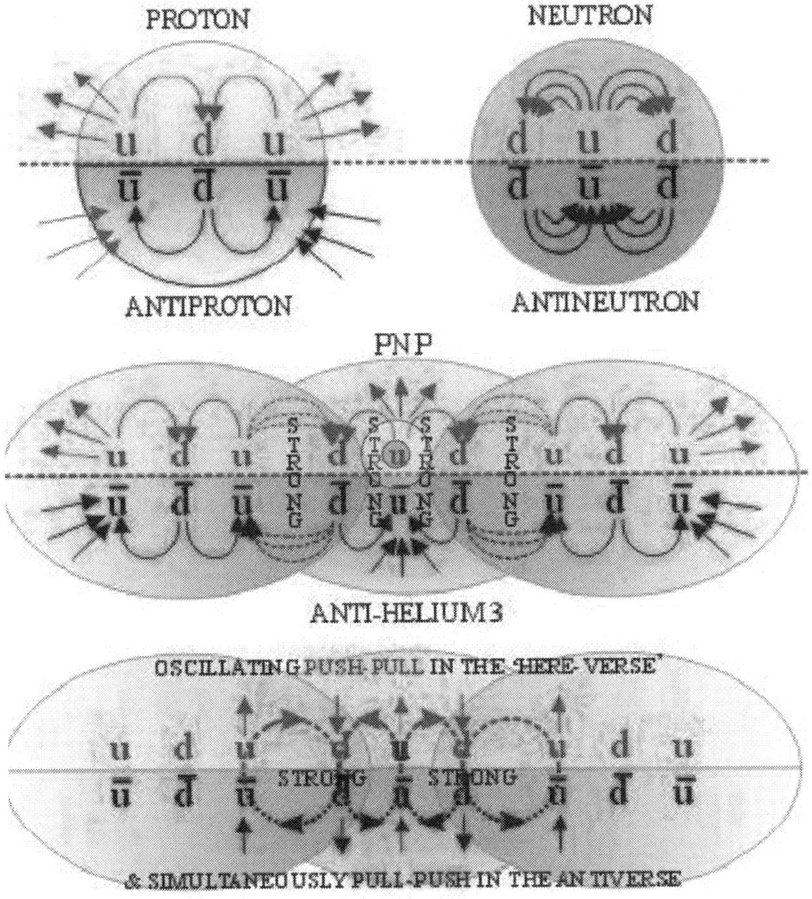

The central back pressure zone around of the neutron is energetic therefore has mass. It oscillates push-pull in the 'here-verse' and simultaneously pull-push in the antiverse. If the proton neutron bond breaks, this energy is released.

The energy of nucleon mid-pressure zones varies with the number of neutrons in the nucleus voiding part of the proton's output to the antiverse.

Science calls these hydraulic energy springs "Gluons"

All nucleons spin

- These diagrams draw attention to the involvement of the quarks with spin.
- They demonstrate zones of varying intensity close to the hub which lead to the polarisation which enables interaction with external charges.

The flip side of an up-quark is its down-antiquark.
The flip side of a down-quark is its up-antiquark.
Quarks are inter-dimensional dipoles.

Deuteron

There are three isotopes of hydrogen which all have only one electron in orbit.

protium $^1_1H = 1\ \text{proton}$, deuterium $^2_1H = 1\ \text{proton} + 1\ \text{neutron}$

and Tritium $^3_1H = 1\ \text{proton} + 2\ \text{neutrons}$

Deuteron is the nucleus of deuterium. The simplest picture of events is that some of the surface quarks of each particle meld.

Deuterons are single positive charges with double mass and are good ammunition for bombarding certain particles.

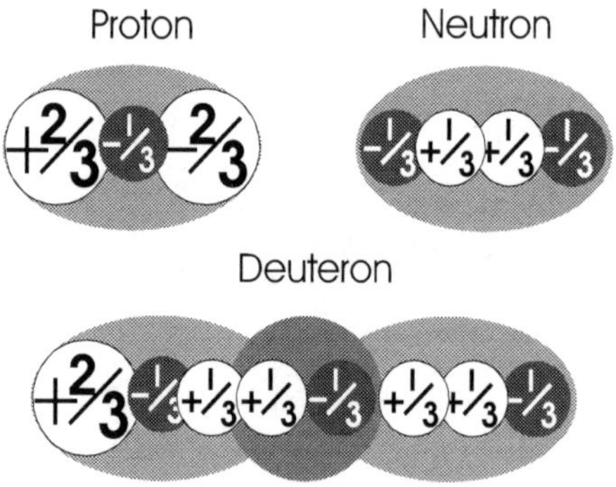

Deuteron quarks cross-couple

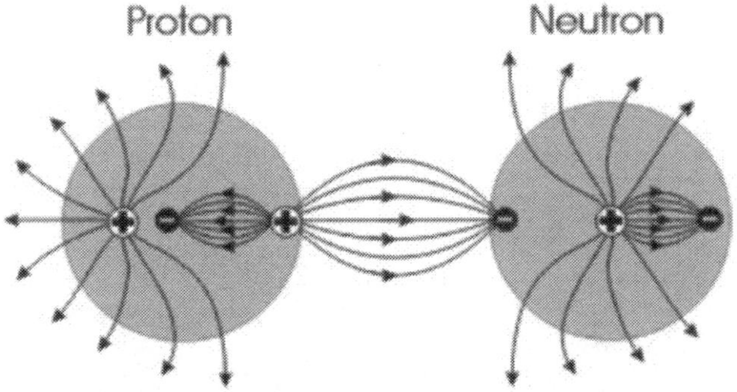

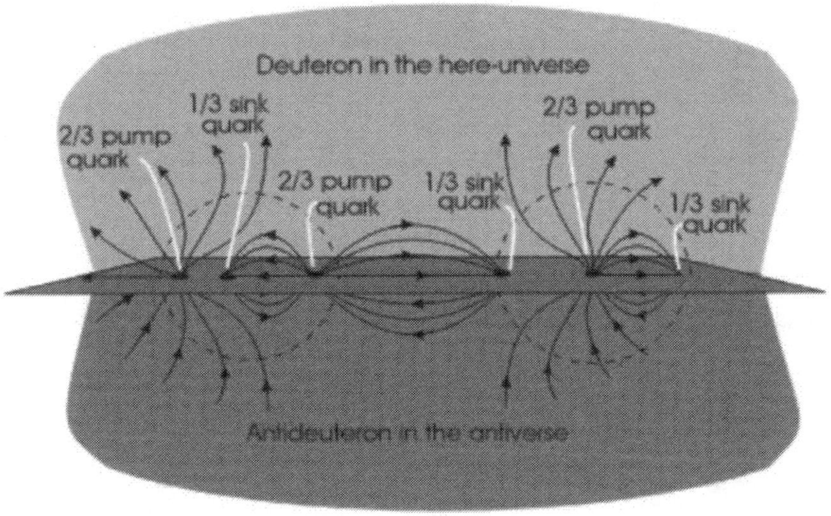

HELIUM-3 ISOTOPE

No stationary proton can get close to another stationary proton without the interposition of a neutron.

Quarks adjusts to pressures in external nucleon quarks which result in their host nucleon becoming partially polarised.

In helium-3, the neutron voids positive pressure from the protons up-quarks.

The neutrons own up-quark now pumps positive energy beyond its host.

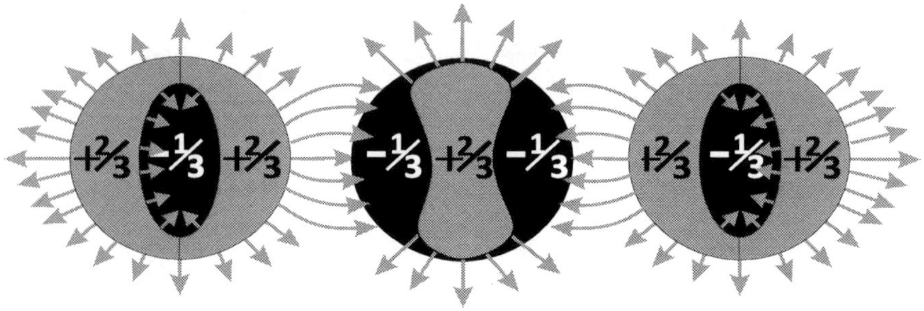

Quark Mass

The mass of a particle is directly proportional to the energy of the cue ball that breaks it out of the pack.

The Mass of an up quark is 0.002 GeV/c^2 (3.56 x 10^{-30} kg). Mass of a down quark is 0.005 GeV/c^2 (8.90 x 10^{-30} kg). Mass of a proton is 0.938 GeV/c^2 (1.67 x 10^{-27} kg).

Two ups plus 1 down = 1/80th the proton mass. The rest of the mass is binding energy/*(gluons??)/(etc????).*

Fluid dynamics propounds that most of this energy is employed circulating the coulomb wind in a loop between the 'here-verse' and the antiverse via the up and down quarks.

An analogy of the proton is a factory with a massive internal air conditioner circulating winds through its underground cellar. Some wind leaks out into the street while the same amount leaks into the cellar from the underground.

A pedestrian only sees the factory and feels the escaping cool breeze.

PI-MESONS & NEUTRINOS

CHAPTER 16

This post follows on from chapter 14; ‘Strong Force Hydraulic Analogue’.

In bi-verse theory all particles in the ‘here-verse’ (our one-verse or universe) are their anti-particles in the antiverse. They are joined at the hip-point. Under certain conditions particles can turn inside out and reveal their antimatter underbelly in the ‘here-verse’. Depending on its mass these inversions require considerable energy which the inverted anti-particle acquires and brings into the ‘here-verse’. This is now negative energy or antimatter which is repulsive to the ‘here-verse’, which as soon as possible forces it to revert, restoring the conservation of energy in both the ‘here-verse’ and the antiverse.

In the following diagrams the dotted centre line represents the bi-verse equator along which lie worm-pin-holes thru which coulomb energy spirals in-up or out-down.

Mesons are a combination of one quark portal and one anti-quark portal bound together by strong enigmatic forces. The only 2-quark combinations allowable have a total charge of +1, 0, or -1.

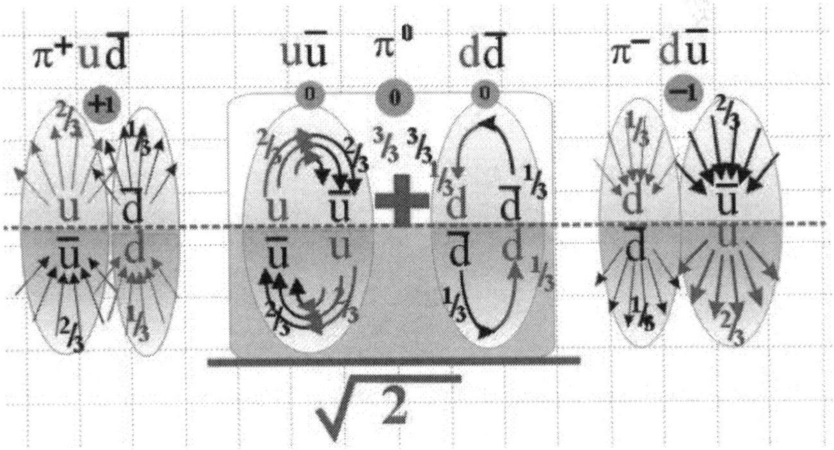

A negative pi-meson or negative pion π^-, which is a down-quark bonded to an anti-up-quark, decays into a negative muon μ^- and an anti-muon neutrino …

or rarely an electron $\mathbf{e^-}$ and an anti-electron neutrino.

$$\pi^- \rightarrow \mu^- + \bar{\nu}_\mu$$

A positive pi-meson or positive pion $\boldsymbol{\pi}$+, which is an up-quark bonded to an anti-down-quark, principally decays into a positive muon μ^+ and a muon neutrino $\boldsymbol{\nu}_\mu$.

The almost non-existent secondary decay mode at lower branching energy, is when a positive pion decays into a positron **e**+ and an electron neutrino $\boldsymbol{\nu}_e$'

$$\pi^+ \rightarrow e^+ + \nu_e$$

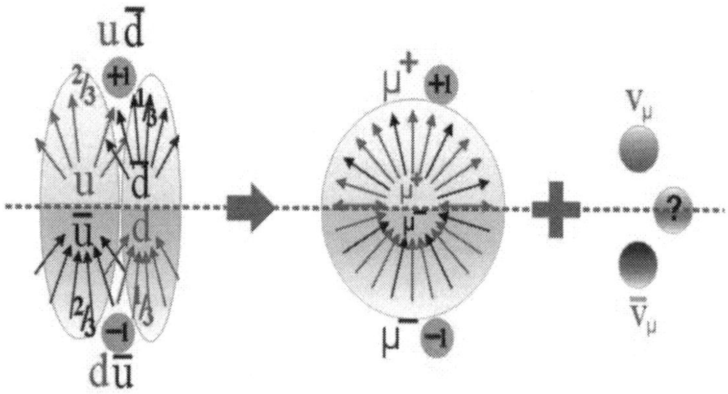

The pion $\boldsymbol{\pi}$+ is a **ud** bonded meson 2/3 up-quark + charge and 1/3 anti-down-quark + charge, and though the total charge is +3/3 = +1, we have to explain why the positive charge elements, +2/3 vs. +1/3, can unite whereas a pair of +1 vs. +1 positrons can't possiblẹ do it (naturally).

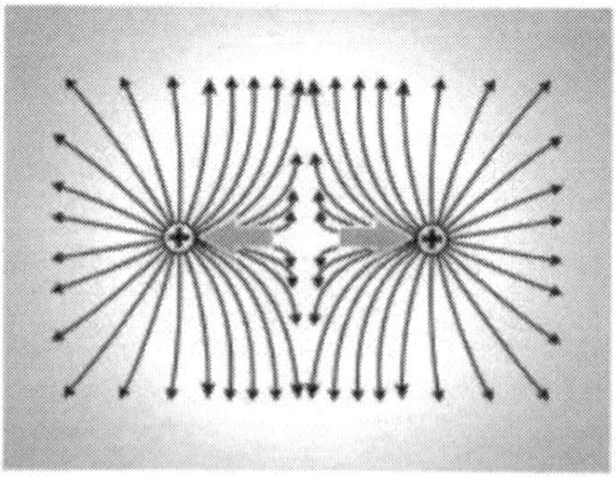

Bi-verse theory holds that instead of flipping back into the antiverse and releasing the energy it inherited getting draged out, the anti-down-quark exercises its other option and pulls itself against the coulomb resistance of the 2/3 up-quark, (which is pulled towards it), and bonds with it for its 10^{-8} seconds or less ot its lifetime.

Thereafter it becomes a positive +3/3 muon in the here-verse, plus the little bit of left over energy not used in bonding which becomes a muon neutrino. Since this energy is not negative energy it is either :

a. A chip off the up-quark or

b. A chip off the anti-down-quark which has flipped back to the antiverse.

Simultaneously to the 'here-verse' events, the polar opposite is happening in the antiverse where an anti-muon and a tiny anti-muon-neutrino are produced.

In the **uū** meson the anti-up-quark has inherited a considerable energy inverting from the antiverse and rather than flip back it finds it easier very briefly, for around 10^{-17} seconds, to bond with the up-quark before they implode into gamma waves i.e. so called photons

$$\pi^0 \rightarrow \gamma_1 + \gamma_2$$

Similarly the polar opposite occurs in the antiverse, producing a gamma wave in it

The **dd** meson has a similar fate to the **uū**, also producing a gamma wave (but a different energy one) in the 'here-verse' and in the antiverse.

The neutral pion π^0 is a composite of the **uū** and the **dd** mesons and therefore produces 2 gamma waves in the 'here-verse' and 2 in the antiverse.

Gamma waves can collide and create electron positron pairs in the here-verse and positron electron pairs simultaneously in the antiverse.

Nutrinos

Neutrinos are fermions imune to electromagnetic and strong nuclear forces but somehow interact with the weak force and gravity. They are thought to be point-like *no-charge* particles with a radius of zero and undetectable mass if any. They travel at close to the speed of light

The Sun produces so many neutrinos *(4 × 10^{38} every second)* that 70 billion of them pass through every square centimetre of the Earth per second and nearly all them heading for Earth passed right through it.

- The probability that it will interact with matter as it passes through is about 1 in 50 trillion.
- The only way science knows that a neutrinos are present is that when they do collide with a nucleus, the nucleus is transformed.
- The bi-products of this transformation identifies the type/flavour of neutrino involved.viz an electron, muon or tau nutrino or their anti-nutrinos.

The neutrino detectors need to be deep underground to protect them from cosmic rays that could interfere with detecting neutrinos and other subatomic particles.

Olympic pool detectors are hundreds of meters underground and patiently wait like frogs for the rare detectable collision with pool atoms.

The Antarez IceCube Neutrino Observatory uses a cubic kilometer of the Antarctic ice sheet near the south pole with photomultiplier tubes distributed throughout the volume.

Neutrinos are portals, entangled with their anti-neutrinos

- They sit on the equatorial fence between the here-verse and the antiverse.
- ***This is the bi-verse interpretation of the theoretical Majorana/Dirac particle which can be its own antiparticle.***

Spin

Portals themselves don't spin. Only what flows thru them does.

The angular spin of Space energy that flows into a potral on one side of the bi-verse, has the the same aangular spin as what flows out ion the other side

Co-parallel flows attract and anti-parallel flows repel, thus the tangent contact zone between two like spinning vortexes repel.

Positrons, or electrons with the same spin be it angular or orbital are strongly repel each other, however something different occurs in the mechanics of a uđ meson where same spins merge.

Perhaps this is because the 2/3 up overrides the 1/3 anti-down or it is some 'here-verse' antiverse spin condition. It clearly does happen because both spins combine into the positive muon spin.

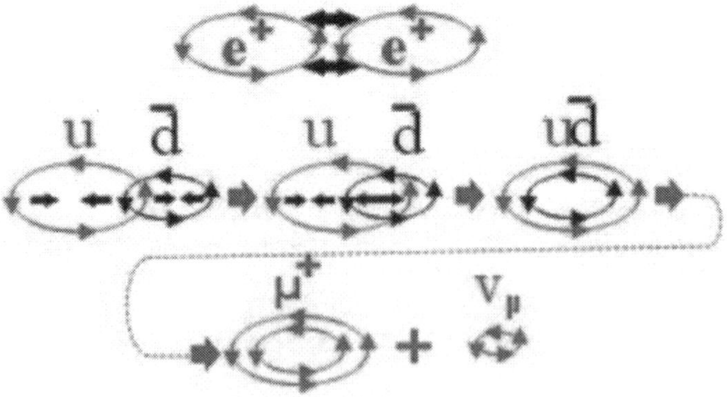

The spins of the uū mesons are opposite therefore co-parallel at their tangent connecting zones as are the spins of the **dd̄** mesons.

Their 10^{-17} seconds almost instant lifespan attests to the contribution of spin to their collapse.

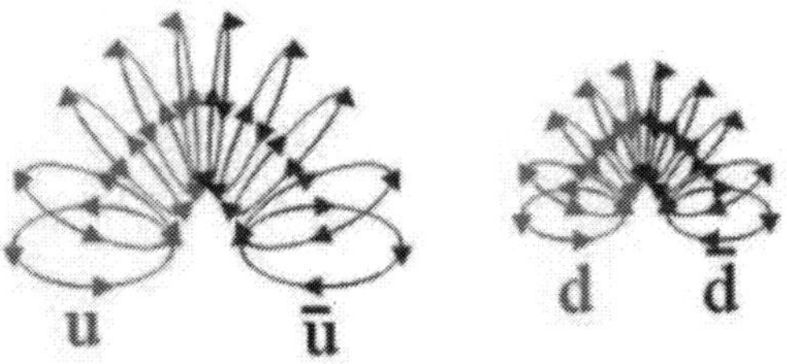

Neutrinos and anti-neutrinos have the same ½ spin as their associated leptons but have zero charge.

Anlogy

An em wave has been described as a packet of pairs of interlinking slinky toroidaal magnetism.

The toroid confines the wave which gives it its imunity.

I see an analogue with particles e.g. neutrinos, where the up vortexes interlink with down vorteexes on each sids of the bi-verse forming a toroid across the interdimensional equator.'

When a neutrino-antineutrino toroid collides with a nucleus, half of the particle say the neutrino, will be absorbed while the antineutrino half is rejected or vice versa.

Perhaps?

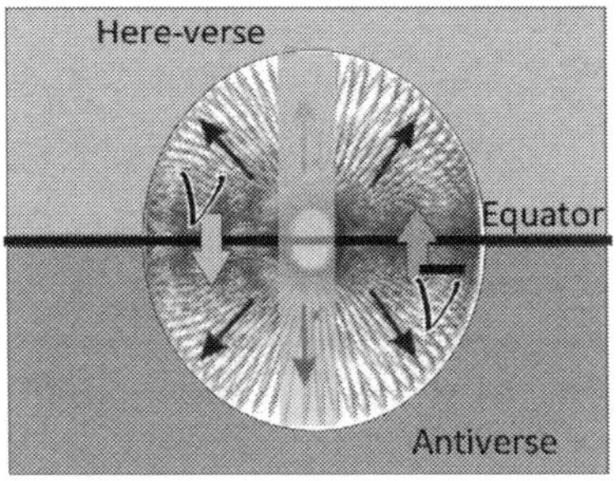

BETA DECAY

CHAPTER 17

The 'Strong force hydraulic analogy showed the neutron in deuterium act as a funnel for hydraulic springs. This caused the neutron to have an oscillating concertina polarization. This could be a permanent inherent vibration which was in tune with the hydraulic springs. *(nothing in the cosmos is static)*

In the bi-verse all particles in the here-verse are their anti-particles in the antiverse.

In some of the following diagrams the fonts for the ups or downs of the quarks is replaced by arrows.

- The dotted centre line represents the bi-verse equator along which lie worm-pin-holes thru which coulomb energy spirals in-up or out-down.
- What spirals down into a pin-hole spirals up from it in the other-verse with the same angular momentum.

Folding the page on the dotted line simulates inside-outs of the bi-verse.

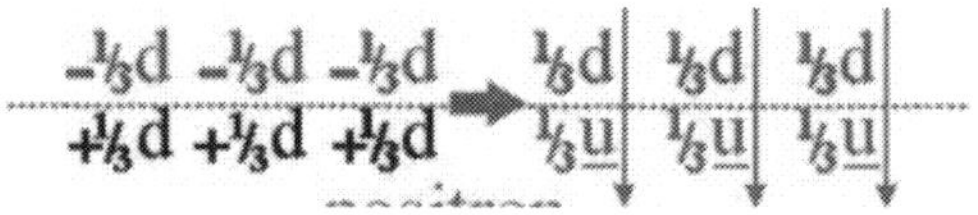

If a neutron is free of the nucleus's restraints it continues to concertina until (after some 15 minutes) it breaks some of its strong neutron double quark bonds (Q-bonds) and is reduced to single type bonds connecting its quarks.

The motivation for this collapse comes from the boson with a charge of minus -1 struggling to escape, which itself contains an electron trying to escape from it.

When the bond's tunnel door breaks open the hydraulic compression blasts out an electron bullet and an anti-neutrino bullet at near-light-speed.

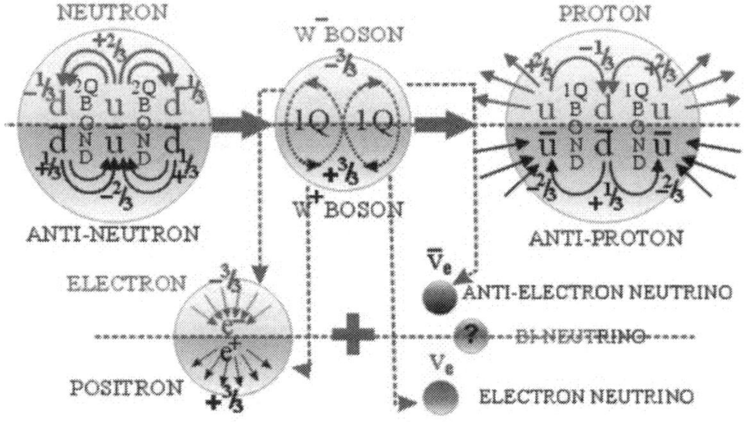

The 'hydraulic analogy' showed suggested models for neutrons and protons which demonstrate their polarizing potential. Here two like charges are at extremities with an opposite charge sandwiched midway.

The conventional images of 3 quark baryons give a triangulated version of events.

In B^{-}decay a [-1/3 down-quark] in the neutron gets transformed into a [+2/3 up-quark] of the proton because?

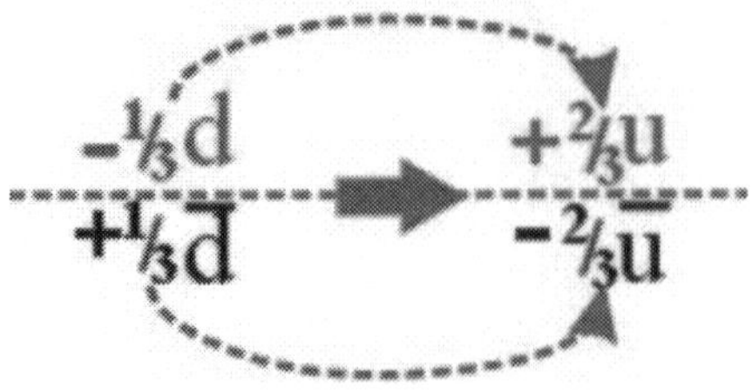

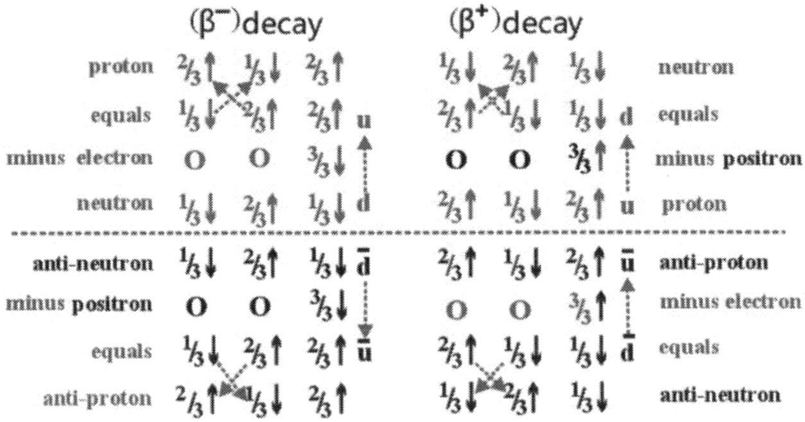

The electron can't be split into separated thirds.

In B⁻decay the Q-bond explosion also delivers an anti-neutrino.

Perhaps because the electron pulled 2/3 of its charge from the antiverse into the 'here-verse', some anti-energy had to be subtracted from the 'here-neutron' to balance the act.??? dk.

MASS BANGS & GRAVITY

CHAPTER 18

The Matter Zoo

Chapter 16 described the zoo of subatomic particles which were in general the fragments of collisions of matter.

Recently Sanford linear accelerator (SLAC) was harnessed to create matter from polarised light.

Polarized light is monochromatic and coherent. Its waves have their peaks and troughs coinciding.

The linear accelerator pushed an electron to near to the speed of light.

When a staccato laser light beam pulse was fired into its approach, the light beam recoiled and occasionally collided with a following light pulse with such violence that an electron positron pair was created.

Convention holds that this was the result of photons colliding.

Fluid dynamics holds that the electron's head-on intake was close to its maximum limit.

The incoming polarized wave was tightened in its approach and bounced back into a partially tightening next pulse.

The conflict generated a pair of inflow and outflow vortexes both in the here-verse and the antiverse

This experiment showed that these collisions of massive energy resulted in positive and negative matter.

Indeed if enough conflicting energy is smashed into one spot there is a veritable menagerie of specks, some coherent, most incoherent and much yet undiscovered.

Lepton Mass

The electron, is 207 times heavier than the muon, and the 3500 times heavier than the tau. All have the same charge but have theoretically have different density Space energy around them.

A muon lives for only about 1.5 millionth of a second and a tau for about 0.3 trillionth of a second. Light travels about a kilometre in 0.3 trillionths of a second.

- The coulomb force on an electron near a proton is about 10^{40} greater then the gravitational force.

Neutrinos

Neutrinos are fermions imune to electromagnetic and strong nuclear forces but somehow interact with the weak force and gravity. They are thought to be point-like *no-charge* particles with a radius of zero and undetectable mass if any. They travel at close to the speed of light

The Sun produces so many neutrinos *(4 × 10^{38} every second)* that 70 billion of them pass through every square centimetre of the Earth per second and nearly all them heading for Earth passed right through it.

- The probability that it will interact with matter as it passes through is about 1 in 50 trillion.
- The only way science knows that a neutrinos are present is that when they do collide with a nucleus, the nucleus is transformed.
- The bi-products of this transformation identifies the type/flavour of neutrino involve.

The neutrino detectors need to be deep underground to protect them from cosmic rays that could interfere with detecting neutrinos and other subatomic particles.9 Ju

Olympic pool detectors Like hundreds of meters underground are poised waiting for the rare detectable collision with atoms.

The Antarez IceCube Neutrino Observatory uses a cubic kilometer of the Antarctic ice sheet near the south pole with photomultiplier tubes distributed throughout the volume.

Gravity

Newton did try to explain that gravitation was caused by differences in *ether* pressure but didn't confirm this as he couldn't find any way of demonstrating it by experiment. When asked what caused gravity he would reply; ***"I make no hypotheses."***

Fatio de Duillier's and le Sage's shadow theory of gravitation.

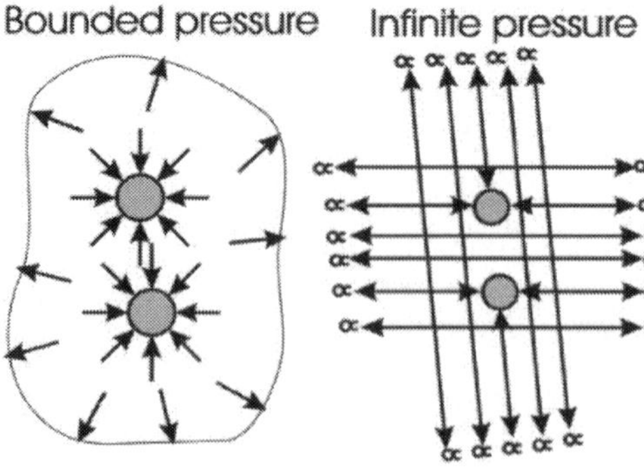

Fatio de Duillier and Goerges-Louis le Sage theorised Space as having pressure vectors coming from infinity then each mass blocks the pressure vectors from other masses. Minus this force directly between them, the resulting offset pressure vectors force the masses towards one another.

Newton thought this was the only theory worthy of consideration

Swedish Nobel laureate Hannes Alfvén claimed that cosmologists placed too much stock in gravity at the almost ignorance of electricity and magnetism.

The Speed of Light

For over a century science has been governed by the conclusions of Michelson and Morley:

1. The speed of light is constant.
2. Therefore there is no ether. (*which is a non sequitur, if A then not B)*

They did not contemplate that a medium could exist which at fixed energy density does not affect the speed of light.

- The speed of light 299792458 km/s becomes slower in denser materials. It is slower by about 90 km/s in air, by 50,000 k/s in water, by 100,000 km/s in glass and even almost stopping is a super-cooled condensate.
- The speed of light is held to be and surely is, relative to the viscosity of the gravity; the greater the mass, the slower the light speed near it.

An outside observer sees light almost stopping as it approaches a black hole. However a person say half way in would still measure it still at 99792458 km/s this since his ruler has shrunk a bit and his clock slowed.(Einstein)

The Genesis

Mankind has applied its devotion to the eternal search for how life begins and ends to Science, and worries how the Cosmos begins and ends.

For over two millennia science has physically and theoretically probed the universe *(one-verse)* looking for its origin and its evolution.

Telescopic eyes looked out into the cosmos while microscopic eyes looked into atoms and their nuclei. These eyes were optical and electric.

In the 20th century reasearch into 'the drop in the ocean' collided with 'the ocean in a drop'.

Dark Matter

The closer a planet is to the Sun the faster it travels. Pluto orbits at 17 thousand kph while mercury orbits 10 times faster.

In 1977 Vera Rubin and Kent Ford took photos of a spiral galaxy where spectrum analysis showed that the speed of outer stars were not much faster than inner ones.

The stars then were not just orbiting around a central focus, there had to be a huge unseen mass present. Also the observed Doppler effect *(red shift)* of light from retreating galaxies has been attributed to the expanding one-verse.

Given their speed and the theoretical age of the uni/one-verse, the fact that these galaxies are still visible means theoretically that some mass must be gravitationally slowing the expansion.

Current theory holds that 27% of matter is dark matter and 68% is dark energy.

(The Milky Way is part of a local cluster of galaxies which include Andromeda Nebula which are attracting each other so their Doppler light is blue shifted. There are many observable clusters and super-clusters in the cosmos)

The Big Bang

In 1929-31 while Edwin Hubble was showing that the Doppler red shifted galaxies are moving away from each other, the Belgium priest and mathematician Georges-Henri Lemaître was theorising that since the number of discrete quanta is ever expanding, then going back in time there would be only one green quanta bottle left hanging on the wall, 'the primordial atom' which would mightily explode dividing over and over like bacteria.

When early sceptics like Einstein and Eddington began tolerated the idea, the 'Big Bang' got legs becoming a major alternative to status quo type theories.

Confirmation for the theoretical big bang came in 1964 when Penzias and Wilson detected background radiation from all directions of the one-verse at 3^0 Kelvin, i.e.a wavelength of 7.35 centimetres. This was close to the predicted cooling of primordial Big Bang radiation.

There were problems. It was generally understood how hydrogen, helium and Space itself were created in the initial nanoseconds but how could the heavy elements form. These could be created in collapsing stars or super nova explosions but if the background radiation was smooth then how could lumps like galaxies form.

In 1978 Alan Guth and Henry Tye tweaked the theory with 'inflation' which fixed the clumping, the apparent flatness of Space, and the needle in the haystack of the predicted magnetic monopoles.

In 1929 Paul Dirac had theorised the necessary existence of massive particles (the size of bacterium) with a magnetic monopole. Grand unified theories supported this, but where were the monopoles? They should be as prevalent as gold.

Guth and Tye inflation eventually tweaked further, after which there was theoretically only one monopole in our known visible uni/one-verse.

Since inflation can not tweak them entirely away the Big Bang has a primordial problem

Antimatter

As well as monopoles Dirac declared that when matter is created an equal amount of antimatter is also created.

In 1932 Karl Anderson detected a positron *(an anti-electron)* from a cosmic ray collision in a cloud chamber, so where was the one-verse hiding the illusive rest of the antimatter?

Swedish scientist and Nobel laureate Hannes Alfvén disputed the Big Bang theory holding that matter was more the result of magnetic fields and plasma clouds. There are equal amounts of matter and antimatter which are separated by super-strong magnetic fields called Leidenfrost layers which repel each other. However if under certain conditions particles cross over they might be observable on Earth or clumps might annihilate in massive explosions causing galaxies to recede from each other.

In 1955 an anti-proton was discovered in the LBL Bevatron and other antimatter have popped up in particle collisions since, but apart from the positron, high altitude balloons have not detected any other antimatter roaming the cosmos.

This negative result exhausted many theories.

Vacuum Energy

The cosmological constant aka Einstein's cosmological constant (symbol lambda: Λ), is the constant coefficient of a term that Albert popped in to his field equations of general relativity to counter the inevitable contraction of the universe by gravity. He favoured a stable universe with which "God doesn't play dice".

When Hubble found the red shift *(i.e. cosmic expansion*) it had to drop out of the equations. After further analysis Albert realised that it need not have been there in the first place; He called this his *"greatest blunder"*.

The discovery in 1998 that the expansion of the universe is accelerating, implied that the constant may have a positive value after all, so it was popped back in.

The current standard model of cosmology known as the ΛCDM model couples the constant to 'dark energy.

Quantum field theory (QFT) has Space as field fluctuations with zero-point energy popping up everywhere throughout the cosmos. Leptons and quarks are also fields of zero-point energy.

The muon and the Tau Leptons

Muons were first detected in in cosmic rays and later in particle accelerators they are 206.7 times more massive than an electron and have a half-life of 1.56 microseconds.

The Tau was accidently detected in particle accelerators. One of the experimental crew had the electric field cranked up to maximum and it dragged the monster in.

It is 3450 times more massive than an electron and has a half-life of 2.9×10^{-13} seconds.

Though they exist unstably for a flicker of time during their life they provide a possible reason for *L*mass.

A significant and nagging back-burner question is

- Why does the muon, tau mass/charge vary?
- Is there a mathematical 1:207:3450 connection?

THE 6D + T COSMOS

CHAPTER 19

If we accept as a priori that all energy circulates then Coulomb energy does not begin at a zero point and perish at another.t

It travels from positive to negative then returns to positive either thru another set of physical dimensions called 'the antiverse' or through the same dimensions in a different time zone; 'the anti-time-verse'.

This work chooses the antiverse option.

- Charges are portals
- They have perpetual coulomb energy Space flowing/pumping thru them.
- It leaves a positive charge under pressure, then flows through ambient Space pressure of the here-verse, and gets squeezed into a negative charge.
- It leaves the negative charge repressurized by its back door, then flows through the ambient pressure Space of the antiverse, to the back door of the positive charge, and get squeezed into it, completing the Coulomb circuit.
- Its journey permeates the surrounding Space, raising it proportionally (1/d) to higher energy Gspace.i.e. $1/\pi d^2$ by volume.
-
- We can see how nanoscopic this is energized by observing the difference between charges..
- Between two electrons the gravity force is $2.3^{(-42)}$ % of the Coulomb force.

The energy that Coulomb forces leak/spread is Gravity.

By increasing the energy of the Space around them charges increase the pressure and density of atoms, molucules mass and Suns

Viscousity

Viscosity is affected by temperature and pressure. For liquids viscosity increases with pressure and/or temperature, whereas for gases it decreases.

In gases, viscosity is caused by the collision and transfer of momentum between gas molecules. The gas molecules move randomly, and as they collide with each other, they transfer their momentum, leading to the resistance of flow.

Water exists between the sates of ice and a gas. At any temperature it is a mixture of low energy and high energy. As a gas it can be heated to a point where there is theoretically only one molecule left in a vacuum.

Correspondingly Space exists between the unattainable states of absolute zero and a perfect vacuum, therefore between being frozen and being at an infinite temperature.

This work holds that Space is comprised of zero point energy in a subspace of an impossibly perfect vacuum.

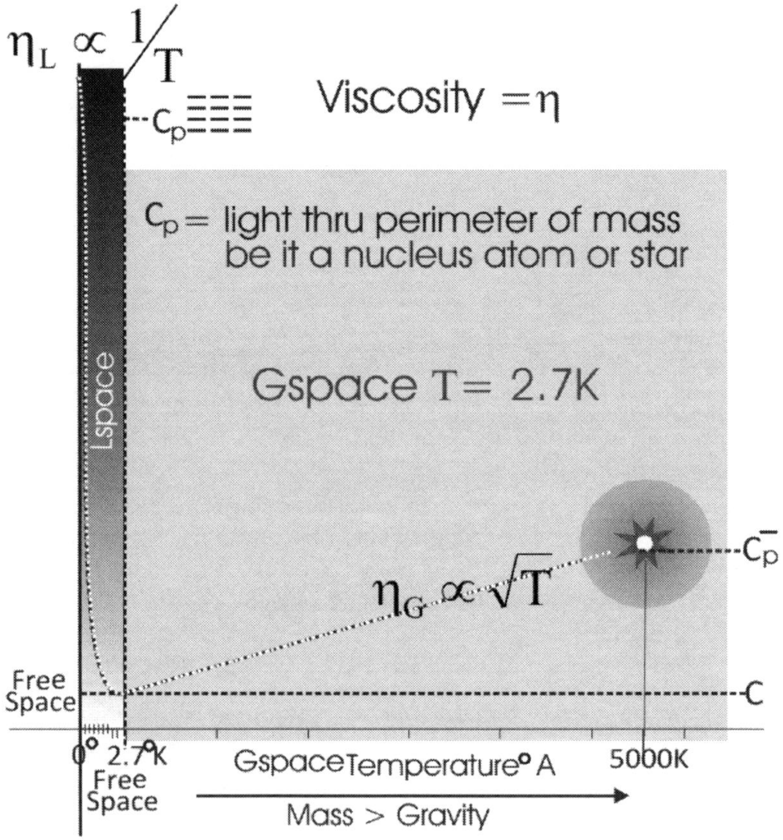

Just as liquids get colder and their viscosity increases, the z-point energy of very low temperature (Lspace) gets slower and more viscous. However just as the thermal energy of gasses increase the friction and consequently the viscosity, then Gspace (Gas Space aka Gravity Space) between the z-points of energy also increases its friction, which increases its viscosity.

This happens in the presence of matter.

Comparing medium viscosity formula:

- As in the Temperature of Gases, Gspace (η_g α √T
- As in Low Temperature Liquids, Lspace (η_L α 1/T)

Between them there is a Space fulcrum point or more likely a fulcrum zone similar to the latent heat of vaporization zone of a liquid.

- The temperature of free space of 2.7^0 K is this point or very close to it.

Above this,

- Space has frictional viscosity(η) similar to that of a gas.
- Gspace increases viscosity (η_g α √T)
- As the temperature increases towards that of a Sun or a neutron star Gspace viscosity slows light.
- A black hole slows light towards a standstill.
- In the above diagram c_p= light passing the perimeter of a mass.

Below this,

- The interZpoint distance decreases which increases viscosity (η_L α 1/T).
- This is the domain of super-magnetism where it stirs the porridge.
- *In my cosmos magnetism flows logically and rationally parallel with its current. Current loops rotate Space.*
- In this domain, in a near 0 ^{0}K condensate, light almost stops moving.

In a Proton and a neutron Coulomb energy circulates cyclonicly via the up-quarks and the down-quarks which leak/spread copious amounts of energy into surrounding Gspace, proportionally increasing local viscosity.

- We can see this increase in viscosity in Bremsstrahlung where electrons zipping very close to heavy nuclei get almost stopped in their tracks by dense viscous Space and have to radiate energy in the form of X Rays.
  (viz. Earth re-entry vehicles generating heat waves)
- We see this in the Shapiro effect when radar pulses passing close to the Sun (short range) were slowing by 1/d, (hyperbola asymptote y=0), and far out (long range) by 1/ln(d), (asymptote y=1)
- We see this in the bending of light rays as it passes by the Sun.
- The lower side of the wave is in a slightly more dense viscous radial zone, which slows it down and bends the light path, just like a glass lens *(Hutgens)*

This is the Gravity lens.

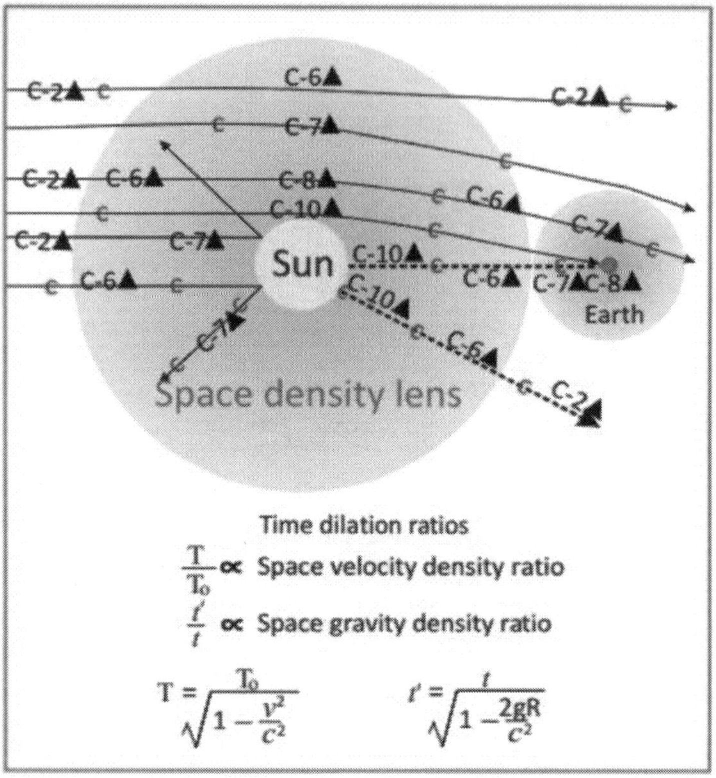

Viscous Gravity

Gspace as a ***viscous*** atmosphere of chaotic energy which is in and around a mass to form a density lens which wanes proportional to $1/d^2$.

Where the density zones of masses intersect the density and viscosity increases

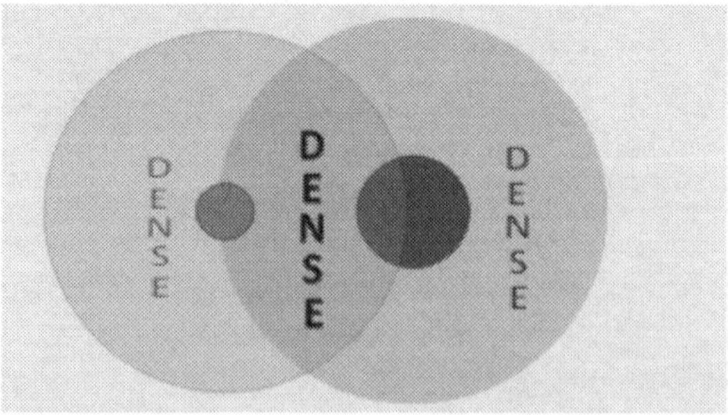

Suns Planets Moons People are sticky honey-pots

Black Holes

Black holes are the ultimate macro-mass,with a macro-density-lens.

There are thought to be humongous holes in the centre of galaxies with the mass of thousands of Suns.

They are so massive that light cannot escape from them, be it originating inside a sphere bounded by what is called the event horizon, or be it if light entered into this sphere from outside.

An analogue its a rocket not having the velocity to escape from earth, only here the density of black hole Space is so great that the initial speed of light appears almost zero to an outside observer.

The horizon acts on incoming light as the no-go-zone acts for returning spacecraft to Earth. The ship either gets caught by the Earth's gravity or it speeds past.

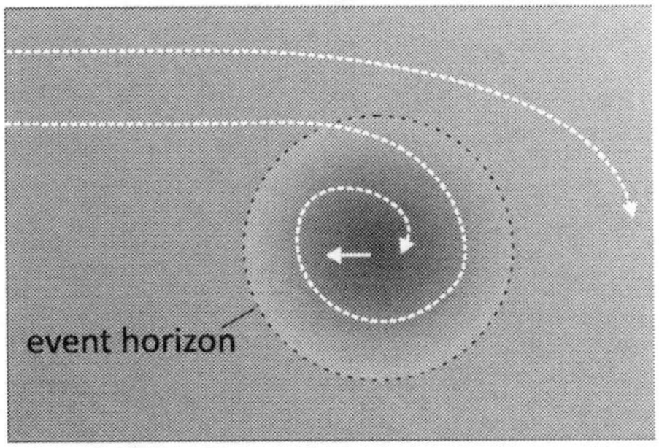

A Space is a density lens slows the nearside of the light wave and bends it. Accordingly Black holes swallow stars and even galaxies

- They swallow thevenergy of of both the here-verse and the antiverse..

Big questions are:

- ***Is there detritus. If so where does the energy go or recycle. (eg Hawkins radiation?)***
- ***Are there white holes?***

Permittivity

The permittivity of a vacuum symbol ε_0 is its resistance to an electric field. In a vacuum on Earth ε_0 is 8.854 ×10^-12 f /m (farads/meter)

Relative permativity of vacuum = 1; of Air = 1.0006

If masses have ether density lenses, Ae-theory predicts that the permittivity of a vacuum near the Sun would be ($\varepsilon_0 + n\Delta$) and in Cosmos free Space ($\varepsilon_0 - n\Delta$)

. If permittivity was zero there would be no resistance to the field, and speed of light would be impossibly infinite.

Since greater density of Space reduces the speed of light, then logically reducing the density increases the speed.

The Space viscosity of free Space is infinitesimal yet we do detect it in the light from distant galaxies.

The permeability constant *(μ_0)*, of free space, is a measure of the amount of resistance encountered when forming a magnetic field in a vacuum. *This is here resistance to Coulomb leakage.*

Quantum Pros &Cons

It is suggested that quantum mechanics has been slipping the nineteenth century luminiferous medium in again, but through the back door, and in disguise."

I see the reverse as well happening, where quantum mechanics is unconsciously illuminating aether theory, and even hurrying it along.

Space ether is the font of all quanta.

Dipole Cons:

Quantum physics extrapolates that a theoretical 'magnetic point dipole' has a magnetic field of exactly the same form as the electric field of an 'electric point dipole'.

The magnetic point dipole is portrayed as a diminutive current loop which has theoretical through lines of force radiating around it.

The electric point dipole is a dipole of opposite charges with almost zero separation which has theoretical lines of force shrink-wrapped around it.

Despite what the iron filing patterns display, 'the map is not the territory'.

Illusions aside, the magnetic dipole is simply a vortex thru which theoretical lines of force describe the eyes of vortlet tunnels generated by the induced magnetic iron filings

Differently an electric point dipole attracts unlike chargers to its unlike poles via its strict (no tunnel) pathway lines.

Graphically they have 'form = parthway' similarities but they have significantly different modus operandi.

It's analogous to uselessly argue that a plane powered by a Ramjet has the same form as a plane powered by Propellers.

- *The Caducius scalar waves are an analogy of turbo props.*

Dipole Pros

Because of its permutivity (its resistance to an electric field), the seeming dialectric property of Space has for over a century fostered speculation that Space is a myriad of dipoles swimming in an ocean of Subspace.

Today the quantum physics parallels this, as it describes its 'quantum vacuum' as having virtual 'bubbles' of electron-positron pairs popping in and out of it.

Recent chemical analysis of Dipole-Dipole Interactions in Macroscopic Systems have shown Variations in the predicted neutrality of random interactions of dipolar molecules .

At room tempersature, dipoles are more probable to form less energetic, onfigurations in accordance with the Boltzmann's distribution.

Only at elevated temperatures do the molecules have higher potential energy.

We can postulate from this that Cosmos energy of dipole Space will be lower at low temperatures of free space and higher in the proximity of elevated energy of a mass.

Dipole Space density is reducing while mass Space density is increasing.

Dipole Space has entropy and Mass Space has negentropy'.

Option A:

- Space is dipolar in the here-verse and in the antiverse.

Option B:

- A positive charge in the 'here-verse' is its inside-out negative charge in the mirror verse and vice versa.
- ***It is an inter-dimensional dipole portal.***
- If space is non-dipolar, and if of the Cosmos is expanding, we can still argue that the just the free Space is expanding while mass Space is contracting
- ***here alwso ; free space has entropy and mass space has negentropy.***

Light Speed Variations

For 175 years the Weber Kohlrausch constant has been SHOUTING from the page

"I AM A HYPERBOLA"

The Weber constant re light is a hyperbola $c = \frac{1}{\sqrt{\varepsilon_0 \mu_0}}$

It shows the Speed of light c will slow if ε_0 or μ_0 is increased.

Similarly, the Speed of light is slowed proportionally by viscosity (η).

$$c \ \alpha 1/\eta_g \text{ and } c \ \alpha \ 1/\ \eta_L$$

alpha (α) substitutes for (is proportional to).

Therefore viscosity $\eta_g \ \alpha \ (\varepsilon\mu) \ \alpha \ \sqrt{T}$

and free Space viscosity $\eta_{(g\,0)} \ \alpha \ (\varepsilon_0\mu_0) \ \sqrt{2.7^0K}$

The Space viscosity of free Space is infinitesimal yet we do detect it in the light from distant galaxies.

• This is the main cause of the red shift if not all of it.

In 1929Swiss astronomer Fritz Zwicky hypothesized 'Tired Light'.Attributed the redshift to light having continuous collisions with dust particles consequently light year images were blurred .

This work attributes the redshift to lightyear light passing through A collective incessant miniscule space density.

- Accordingly the speed of light is variable as must be permittivity (μ) and permeability (ε);
- and since (μ,ε) are measurements of events in a vacuum then the vacuum itself (i.e.Space) must be variable.
- and because of black holes, we know that as c approaches zero (μ,ε) must approach infinity.
- Thus (μ,ε) (aka Space density) at a radius R of a mass is proportional to its mass/volume aka mass density. *(plus the miniscule* Δ*-density* δ *of free Space)*
- The Weber hyperbola shows that the speed of light is proportional to the density of ether.

If a Space ship could thin Space (i.e. expand Space) it could travel faster than ambient Space density light speed.

Similarly the 'Alcubierre drive' predicts warp drive by contracting Space energy ahead and expanding Space behind.

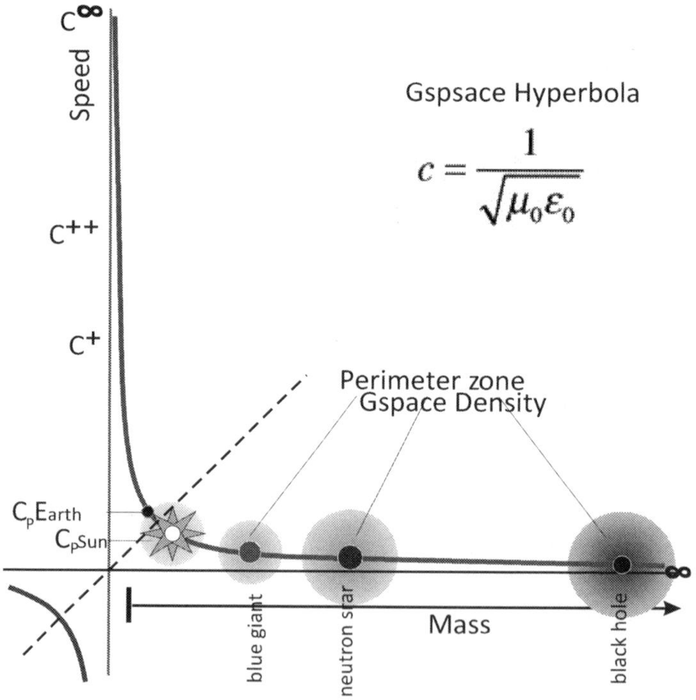

What Weber/Kohlrausch measured was the permittivity of a vacuum on Earth which is constant anywhere at sea level. If they took their Leyden jar in a 'miraculous Earth fixed Fitzgerald contraction cage' to sea level on Uranus and timed its discharge, it would take longer than Earth time.

Light leaving the Sun gets faster with distance as the Sun's Space density decreases; then it gradually slows as it encounters the Earth's increasing Space density.

If its speed was measured at any Space density point on its journey using local devices it would still be the constant 299 792 458,m/s

So where is Earth on the W/K hyperbola?

- We can speculate that since Eddington was able to detect some minor deflection of light passing through the Space density lens of the Sun, then the Earth and the Sun should lie on the acute curve near the transverse axis of the hyperbola.
- If was further out along the 'y' axis it would be near impossible to measure any mini-minuscule variation

THE BIVERSE Gspace HYPERBOLA

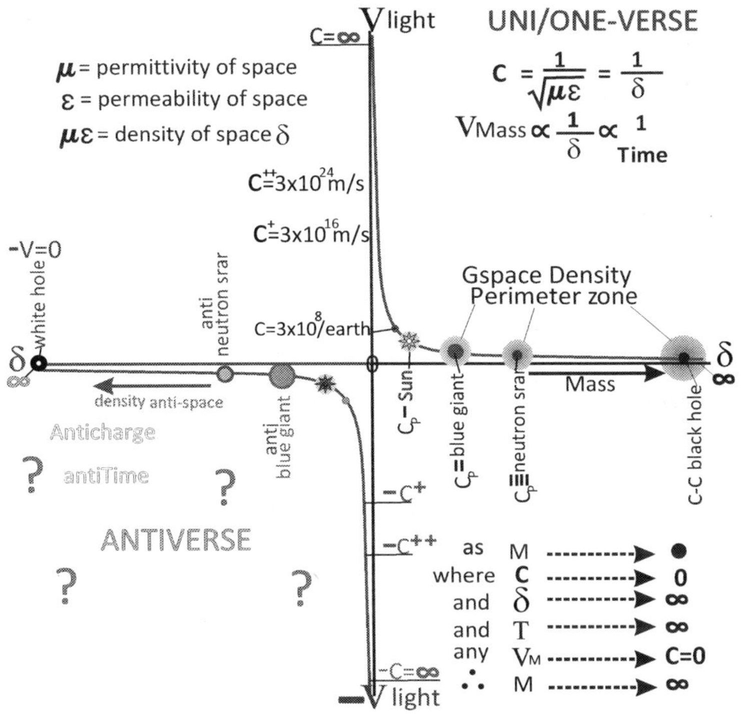

The greater the mass, the denser the Space energy in it and around it and the slower the light through it.

Conversely the thinner the Space density, the faster light the light through it.

If the Space density approaches zero then the speed of light approaches infinity.

Relativity and Frame Dragging

A light wave front passing the Sun has wavelets closer to the Sun in denser Space, where they slow and bend the light à la Huygens diffraction.

Light leaving the Sun gets faster with distance as the Sun's Space density decreases, then it gradually slows if it encounters the Earth's increasing Space density.

However if its speed was measured at any point on its journey at local Space density, using local devices (e.g. a shrunk tape measure), it would be the constant c = 299 792 458 m/s.

In denser Space wavelengths increase as frequency red shifts.

A satellite orbiting at 14,000 km/h encounters head on increased Space density which slows its clock by 7 microseconds / day compared to an Earth clock. Because it is orbiting 20.000 km above in the Earth's less dense lens zone, its clock gains 45 microseconds / day. On-board GPS systems compensate by slowing its clock by 38 microseconds / day.

As a current coil drags around its me-flow vortex, orbiting masses drag around Space

Viscous Space causes frame dragging.

Events with the biverse

Since electron is a Zero point charge, Space itself must be spatially dimensionless in both the 'here-verse' and the antiverse, to allow it passage through a portal.

- The viscous friction energy of Gspace at 2.7^0 K 0is about one joule per cubic kilometer.
- *Rest mass is a measure of the total compressed Space-energy to infinity relative to/above the ambient pressure of free Space (Casimir)*
- The greater the pressure, the greater the Space energy density and therefore the greater the mass.

There are theorised a menagerie of multi-verses but accordingly here, they must be multi-bi-verses.

Each 'mirror-verse' is the domain of anti-charge not antimatter

Anti-charge is Not Antimatter

Matter is defined any substance that has mass and takes up Space by having volume.

Antimatter has to me always been a misnomer, as it erroneously implies some sort of anti-mass. yet examples of it are only anti-charge.

A huge concentration of energy might be able to flip a particle from the here-verse to the antiverse and vice versa, but these only survive for say 10^{-8} seconds.This is anti charge not anti mass.

Coulomb energy flows from positive pressure into negative pressure (which it attracts), and some how returns via another set of mittor dimensions, possibly by some amorphic subspace, like Brane theories' predicted fourth dimension, or or thru another time zones; or as I predict and prefer via a more direct route, thru the antiverse where time is mutual to the here-verse.

- An electron in the 'here-verse' moving north causes Space positive charge / aka me-flow to move south.
- Simultaneously its antiverse underbelly positron moves north with its Space positive charge/me-flow also moving north.

In the bi-verse:

- ee-flow force between two charges is a combo of the 'here-verse' and the antiverse forces.
- Similarly the me-flow force of a current conductor on an electron is the sum of 'here-verse' and underbelly antiverse positron where the current flows in the conductor in the opposite direction.
- ***In the Bi-verse charges and their total ee-flow and/or me-flow have potential enrrgy but zero kinetic energy.***

Transparent Energy

Since light is travelling through ambient Space with its ambient energy, it will be progressively affected and indeed weakened.

As the speed stays constant in constant density the only property that can be impaired is its frequency; the wave length gets longer.

Part or all of the non-cluster galactic Doppler effect could be attributed to transparent Space energy.

Space is transparent energy
Its transparency is one of the greatest of all natural illusions.

Moving Mass

A mass in motion thru Space has a permanent constant velocity which is one of the principle objections to the existence of Space as a resistance-less medium.

We know the speed of light in transparent matter is more or less inversely proportional to the density of that matter. It is1·000003% slower in air, about 30% slower in water, 60% slower in diamond and almost zero in condensates.

Similarly its speed is proportional to ether density.

In a fixed density the elements of chaotic light speed Space cannot exceed or reduce their light speed fixed velocity; so they can not gain or lose energy from collisions with a mass in motion nor each other.

Portals moving thru Space adjust their intake *pull* and output *thrust* and therefore appear frictionless until they ac-de-celerate, (Inertia,Induction) .

- A mass moving at a constant velocity through ambient energy free Space is equivalent to an assembly of stationary charges in a steady me-flow.
- A mass at constant acceleration through free Space is equivalent to an assembly of stationary charges in a steady ee-flow.

- **At relativistic speed**
- The head-on me-flow and/or head-on free Space increases the masses' own head-on Space density.
- This is longitudinal mass = *Lmass in the direction of propagation*

$$Lm_v = Lm_0 \sqrt{1/(1 - v^2/c^2)}$$

- Its rear mass -*Lmass in the direction of propagation,* decreases proportionally.

- Transverse mass *Tmass* is unchanged by variations of longitudinal velocity.

Matter-waves.

The matter-wave side of quantum duality superposition describes the relationship between momentum and wavelength.

- The wavelength is inversely proportional to the momentum (mass and velocity) of the particle). λ=h/mv
- The smaller the wavelength of the matter wave, the faster the particle moves.
- The De-Broglie wavelength increases as the particle become lighter.
- They are probability waves.

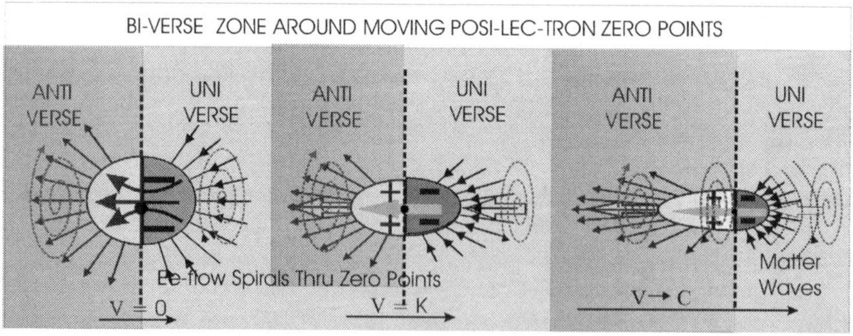

A portal like an posilectron moves with its zone of compressed Space.

At relativistic speeds, even though not accelerating the portal can't cope with the head-on intake from ambient Space, which it rejects as a matter wave.

Electromagnetic waves are not matter waves.

- They travel at the set speed of light per medium.Bi-verse Mass
- Different wavelengths have the same velocity.

The Cosmos Bi-verse Cycle

Dirac held that when matter is created an equal amount of antimatter is also created.

This work holds the secular view that energy has no origin of creation:

Energy simply is, and It ciculates.

Above absolute zero all energy circulates.

This denies the Big Bang of creation

Accordingly, even when traveling to infinity, it must somehow find a way home thru other dimensions.

Infinity then is the virtual vertex of infinite energy's inter-dimensional loop.

It's uncanny how John Wallis's infinity symbol: '∞' purports vertex entanglement.

Be it local or infinite, energy is inter-dimensionally entangled with its mirror energy.

This unravels the double-slit 'which-way' enigmas.

This work theorizes that the aether itself is expanding the Cosmos, and Mass attraction which is restraining/ contracting it.

Aether has entropy; Mass has negentropy.

APPENDIX 1

Ferro-Magnetism

In ferromagnetic material, *usually iron, cobalt or nickel*, inter-atomic forces cause the magnetic moments of neighbouring atoms to align in the same direction. These become domains that resemble the assorted sized, snug fitting blocks of stone in Inca walls.

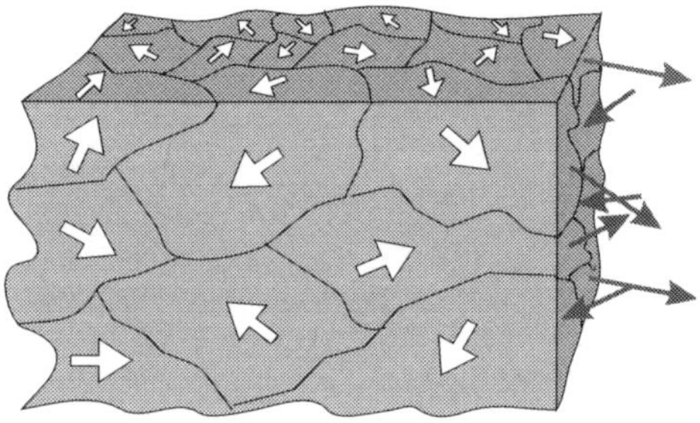

- Each domain has an inherent but random magnetic orientation when the bar is not magnetised.
- If finely ground iron oxide is dusted onto a bar magnet it will collect on the borders between the domains. These can just be seen with the naked eye, but are clearly visible through a magnifying glass.
- A current flowing through a coil generates magnetism. When the coil is wrapped around a ferromagnetic bar, the domains with a sympathetic magnetism to that of the coil grow at the expense of those not simpatico which shrink. The bar becomes directionally magnetised, increasing the overall magnetism as much as 10,000 fold that of the isolated coil.
- When the coil current is turned off the coil's magnetism collapses but many of the domains in the bar do not revert to their pre-magnetised condition. These directionally magnetised domains linger for a long time. The harder the steel is, the longer they linger and the more permanent the magnet.
- Various environmental disturbances such as heat can cause magnets to revert to their random domain alignment.

The Curie point is the temperature at which heat breaks magnetism down. The Curie point for iron is 770° C. Combo iron-neodymium-boron magnets are more resilient and more powerful.

Currents and Iron Discs

An electric current will magnetise a piece of ferromagnetic material near it.

When a coin sized iron disc is placed near a straight conductor, the stronger me-flow near the conductor encourages sympathetic magnetic dipole moments in small locations of the disc which are called domains.

Simultaneously it discourages non-simpatico moments in other domains therefore the simpatico moments dominate the disc.

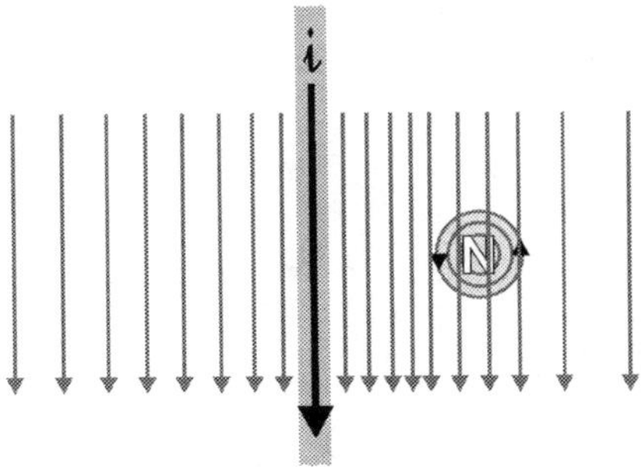

Taken as a whole, the total moments approximately mimic the magnetism which would be generated by a tiny electric current circulating the perimeter of the disc.

This mini-current would generate a vortex of co-rotating me-flow whose axis is perpendicular to the disc and through its centre

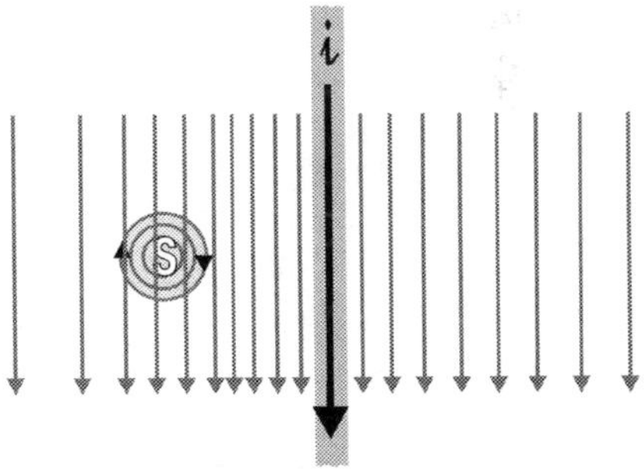

The kick-started domains now radiate energy back into the system.

- They increase the me-flow energy between disc and conductor causing attraction.
- and retard the far-side flow energy beyond the disc thereby pushing the disc towards the conductor.

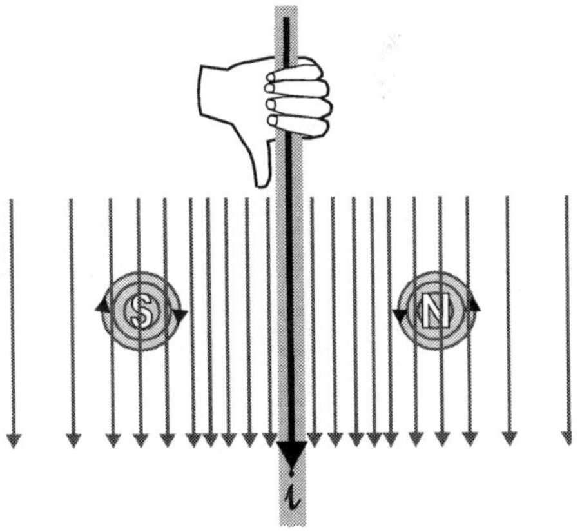

- The right hand rule reveals the conventional polarity on the disc.
- The face of an ant-clockwise me-flow vortex is a north pole.

Pivoting Discs

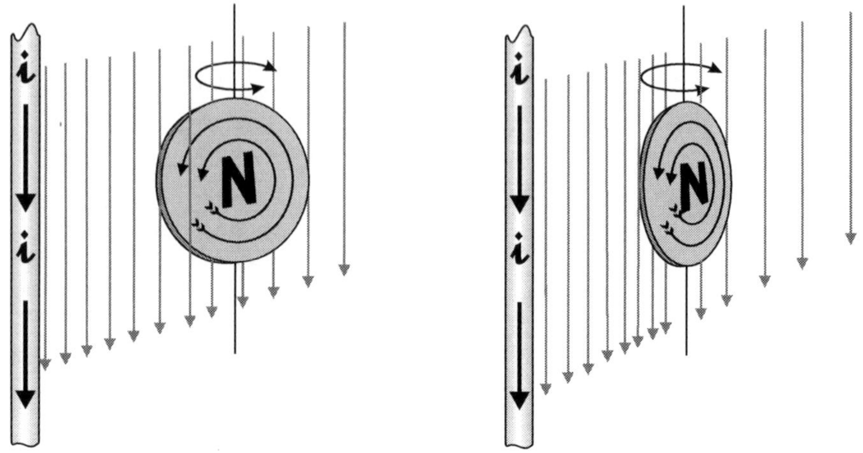

If a disc is allowed to pivot it will rapidly seek an alignment where the difference between the near-side force and the far-side force is greatest. This happens when the disc is aligned along a radius line from the conductor. When aligned, the disc's vortexes contribute their maximum towards increasing the me-flow between the disc and the conductor.

This is how and why compasses orientate about a conductor.

Discs placed around a Current Ring

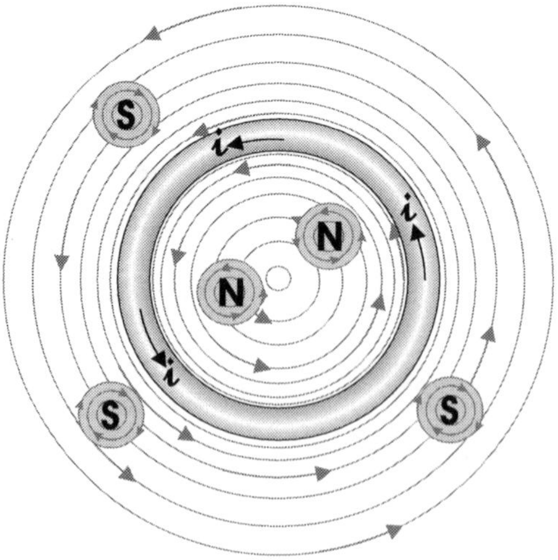

Perpendicular Conductors

Perpendicular conductors do not affect one another which confirms that their *me-flows* run perfectly parallel right across their planes of motion.

Three conductors

- Forces are shown acting on a third conductor (left), by pairs of co-parallel or anti-parallel conductors (right), and respectively placed current rings.
- Associated parallel me-flows are not shown.
- The resultant forces shown acting on the corresponding current rings, are those generated by each ring's me-flow vectors which are parallel to the straight conductor.
- Forces acting on and within the pairs are not shown.
- F_R= resultant force.

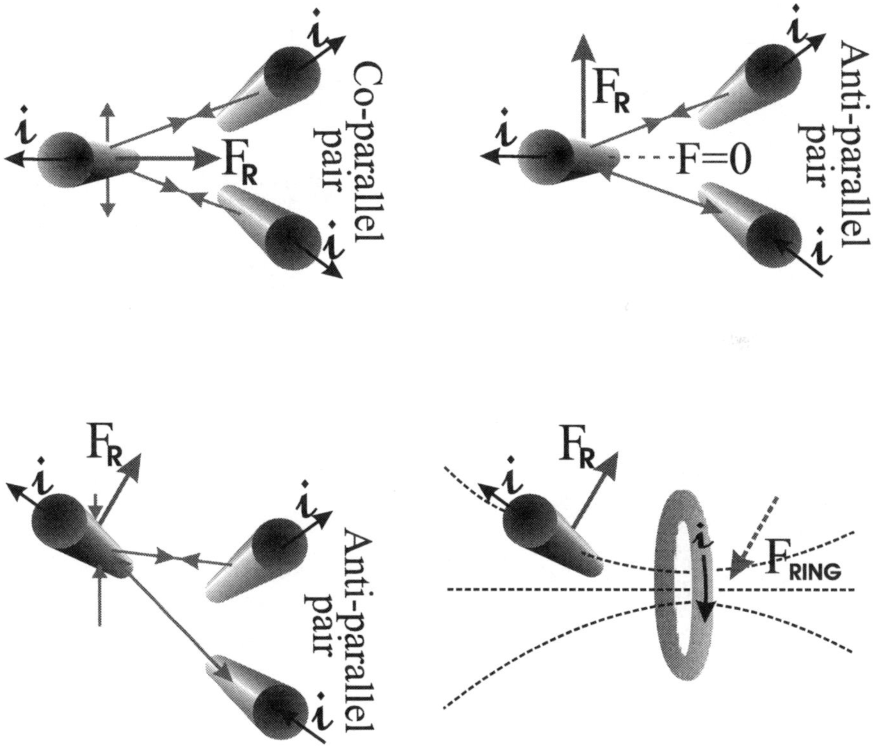

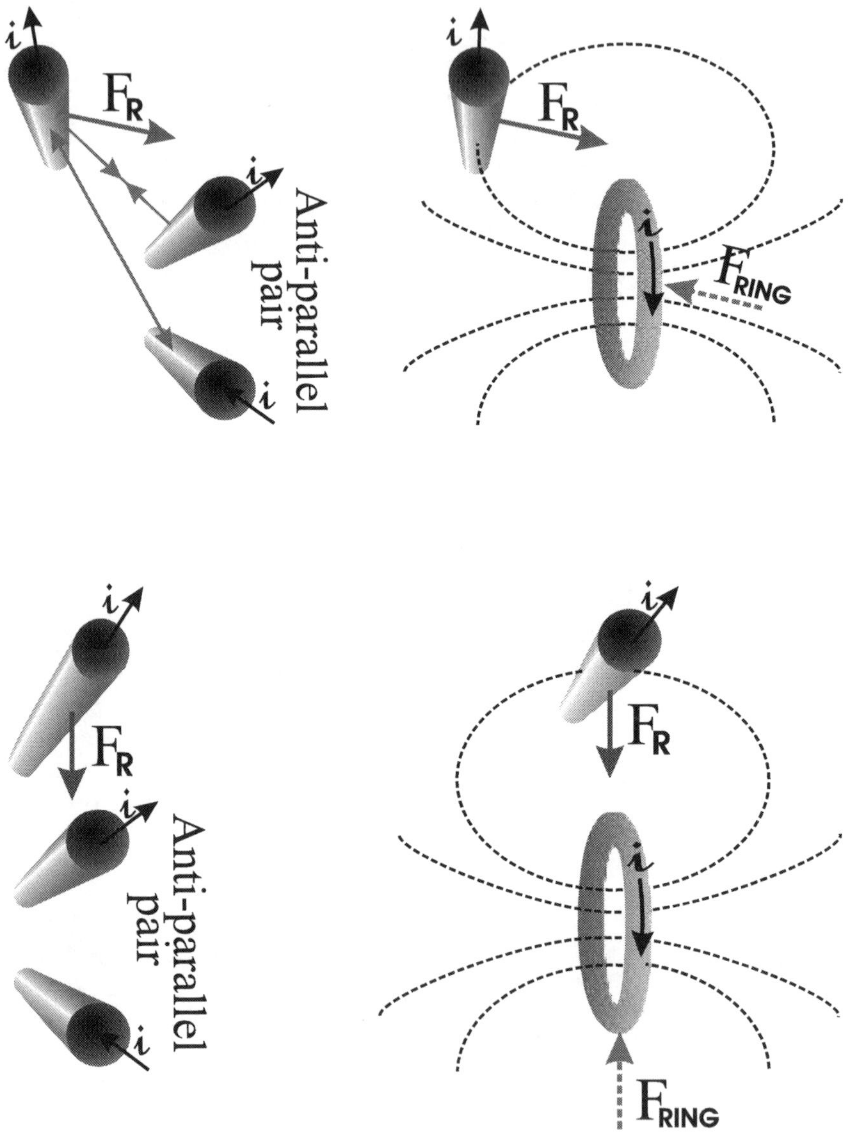
i
F_R
Anti-parallel pair
i
F_R
F_RING
i
F_R
Anti-parallel pair
i
F_R
F_RING

Electromagnetic diagrams

When seen from above, a half way cross section of a pivoting iron disc submerged in an external me-flow, is represented by cross connected to a dot in a circle.

- The cross, symbolising the rear view of the tail feathers of a departing archer's arrow, represents the direction into the page of the overall induced me-flow vortex on that side of the disc.
- The disc's me-flow out of the page is represented by a dot in a circle symbolising the head on view of an approaching archer's arrow.

Compass Needless

The relationship between conventional north pole and a disc's magnetism is:

- A conventional bar magnet has an N for north pole stamped on one end and a S for south pole on the other.
- When the bar magnet is suspended by a thread around its waist, the N-pole swivels until it points to the Earth's South Pole.
- The N stamped on a compass needle points to the Earth's North Pole. It was so called because it pointed to the north star, however to do this it must be a south pole. (Hikers with improvised compasses beware.)

In the next two diagrams:

- The plane of the page bisects the current rings.
- Light grey background shows me-flow out of the page. (the lighter it is the greater its flow energy).
- Dark grey background shows me-flow diving into the page. (the darker it is the greater its flow energy).
- Fine white and black lines are selected theoretical lines of force.
- Swivelling iron discs are shown orientated to directions they would naturally adopt were they placed on some selected imaginary iron fillings lines of force.
- Compasses would point along the white lines with their N-poles directed towards the ring's north pole.

Current Rings

All the magnetic energy flowing in and around a current ring rotates as a vortex whose axis is the axis of the current ring. The flow is strongest beside the ring.

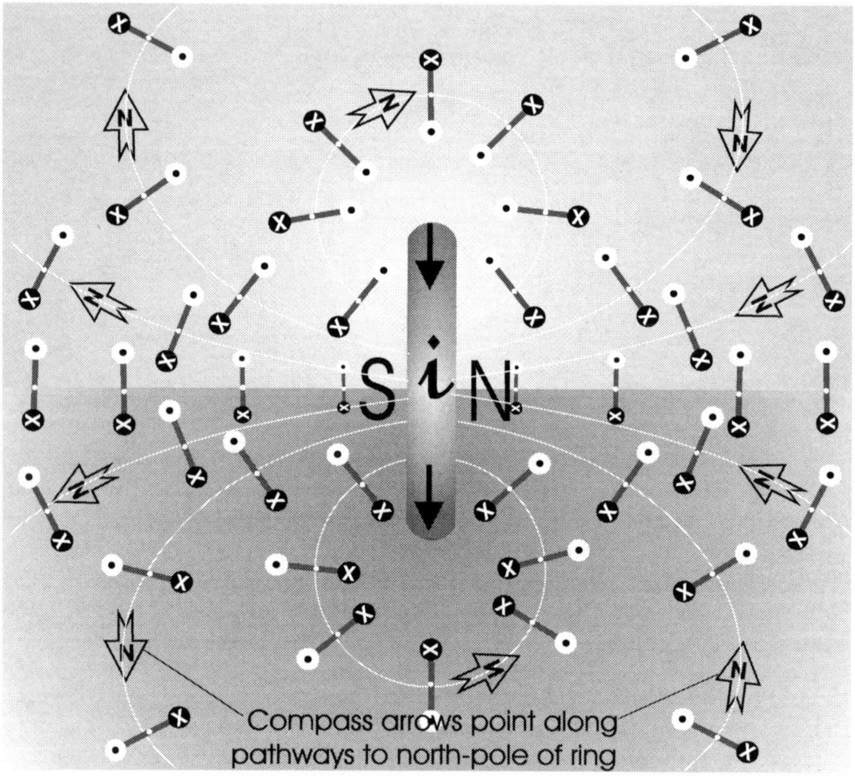

Cylindrical Coils

Several linked current rings become a coreless cylindrical coil, stretching the events which occur in a single current ring.

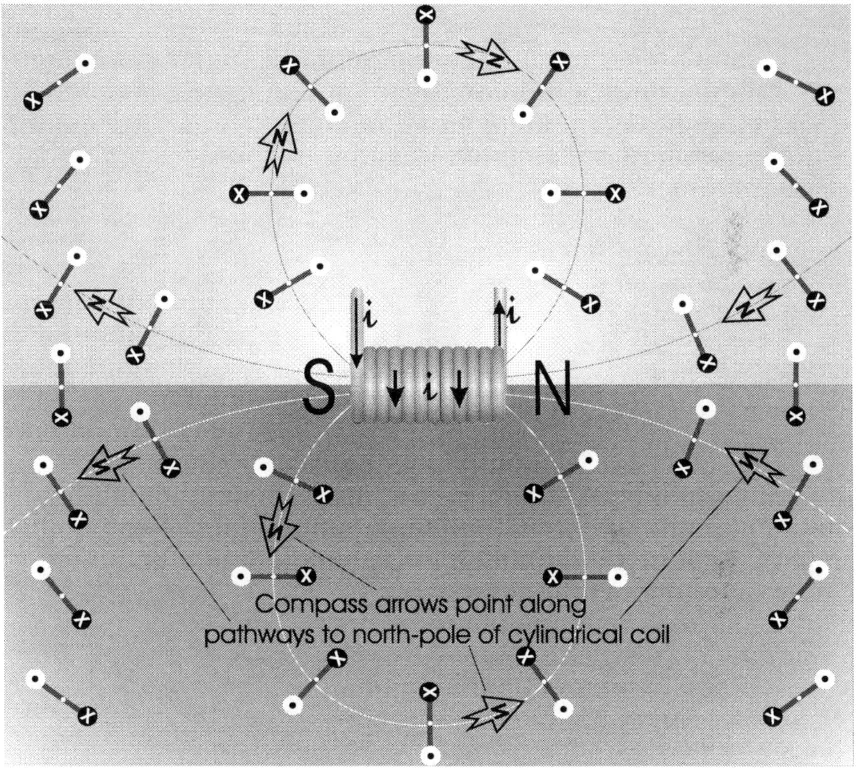

APPENDIX 2

Why Patterns Form

- Filings become miniature magnets and adopt positions where the difference in the me-flow on their near sides to that on their far sides is greatest.
- These filing generate their own mini-vortexes which combine with vortexes of the cylindrical coil or magnet.
- Friction between iron filings and the page lightly holds them in place.
- When a filing falls straight down onto another in-situ-filing, it has a matching me-flow rotation which it picked up before or during the descent.
- It will land by attraction either fore or aft of the in-situ-filing along its magnetic axis, or land by repulsion almost parallel to it but a short distance to either side.
- With the rain of more filings (and some tapping), long chains of filings are formed whose magnetism rotates in the same direction.
- These chains can be viewed as near parallel long magnets which are mutually repulsive, therefore the chain lines are spaced apart.
- These visible lines resembling contour lines are what Faraday believed were 'lines of force' travelling from one pole to the other. They are the rationales for the great magnetic illusion.

The diagram here shows a hollow cylindrical coil with its magnetism coming out of the page above and diving into the page below.

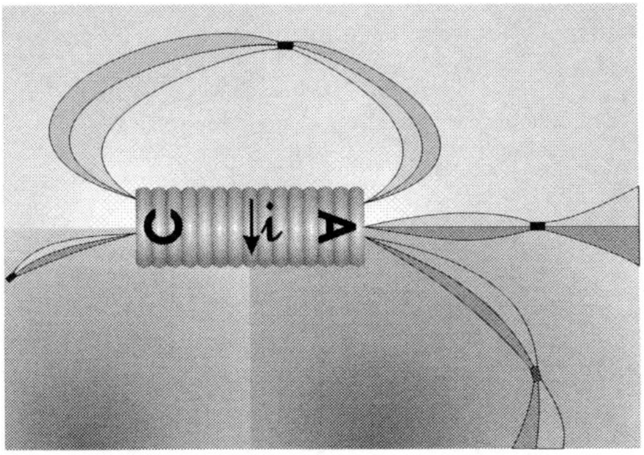

- A few iron filings have landed on the page
- They are the source eyes of mini-vortexes which connect to the cylinders magnetic poles

note

- If the coils were replaced with a bar magnet, and if its mini-vortexes had melded then the filings would again establish their own mini-vortexes
- If the bar magnets mini-vortexes were not melded then the filings would align along the established eyes

Magnetism Graphics and their Filing Maps

The following diagrams show coreless cylindrical coils with their magnetic vortexes, and use magnets with filing maps to corroborate events.

To my mind these validate that magnetism rotates around the long axes of magnets and cylindrical coils.

The jury is still out as to whether or where mini-vortexes meld.

The forces of attraction and repulsion here between cylindrical coils or magnets are not shown.

Co-linear Co-polar Coreless Cylindrical coils

The me-flow vortexes from opposite poles of co-linear cylindrical coils enhance one another increasing their rotational energy, *(crowding their streamlines)* thereby reducing the axially directed pressure between them. The cylindrical coils attract each other, as do the magnets in the filing map below.

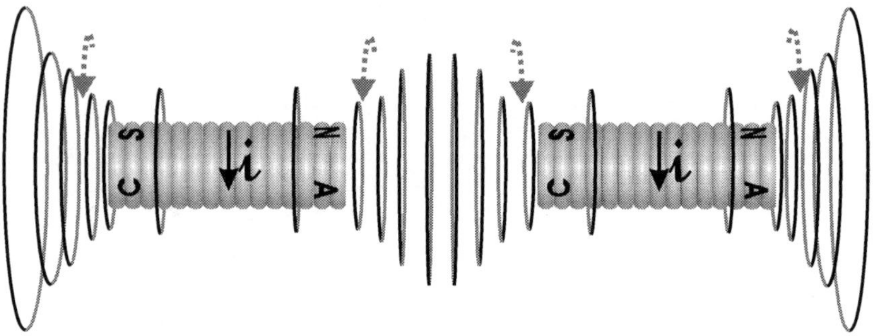

Co-linear Co-polar Bar-magnet Filing Map

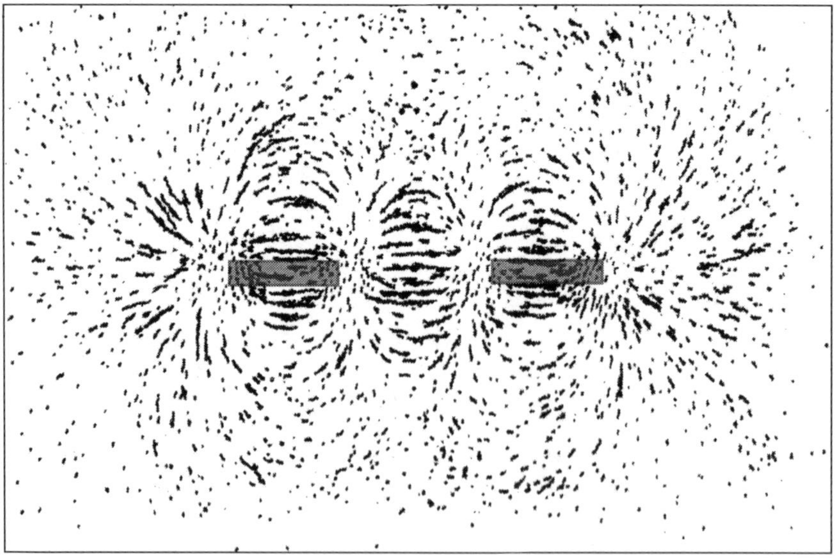

Co-linear Anti-polar Coreless Cylindrical Coils

The converse occurs when the current in one cylindrical coil is reversed. The me-flow vortexes between the like poles now counteract one another increasing the axially directed pressure between them. The cylindrical coils repel each other, as do the magnets in the filing map below.

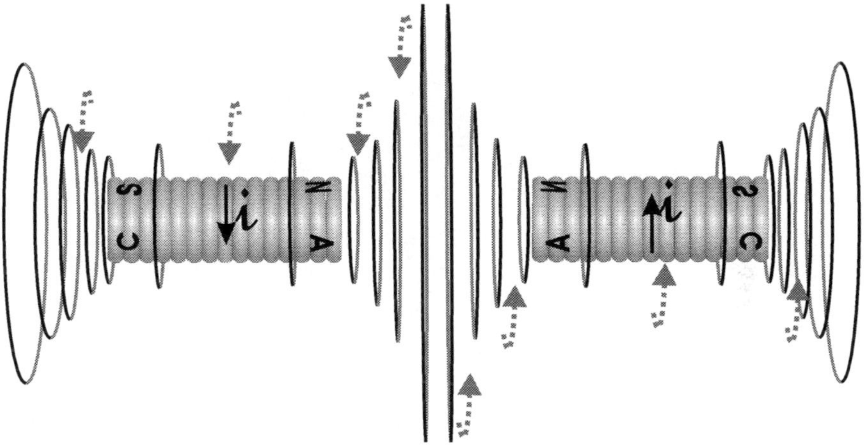

Co-linear Anti-polar Bar-magnet Filing Map

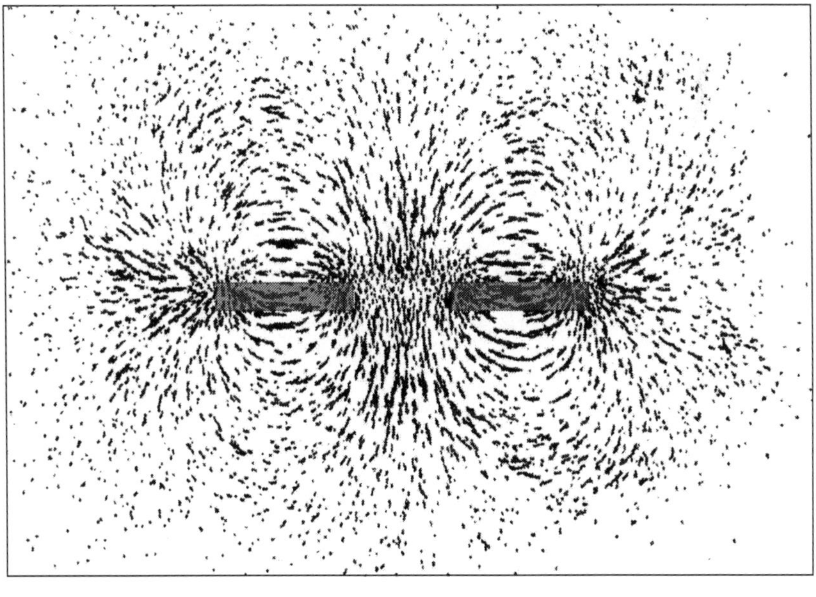

Co-parallel Co-polar Cylindrical coils

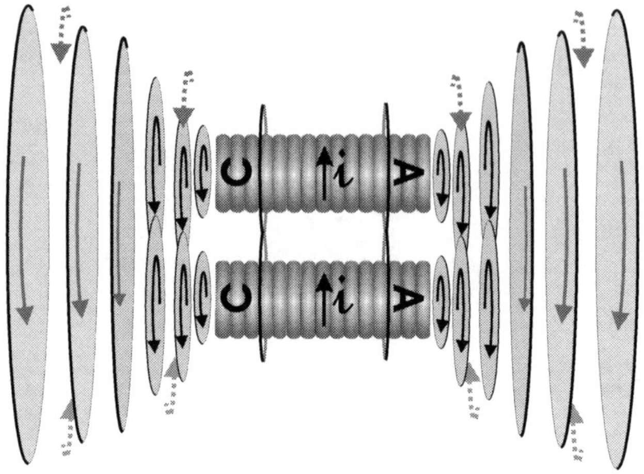

The filling map here appears to support the hypothesis that me-flow vortexes meld after a short distance.

Co-parallel Co-polar Bar-magnet Filing Map

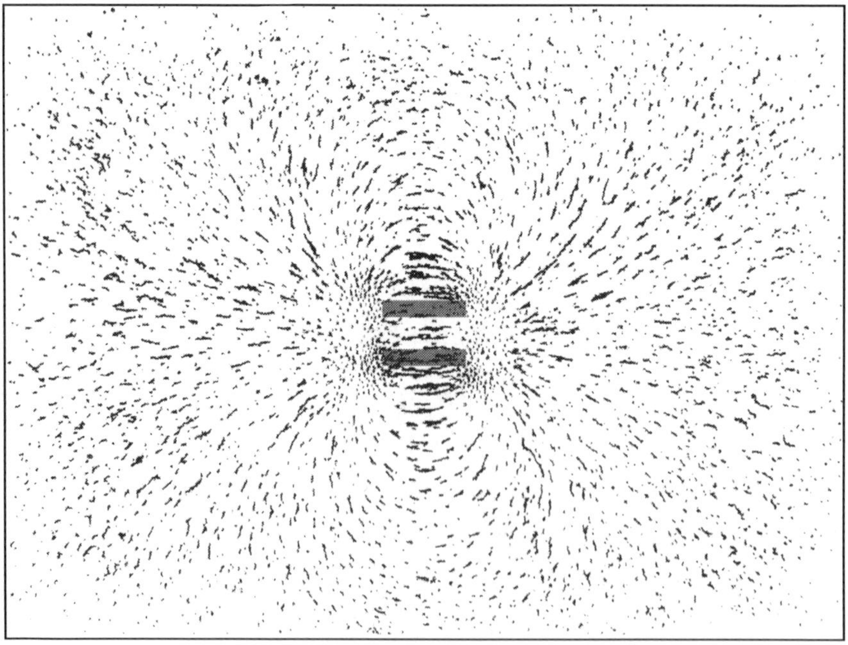

Cylindrical Coils and Bar-magnets Moving Through 180°

Most of the following drawings on the left of page show the interactions of the central vortexes from each cylindrical coil.

The light grey section of a vortex represents me-flow rolling out of the page while the dark grey section represents me-flow rolling into the page.

Each drawing is accompanied by its corresponding filing map.

When the polarities of the cylindrical coils are reversed they create the same shaped but reverse spinning vortex, while both magnets reversed create the same filing maps.

ATTRACTION

or

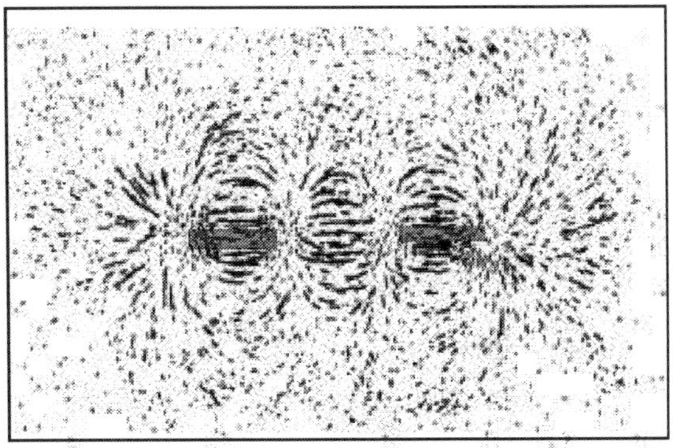

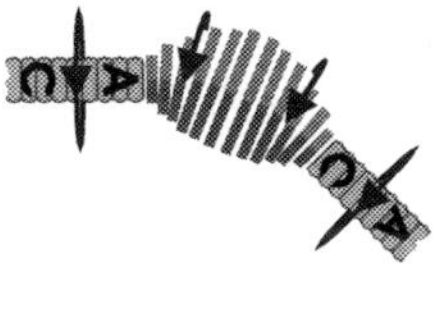

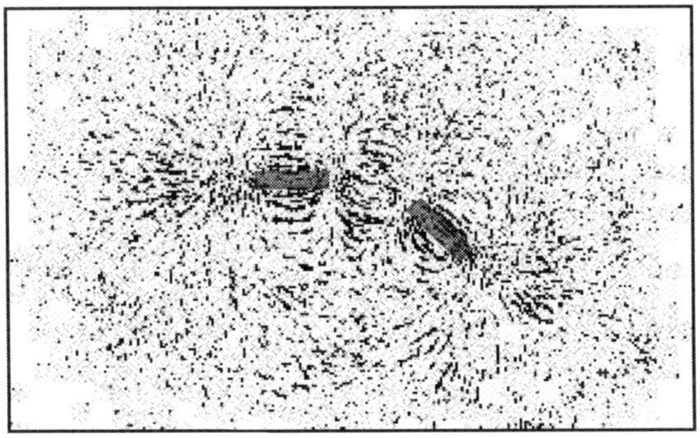

Rotation generated across the poles of anti-parallel magnets is much stronger than that from around their waists.

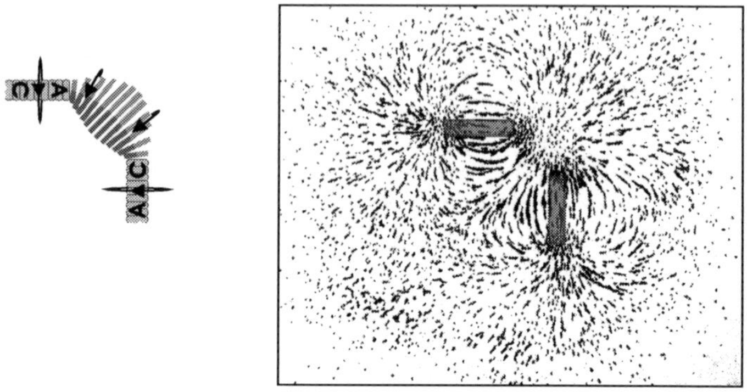

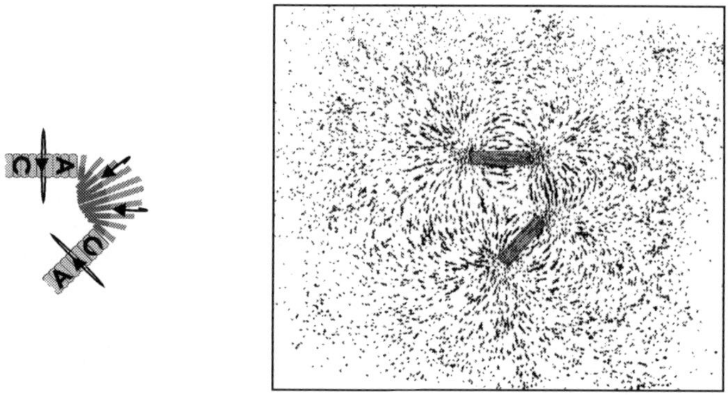

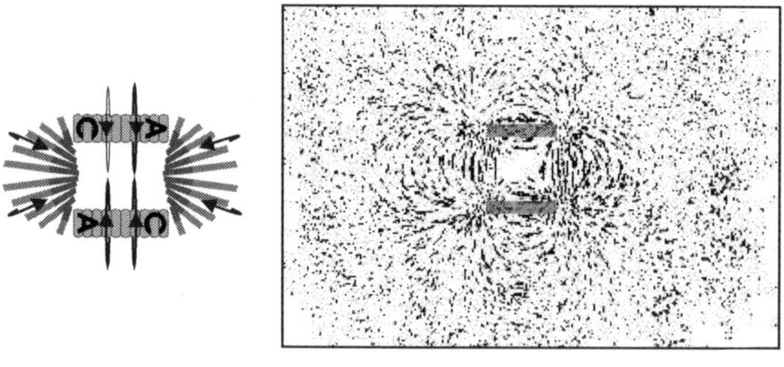

REPULSION

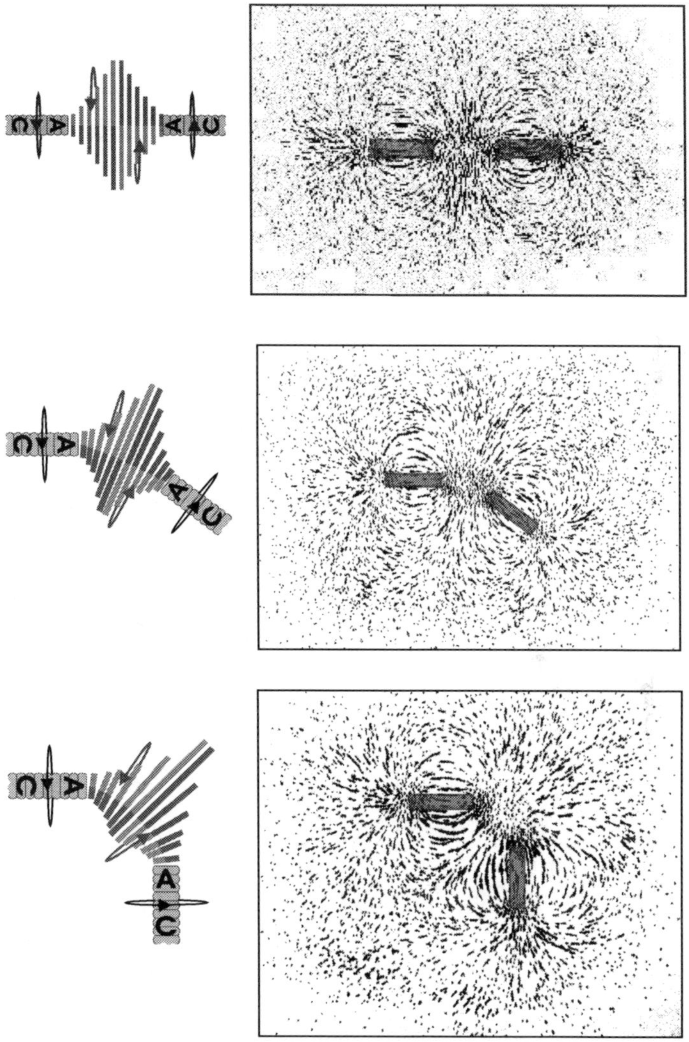

Melded and unmelded vortexes are included here.

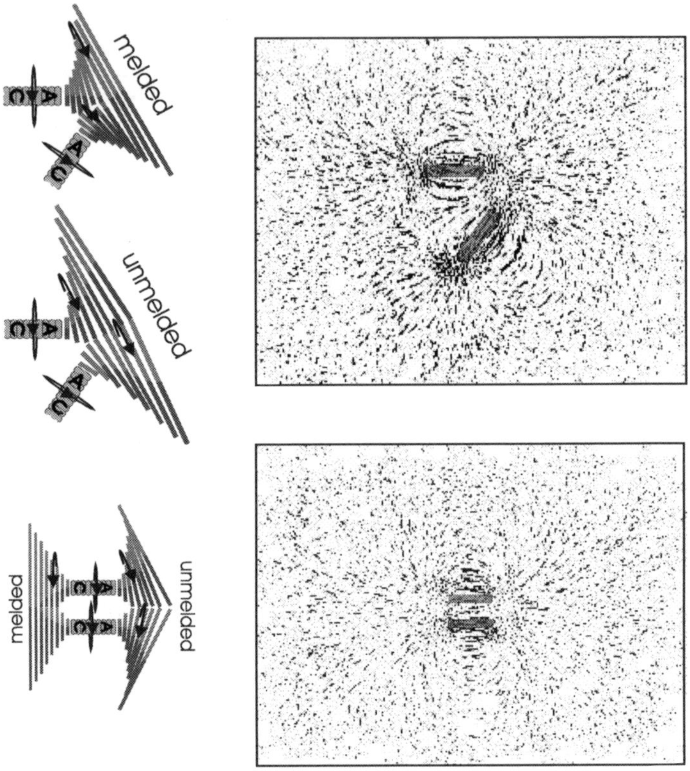

Co-parallel Co-linear Pairs of Cylindrical Coils

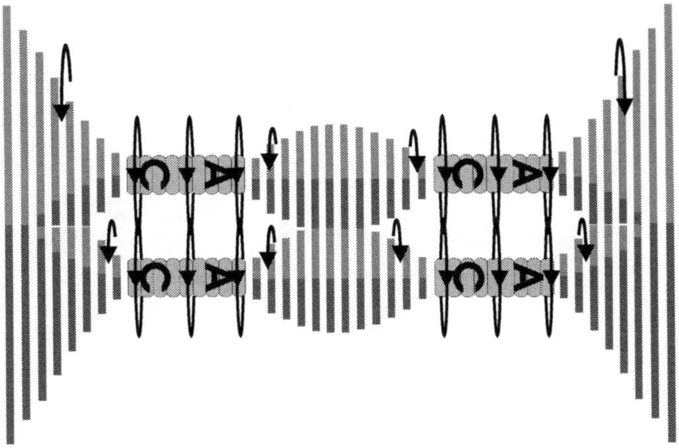

Co-parallel Co-linear of Pairs of Bar-magnets Filing Map

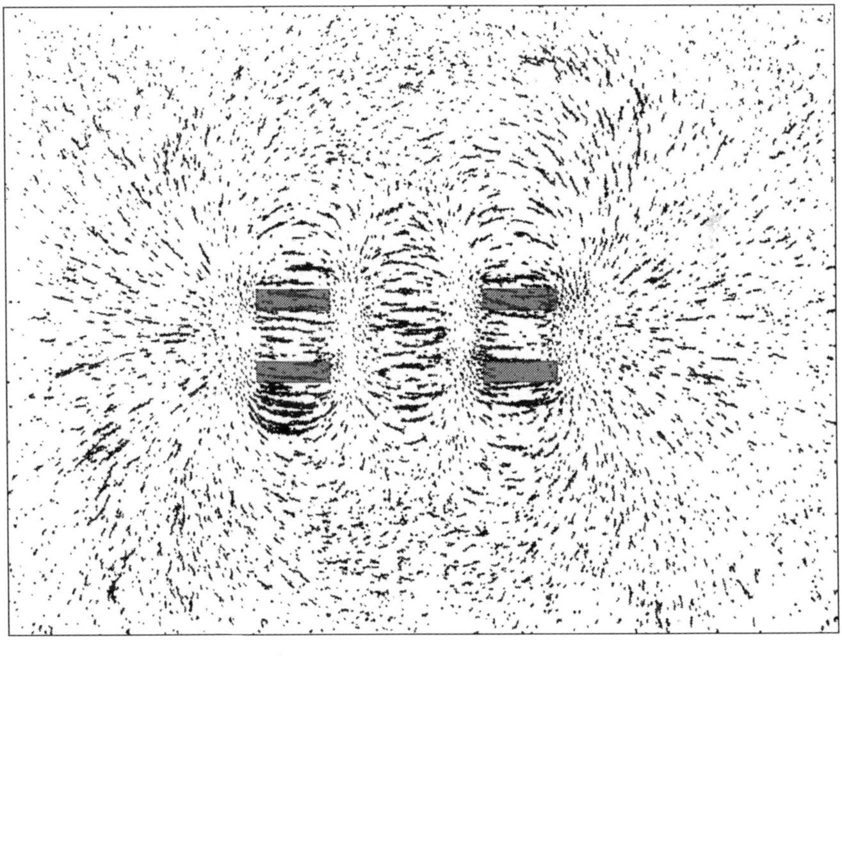

Sliding Co-parallel Co-polar *(like spin)* Magnetism

- The left coil or magnet is fixed.
- The right coil or magnet slides to fixed parallel positions along a perpendicular axis track and is prevented from twisting.
- 'F' refers to axial forces.
- 'T' refers to torsion forces trying to twist the coils or magnets.
- The lengths of arrows associated with 'F' and 'T' in the diagrams are guides to the ratio of 'F' to 'T' per diagram. (not necessarily from one diagram to the next)

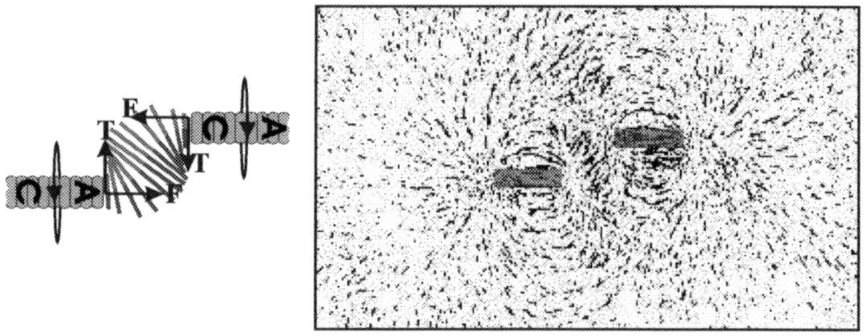

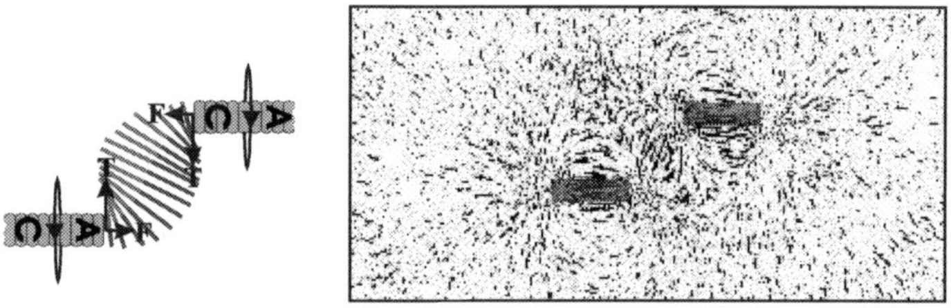

At the meridian point

- There is only torque. 'F' = 0
- From here if the right side coil or magnet is moved up, it experiences a force 'F' to the right. If it moves down it experiences a force 'F' to the left.

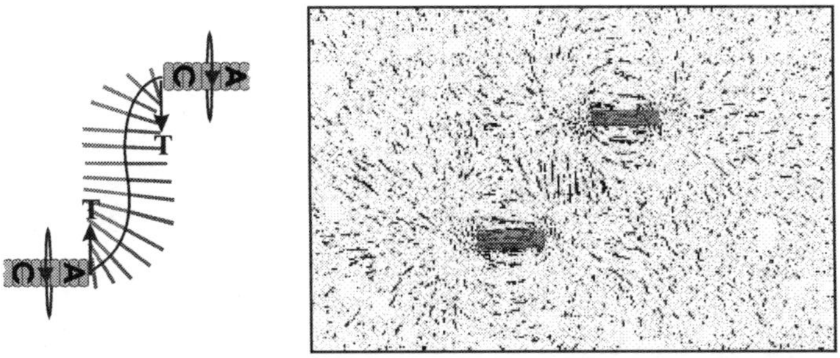

Beyond the meridian

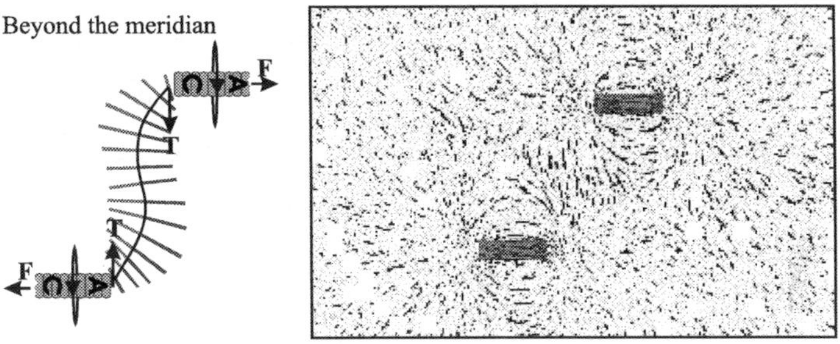

Straight Conductors through a Me-flow Vortex

A current flows along straight conductor passing through a me-flow vortex generated between two coreless cylindrical coils. Wherever the straight conductor's me-flow and the coil's me-flow enhances one another, the pressure is reduced causing attraction. Conversely wherever they hinder one another the pressure is increased causing repulsion.

The force upwards on the conductor equals the sum of the reciprocal forces downwards on the cylindrical coils. *The attractive force between the cylindrical coils is not shown.*

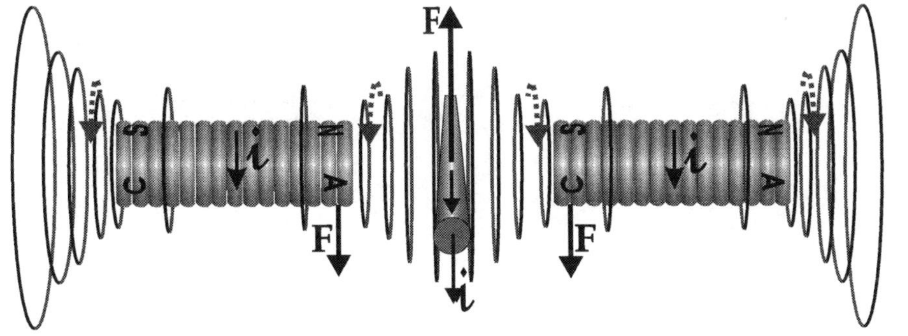

Only the parallel-to-conductor vectors 'Ve' of the coil's me-flow vortex operate.

- Co-parallel flows attract the straight conductor.
- Anti-parallel flows repel the straight conductor.
- When the current reverses the forces reverse.

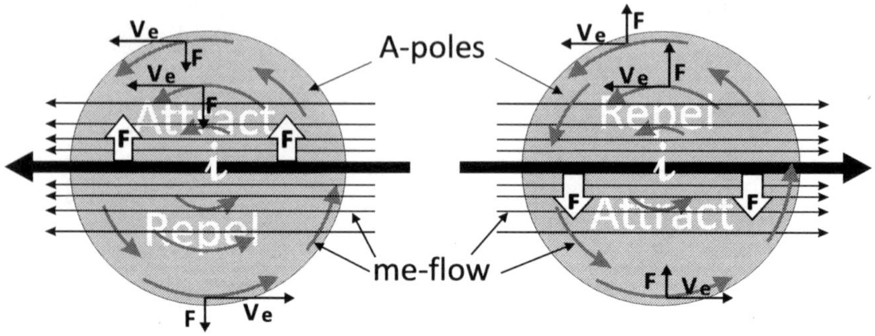

No illusion, no right hand rule, no fingers or thumbs,
No Perpendicularitis!

Multi Co-linear Co-parallel Cylindrical Coils

When opposite poled magnets co-linear end to end are close to each other, the columns jump the gap.

- They likely merge with the opposite columns.
- However they might remain independent and just crowd the gap.

Outside vortexes here are shown melded.

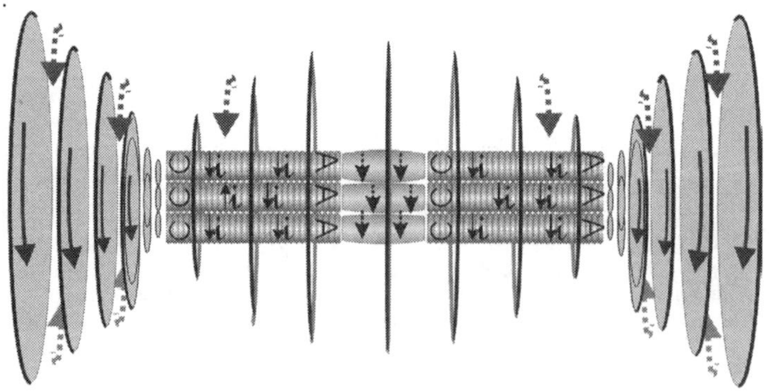

APPENDIX 3

CIRCUITS

Resistors

A resistor is to a conductor what a tap is to a water pipe. Both can restrict their respective current flow. Both in their own way create noise as they restrict their flows.

The symbol for a resistor is: 

Ohm's Law states that the amount of current passing through a conductor is directly proportional to the voltage across the conductor and inversely proportional to the resistance of the conductor.

$$R \propto V/i$$

Resistors in series in a circuit:

$$R_{total} = R_1 + R_2 + R_3 + \ldots..$$

Resistors in parallel in a circuit:

$$R_{total} = \frac{1}{\frac{1}{R_1} + \frac{1}{R_2} + \frac{1}{R_3} +}$$

The unit of resistance is the ohm, symbol: Ω

Resistivity

Resistivity is the electrical resistance of a substance of a standard length and cross section, also called specific resistance.

The symbol for resistivity is Rho = σ measured in ohms/meter

Raising Resistance

Free electrons in conductors continually crash into their host atoms which are constantly vibrating about their equilibrium position in the lattice.

If the conductor is cooled, this random vibration slows and the roving electrons find smoother passage as they thread their way about the lattice.

At absolute zero i.e. -273°C these collisions theoretically stop. *(Theoretically because like infinity this point can never be reached.)*

Conversely as the temperature rises so does the random vibrations of the ionic cores and so do the collisions with the free electrons, and therefore so does the resistivity.

For copper the resistivity temperature graph line is almost straight.

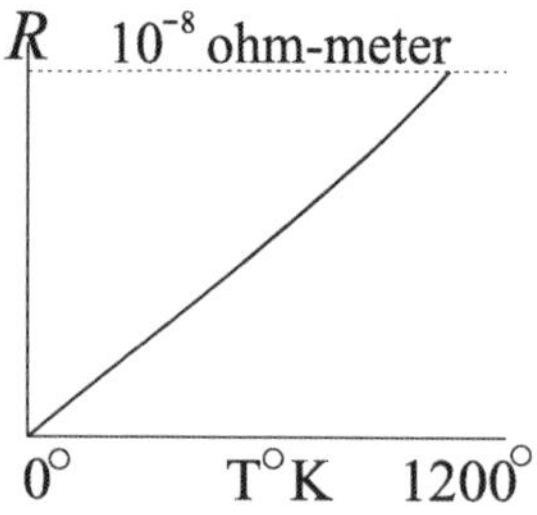

Collisions can also be increased in conductors by:

a) Blending impurities into their structure.

b) Drawing the conductor through dyes which increases imperfections in the lattice structure.

Capacitors

A battery is a storehouse for electric charges. These are held in cells which contain positive ions in one place and negative ions in another. The ions swim in a solution called an electrolyte. Posts connected to the positive and negative zones protrude from the cells. Throughout the battery including the posts, electrostatic charge flows from positive ions and its post into the negative ions and its post.

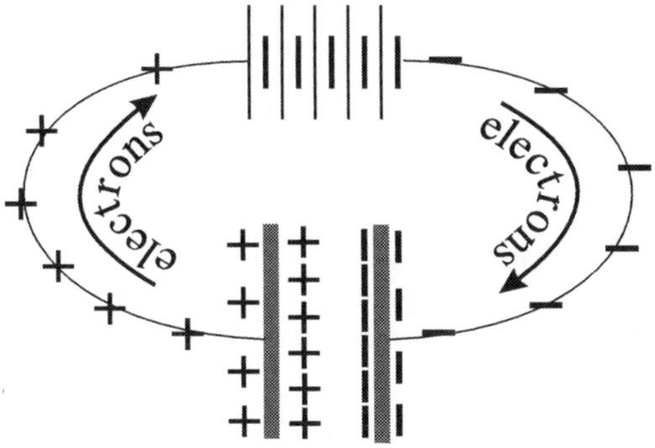

- A basic capacitor has two metal plates connected via conductors to the positive and negative terminals respectively of a battery.
- When the plates approach each other some of the ee-flow across the battery is diverted to flow across the divide between the plates.
- This causes free electrons in the positive plate to migrate upstream leaving that plate positively charged, i.e. electron deficient.
- Simultaneously free electrons in the negative plate are drawn towards the positive plate. These pile up on the gap side of the plate. Their vacated zones are refilled from the battery.
- It is a complete system with all parts dependent on each other. If a switch anywhere in the circuit is turned off, all movement stops.
- When the plates are opposite each other and the switch is turned on electrons begin their migration.
- Initially this requires little energy, however as they build up on the negative plate it becomes increasing harder for more electrons to flow in.
- Simultaneously the increasing attraction of the positive plate makes it ever harder for its electrons to migrate away.
- Eventually the process slows to a halt.
- It has reached capacitance.

This charging of the plates is a timed process whose duration depends on the size of the plates. Quadrillions of free electrons have to drift across the plates as well as down or up the thin conductors.

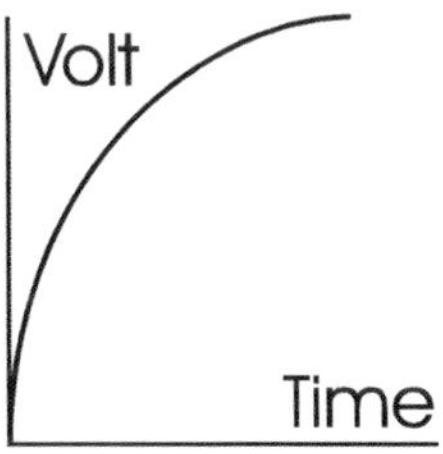

- Because there are quadrillions of them they only have to drift microns to totally charge the capacitor.

- The plates or any other shapes that are likewise chargeable are called capacitors. Another name for them is condensers.
- If the plates are now brought closer together the ee-flow across the plates and therefore in the circuit increases as it does when oppositely charge objects approach each other.
- The capacitance increases again until it reaches a new higher capacity.
- If the plates are draw apart, this reverses the process.
- If the plates are brought into contact with each other a dead short occurs and charge flows freely until the battery is exhausted or flat.

The unit of capacitance is the farad; symbol 'F'.
A microfarad symbol is 'μF'

$$C = \frac{q}{V} \quad \text{farad}$$

A capacitor has a capacitance of 1 farad if it carries one coulomb of charge when 1 volt is applied across it.

Discharging a Capacitor

Once a capacitor is 'charged' if it is pulled away from the battery, by an insulated hand, it will remained charged for some time, until electrons find a way from the negative plate to the positive plate, possibly by hitching a ride on atmospheric water vapour molecules.

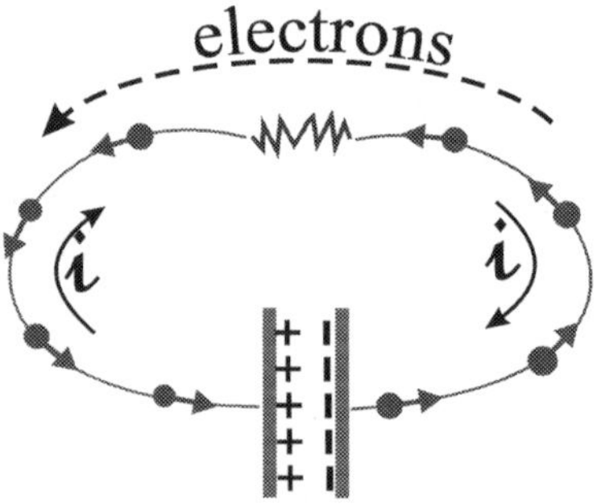

When the conductor tails of an isolated charged capacitor are brought together there is often a spark just prior to contact. The free electrons move as a spark to de-crowd the negative plate and neutralise the opposing positive plate and its conductor.

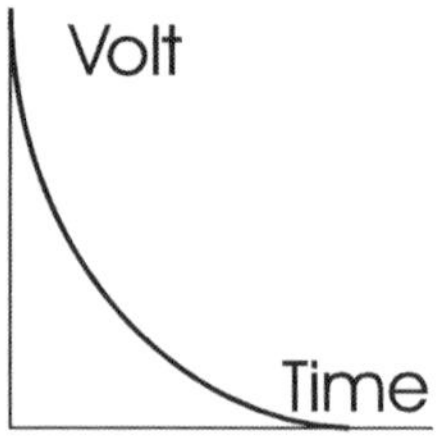

Discharging a capacitor also starts off easily then slows as the potential difference or voltage across the capacitor falls.

Dielectrics

If the plates are brought too close together, a spark or sparks (which are streams of electrons) will jump the gap until the power of the battery is to low to energize their jumps.

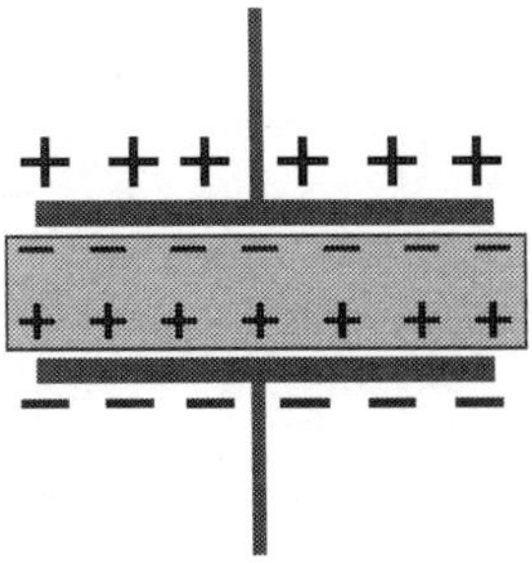

- Sandwiching an insulator between the plates does not disturb the ee-flow but it does prevent the sparks from shorting across the gap.
- Now the plates can be brought very close together which increases the plate's capacitance.
- This insulator is called a dielectric.

Many molecules are polarised. They have one end more positive than the other which is consequently more negative.

When these get sandwiched between oppositely charged metal plates the positive ends get drawn to the negative plate while the negative ends get drawn to the positive plate.

This requires energy so more energy can be stored in the combined system of plates and dielectric.

The ability of a material to polarise in this way is measured in terms of its dielectric constant.

Approx Dielectric Constant	
Vacuum	1.0
Air	1.00054
Glass	4.0
Fired Clay	6.0
Neoprene	7.0
Water	78.0

Water, which is 'polar', has a staggering dielectric constant of 78.0, which is why the Leyden jar has such clout.

The ee-flow extends beyond the ends of a capacitor into a proud dielectric.

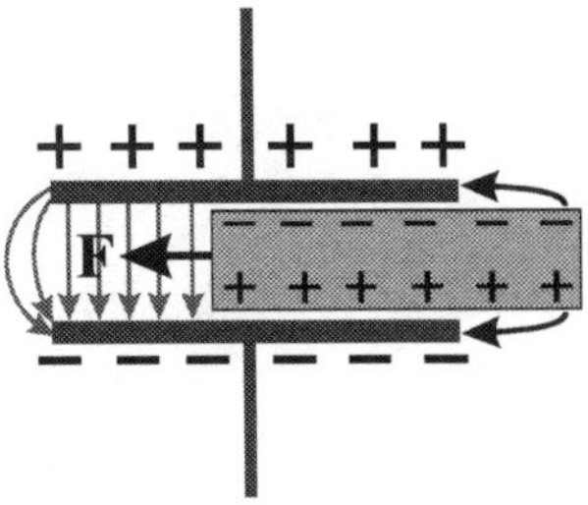

- This aligns polarised sections of the dielectric which are proud but still near the capacitor.
- If free to move the dielectric is sucked into the capacitor.

The Greater the Area the Greater the Capacitance.

In a capacitor, the free electrons mainly accumulate on the inside of the negative plate and consequently those in the positive plate are pushed away leaving the near side of it positive.

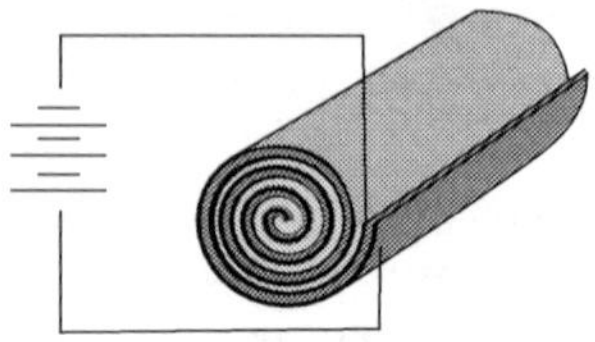

The capacitance of the plates is dependent on their area.

When a strip of plastic insulator *(doubling as a dialectic)* is sandwiched between matching strips of metal foil, the total can be rolled into a compact cylinder, having a large plate area, therefore large capacitance.

One sees a myriad of these in circuit boards.

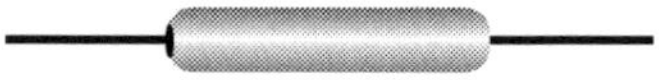

The capacitance of a parallel plate conductor filled with dielectric is given by:

$$C = \frac{k \varepsilon_0 A}{d}$$

k = *dielectric constant*
ε = permittivity constant
A = *area or the plate*
d = distance apart

Me-flow And Capacitors

While a capacitor is charging or discharging an electric current is said by convention to flow across the gap between its plates, even though there is no transfer of electrons or holes across them.

When a capacitor is partially or fully charged there is always electrostatic energy (ee-flow) across the plates.

When the capacitor is fully charged the electrostatic energy does not have any associated magnetism either inside or beside it, or beside the connecting battery wires.

However when the charge is increasing or decreasing there is associated me-flow, not only as companion to the movement of electrons in the battery wires but also as companion to the altering ee-flow across the plates.

Here this concentrated me-flow is co-parallel to the altering ee-flow.

Beyond the ends of the capacitor some altering ee-flow crosses via a waist proud arc. Associated me-flow here is tangent to the arc.

Just as there is pressure between a conductor and its me-flow, similarly there is pressure between the altering ee-flow and its companion magnetism.

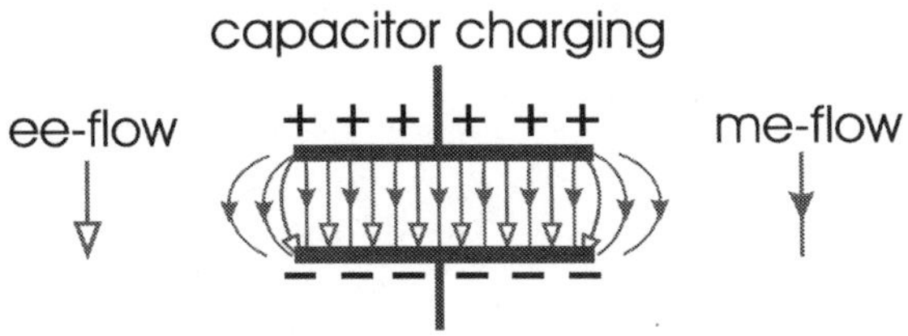

Electromotive Force

Devices like batteries and electric generators are able to keep up their voltage or maintain their potential difference between their terminals while they are being used.

Such devices are called seats of electromotive force (emf); Symbol

RC Circuits

In the diagram, when the switch is thrown to *'a'* the current flows through the resistor as the capacitor charges. Simultaneously quadrillions of free electrons have an average movement of microns in the opposite direction.

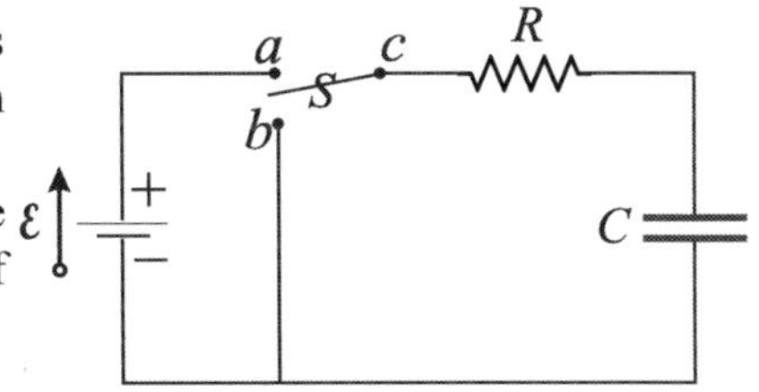

- A surfeit of quadrillions of free electrons builds up on the negative plate while a corresponding deficit occurs on the positive plate.
- The rate at which the capacitor charges is governed by the resistor and the energy used is recorded as joule heat in it.
- When the capacitor is fully charges the current stops.

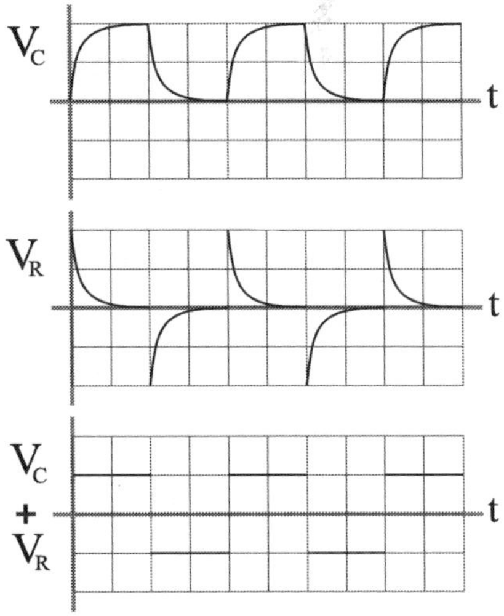

The graph shows the oscilloscope screen pattern of the circuit switching from *a* to *b* charging, then *b* to *a* discharging, then *a* to *b* to *a* to *b* to *a* etc.

- Initially the first electrons drift into one plate and out of the other with almost no effort, so the initial state can be thought of as having an all-but short circuit across the capacitor.
- Consequently the voltage drop across the resistor is the same as the voltage drop across the battery.
- As the charge builds up in the capacitor it gets progressively harder for the free electrons to drift into the increasingly negatively charged plate. Simultaneously it gets proportionally harder for their opposites to leave from the positive plate.
- Therefore the voltage drop across the resistor falls.
- When the capacitor is fully charged the current stops so the voltage across the resistor becomes zero, while the voltage drop across the capacitor is now that of the battery.

The process reverses when the switch is turned to *b* and the battery is now bypassed.

- The capacitor discharges initially using all-but minimum energy so the voltage drop across the resistor is the same as the voltage drop across the capacitor.
- This drop is in the opposite direction to the initial drop in the *a*-circuit.
- The discharge progressively slows as the circuit approaches neutral ground then stops.

LR Circuits

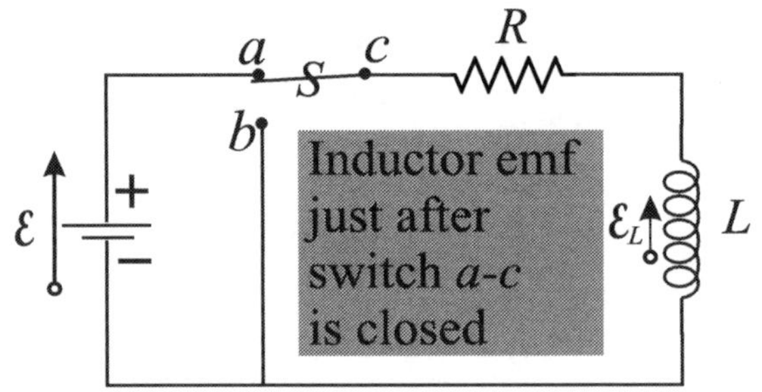

Except for the combined voltages *(bottom graph)*, the oscilloscope screen patterns show the *LR* circuit voltages are the reverse of the *LC* circuit voltages.

- Initially as the switch is thrown to *a* there is an all-but dead short and the voltage drop across the resistor is zero.
- Simultaneously the current has very little trouble winding up the *magnetic energy vortex* in and around the inductor.
- With time it gets progressively harder to wind up the me-flow *spring*.

This is because:

a) A back emf has been generated in the inductor.
b) The volume being wound is getting progressively larger.

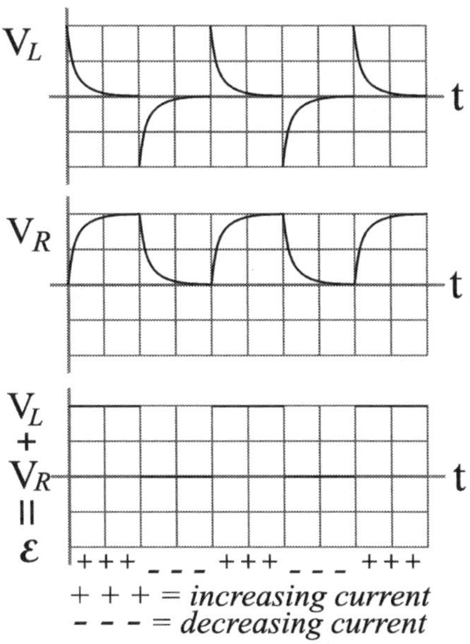

- The current soon slows then stops.
- The voltage on the positive side of the resistor is now the voltage of the positive battery terminal.
- Since the current isn't going anywhere the voltage on the far side of the resistor is zero.
- Now the voltage drop across the resistor is that of the battery.
- Simultaneously the voltage on both ends of the inductor is zero so the voltage drop across it is zero,

When the switch is thrown to *b* the me-flow *spring* unwinds quickly in the beginning sending a current, which now bypasses the battery, back through the resistor.

This unwinding progressively slows as the circuit approaches neutral ground then stops.

Made in the USA
Monee, IL
05 November 2026

67811373R00186